Olav Giere

Meiobenthology

The Microscopic Fauna in Aquatic Sediments

With 102 Figures and 20 Tables

Springer-Verlag

Berlin Heidelberg New York
London Paris Tokyo
Hong Kong Barcelona
Budapest

Prof. Dr. Olav Giere
Universität Hamburg
Zoologisches Institut und
Zoologisches Museum
Martin-Luther-King-Platz 3
D-20146 Hamburg, Germany

ISBN 3-540-56696-1 Springer-Verlag Berlin Heidelberg New York
ISBN 0-387-56696-1 Springer-Verlag New York Berlin Heidelberg

Library of Congress Cataloging-in-Publication Data. Giere, Olav, 1939–Meiobenthology : the microscopic fauna in aquatic sediments / Olav Giere.   p.   cm. Includes bibliographical references (p..) and index. ISBN 3-540-56696-1 (acid-free paper : Berlin). – ISBN 0-387-56696-1 (acid-free paper : New York)   1. Meiofauna. 2. Benthos.   I. Title.   QL120.G54   1993   592'.092 – dc20   93-14010

© Springer-Verlag Berlin Heidelberg 1993
Printed in Germany

The use of general descriptive names, registered names, trademarks, etc. in this publication does not imply, even in the absence of a specific statement, that such names are exempt from the relevant protective laws and regulations and therefore free for general use.

Typesetting: Thomson Press (India) Ltd., New Delhi

31/3145/SPS–5 4 3 2 1 0 – Printed on acid-free paper

To GABY
to whom I owe it all

Preface

Also bestimmt die Gestalt die Lebensweise des Thieres,
und die Weise zu leben sie wirkt auf alle Gestalten
mächtig zurück.

So, the shape of an animal patterns its manner of living,
likewise their manner of living exerts on the animals' shape
massive effects.

GOETHE 1806: *Metamorphose der Thiere*

Studies on meiobenthos, the motile microscopic fauna of aquatic sediments, are gaining in importance, revealing trophic cycles and assessing impacts of anthropogenic factors. The bottom of the sea, the banks of rivers and the shores of lakes contain higher concentrations of nutrients, more microorganisms and a richer fauna than the water column. Calculations on the role of benthic organisms reveal that the "small food web", i.e. microorganisms, protozoans and smaller metazoans play a dominant role in the turnover of organic matter (KUIPERS et al. 1981). New animal groups – even of high taxonomic status – are often of meiobenthic size and continue to be described. Two of the most recent animal groups ranked as phyla, the Gnathostomulida and the Loricifera, represent typical meiobenthos.

A textbook introducing the microscopic fauna of the sediments, their ecological demands and biological properties, does not yet exist, despite the significance of meiobenthos indicated above. The recent book entitled *Introduction to the Study of Meiofauna* (eds. HIGGINS and THIEL 1988) gives valuable outlines for practical investigation, and the *Stygofauna Mundi*, a monograph edited by BOTOSANEANU (1986a), focusses on zoogeographical aspects mainly of freshwater forms, but neither of these fulfils the needs for a comprehensive text on the subject of meiobenthology.

It is the intention of this book to provide a general overview of the framework and the theoretical background of the scientific field of meiobenthology. The first of three major parts describes the habitat of meiobenthos and some of the methods used for the investigation. The second part deals with morphological and systematic aspects of meiofauna, and the third part reports on the meiofauna of selected biotopes and on aspects of community and synecology of meiobenthos. However, a monographic text cannot include an adequate survey of general benthic ecology, or be a textbook on the zoology of microscopic animal groups. The primary purpose of this text is to give an ecologically oriented scientific basis for meiobenthic studies. Further advice for practical investigations is found in important compilations by HIGGINS and THIEL (1988) and HOLME and McINTYRE (1984). Hence, aspects of sampling procedures and strategies, statistical treatment and fauna processing will be treated here in brief only. In these fields, the present work should be considered a supplement to the books mentioned above and will rather focus on some critical hints, methodological limitations, and a few neglected practical aspects.

Writing this book was particularly difficult because the literature on meiofauna is so widely dispersed in journals and congress proceedings and has so rapidly increased in volume that a complete coverage is impossible. Regardless of my efforts, therefore, there is no pretense of this text being absolutely comprehensive. Where it is important for the general context, the major chapters of the book may contain some overlap in details of information. This is deliberate, and intends to provide the reader with chapters complete in themselves and to avoid too many cross-references. While the scope of this book does not allow for details on all aspects, the literature given in the text should enable the reader to inquire further for more specific information. For this approach a fairly extensive reference list is required. However, in order to maintain a readable, coherent style, citation of specific references had to be restricted. Thus, the Reference List of this text does not represent all the sources drawn upon in the production of the book. Where appropriate, references compiled in a paragraph "More detailed reading" are given at the conclusion of many sections and chapters. They will serve as supplementary information and, hopefully, will compensate for my own subjectivity.

The selection of topics and the emphasis given to them is admittedly subjective. In particular, the brief treatment of freshwater meiobenthos (Chap. 8.2) by no means reflects the exhaustive achievements and importance of this field of meiobenthology. This book does not include the nanobenthos, since this represents a microbiota completely different from the meiobenthos in its size range, methodology and taxonomical composition (mainly prokaryotes, often autotrophic protistans and fungi). Should there be incorrect or misunderstood data reported in the text, I would be most grateful for information.

This book resulted from a series of lectures for advanced students given by the author over a period of several years at the University of Hamburg. Studying the tiny organisms living in sand and mud fascinated many of my students and provided the encouragement and persisting stimulus needed to write this book. It will achieve its goal if it further promotes interest in the diverse and cryptic microscopic world of meiobenthic animals, emphasizes their ecological importance, both from a theoretical and practical view, and contributes to the awareness that small animals often play a key role in large ecosystems which are becoming increasingly threatened.

Acknowledgements. I am deeply obliged to Dr. Robert P. HIGGINS (Washington, D.C.), who critically reviewed the entire text not only for linguistical flaws. My thanks go to my graduate students who supported me in selecting figures and designing graphs. I am grateful to several of my colleagues for their valuable comments on parts of the text, also for providing me with manuscripts sometimes still in press and for other helpful hints. It was my intention to include only originals or redrawn figures. This was possible through the patient work of A. MANTEL and M. HÄNEL (both Hamburg) for which I am most grateful.

Hamburg, July 1993   OLAV GIERE

Contents

CHAPTER 1

Introduction

1.1 What is Meiofauna? Definitions

The study of meiofauna is a late component of benthic research, despite the fact that meiobenthic animals have been known since the early days of microscopy (meiofauna and meiobenthos are largely used in this book as synonyms). While the terms macrofauna and microfauna had been long established, it was not until 1942 that "meiofauna" was used by MARE to define an assemblage of mobile or hapto-sessile benthic invertebrates (meiobenthos) distinguished from macrobenthos by their small size. Earlier, most researchers had referred to typical meiobenthic animals as microfauna. Derived from the Greek μειος meaning smaller, members of the meiofauna are mobile, sometimes also hapto-sessile benthic animals, smaller than macrofauna, but larger than microfauna (a term now restricted mostly to Protozoa). Today, the size boundaries of meiobenthos are based on the standardized mesh width of sieves with 500 μm (1000 μm) as upper and 42 μm (63 μm) as lower limits: all fauna passing the coarse sieve, but retained by the finer sieve during sieving is considered meiofauna. In a recent move, a lower size limit of 31 μm has been suggested by deep-sea meiobenthologists in order to quantitatively retain even the smallest meiofauna organisms (mainly nematodes). The significance of what began as a subjective defined size-range of benthic invertebrates has since been supported by more extensive studies on the size spectra of marine benthic fauna (SCHWINGHAMER 1981a; WARWICK 1984; WARWICK et al. 1986a), which will be described later. These studies infer that the (marine) meiobenthos represents a separate, biologically and ecologically defined group of animals, a concept long accepted in the case of the (interstitial) meiofauna of coarse sands. For the interstitial fauna, REMANE (1933) recognized this community concept on the basis of common and unique adaptations to their special habitat (see Chap. 1.2). There is no answer to the question why meiobenthos was not earlier recognized conceptually as a valid intermediate between the micro- and the macrobenthos. It seems inconsistent with the fact that in the water column, the microscopic fauna had long since been considered an equally discrete faunistic assemblage. Personally, I believe that clean sand bottoms, beaches or often odiferous muds were considered unlikely habitats for a diverse fauna of minute dimensions.

More detailed reading: WARWICK (1989)

1.2 A History of Meiobenthology

Zoological investigations and taxonomic descriptions of minute benthic animals were being published by the mid-19th century. One of the first of these was on the discovery of the Kinorhyncha by DUJARDIN in 1851. In 1901, KOVALEVSKY studied Microhedylidae (Gastropoda) in the Eastern Mediterranean, and in 1904, GIARD described the first archiannelid, *Protodrilus*, from the coast of Normandy. He even found the meiofauna to be so rich "that it would take years to study them." But these pioneers of meiofauna studies considered only isolated taxa, often the exceptional species of known invertebrate groups, not the ecological grouping and the community aspect.

Since then, emphasis of field investigations has been biased towards the commercially more interesting macrofauna. Consequently, a suitable methodology for sampling specifically the smaller benthic animals had to be developed. It was REMANE who first used fine-meshed plankton nets for filtration of beach "groundwater", and for equally pioneering studies of the microscopic fauna of (eulittoral) muddy bottoms ("pelos") and of the small organisms associated with surfaces of aquatic plants ("aufwuchs" or "phyton") using dredges with a sac of fine gauze. With the development of effective grabs (PETERSEN 1913) and dredges (MORTENSEN 1925) for sampling subtidal bottoms, the abundance and complexity of the smaller benthos became much better realized. REMANE summarized his work in a monograph *Verteilung und Organisation der benthonischen Mikrofauna der Kieler Bucht* (1933), where he first used the word "*Sandlückenfauna*," later termed by NICHOLLS (1935) "interstitial fauna". Aside from the important descriptions of new kinds of animals, the significance of REMANE'S work is reflected in his contention that the mesobenthic fauna of sand is not merely a loose aggregation of isolated forms, but "a biocoenosis different not only in species number and occurrence, but also in characteristics of form and function" (Figs. 77,78). In his 1952a paper he embodied this concept in the word "Lebensformtypus" which has become incorporated into the terminology of general ecology.

From REMANE's school emerged numerous German scientists of considerable influence on meiofauna research, e.g. AX, GERLACH, NOODT, just to name a few. Through their work REMANE's stimulus even proliferated to further generations of meiobenthologists in Germany. With improved methods (e.g. MOORE and NEILL 1930; KROGH and SPÄRCK 1936), studies on the small benthos soon emerged from many parts of the world.

From Britain, MOORE (1931), NICHOLLS (1935) and MARE (1942) should be mentioned. BOADEN and GRAY in the beginning of the 1960s belong to the first to perform experiments with marine meiofauna. In 1969, McINTYRE compiled the first review on *The Ecology of Marine Meiobenthos*, still a valuable source of information paticularly referring to data on meiofauna from tropical areas.

Studying the fauna of the Normandy coast of the Channel, the Swedish researcher SWEDMARK focussed attention on the rich interstitial fauna, and described many hitherto unknown species. His review on *The Interstitial Fauna of Marine Sand* (1964) is considered a classic among the meiofauna literature.

Working along the shores of the Mediterranean, DELAMARE DEBOUTTEVILLE concentrated his research on the brackish transition areas between marine and freshwater meiobenthos. He was the first to conduct meiofauna research along the African shores. His book *Biologie des eaux souterraines littorales et continentales* (1960) is another much-esteemed compendium of meiofauna research. The intertidal zone of the French Atlantic coast was the main investigation area of RENAUD-DEBYSER and of SALVAT, who, beginning in the early 1960s compiled comprehensive accounts on the interstitial meiofauna and their abiotic ecological factors.

What about North America, now one of the centres of meiofauna research? The early meiofauna studies were linked to just a few names, e.g. PENNAK, SANDERS and ZINN, who discovered important new crustacean groups. Additionally, some European scientists during stays in the USA contributed to the further development: the studies of the Austrians WIESER and RIEDL in the 1950s and 1960s stimulated meiobenthologists to work in this young field of research along the American coasts. The 1960s saw the beginning of American investigations directed primarily towards ecology and which continue to be the major thrust of American meiobenthology, mostly concentrated along the Atlantic coast of the United States. Beginning in the 1970s, the school of COULL investigates the soft bottom meiofauna often using field experimental methods. Its impact drew the attention of general marine benthologists to meiofauna.

The development of meiobenthology in the marine and freshwater realm went separate ways, used different methods, and even developed a separate nomenclature. Still now, research on freshwater meiobenthos is not too well coupled with its marine counterpart, although both REMANE and DELAMARE DEBOUTTEVILLE often emphasized the connections between marine and freshwater meiofauna, especially those of a zoogeographical and evolutionary nature. Freshwater meiobenthology started and developed independently with the Russian SASSUCHIN and colleagues (1927), who sampled at a river shore. They first described the "psammon", i.e. the fauna and flora of sand. This term is today specified as "mesopsammon", the fauna between sand grains (= interstitial fauna), in contrast to the mostly macrobenthic "epipsammon" (i.e. species that live on the surface of the sand), and "endopsammon", which live burrowing in the sand. WISZNIEWSKI (1934) conducted similar studies in Polish rivers and lakes, emphasizing the important role of rotifers. He introduced a nomenclature for various shore regions (e.g. "hygropsammon") still in use mainly in freshwater meiobenthology. While in England, Germany, France and Belgium, early papers on the freshwater psammon remained rather isolated and mainly taxonomic in nature, it was the American PENNAK who included a wider faunal spectrum in his ecological and faunistic considerations. His monograph (1940) on some Wisconsin lake beaches is one of the classic publications in freshwater meiobenthology. His ecological comparison of freshwater and marine interstitial fauna (1951) gave valuable insights into the characteristics of these two biomes.

Related to the research of DELAMARE DEBOUTTEVILLE were the investigations of ANGELIER (1953) on the river shores and banks in the south of France exposed at the dry season. Detailed granulometric and physiographic description of the biotopes are a characteristic of this outstanding work. The importance of the hydrological regime was the subject of the meiobenthos studies by RUTTNER-KOLISKO (beginning in 1953) in Austrian mountain streams and rivers.

The Swiss CHAPPUIS started a series of investigations (beginning in 1942) on the fauna of the groundwater. He found the "stygobios" to be a distinct faunal element, different from the "hyporheic" or "phreatic" fauna in sediments where the interstitial milieu is more or less exposed to the waterflow (see Chap. 8.2.1). These biotopes beneath streams and rivers were the research domain of KARAMAN (1935 ff), ORGHIDAN (1955) and collaborators. These researchers were attracted by the interesting subterranean fauna of karstic rivers in the southeast of Europe, and contributed much to the early knowledge of cave meiobenthos, today also termed "troglobitic" fauna.

SCHWOERBEL (mainly 1961, 1967), by means of faunistic elements, tried to distinguish between the hyporheic biotope and the phreatic groundwater zone, a separation which is hardly to be generalized (DANIELOPOL 1976). Since the 1960s, DANIELOPOL has worked intensively on hyporheic and lacustrine meiobenthos, mainly in Austria. Although specializing on ostracodes, he focussed also on general evolutionary aspects discussing the pathways of colonization of subterranean habitats.

Today, several hundred scientists are working to expand our knowledge on meiofauna from alpine lakes to the deep-sea floor, from tropical atolls to antarctic sea ice. However, wide areas in Africa, South America, Asia and Australia remain terra incognita in the field of meiobenthology; their coastal and inland meiobenthos still awaits description. Most meiobenthologists are members of the International Association of Meiobenthologists, and thus receive its newsletter PSAMMONALIA for information on actual points of interest and recent literature. The triannual Conferences of the I.A.M. are important occasions for mutual exchange of results, experiences and developments. Thus, it can be hoped that in the future the relevance of meiofauna will be further elucidated and the historical separation between marine and freshwater meiofauna research will be significantly reduced.

More detailed reading: REMANE (1933); PENNAK (1940); SWEDMARK (1954); DELAMARE DEBOUTTEVILLE (1960); AX (1966); SCHWOERBEL (1967); MCINTYRE (1969)

CHAPTER 2

Habitat, Habitat Conditions and Their Study Methods

2.1 Abiotic Habitat Factors (Sediment Physiography)

2.1.1 Structure of Sediment Pores and Particles

When describing meiofauna habitats, grain size is a key factor which directly determines spatial and structural conditions, and indirectly determines the physical and chemical milieu of the sediment. Poorly sorted sediment particles (e.g. sand mixed with gravel and silt) become packed tighter and the interstitial pore volume is often reduced to only 20% of the total volume. Well-sorted (coarse) sediments contain up to 45% pore volume. According to RUTTNER-KOLISKO (1962), most field samples of unsorted freshwater sand have a 40% pore volume.

Beside pore volume, also the size of the internal surfaces of the sediment particles is an important determinant for meiobenthic life. It defines directly the area available for the establishment of biofilms (bacteria, fungi, diatoms, mucus secretions) which form a substantial biotic parameter for microscopic animal life (see Chap. 2.2). This internal surface is unbelievably large: for $1 \, m^3$ stream gravel it has been calculated to amount to about $400 \, m^2$. One gram of dry fine sand with an average particle diameter of 10 to 50 μm may have a total surface of 3 to $8 \, m^2$ (MAYER and ROSSI 1982).

However, this clear relation is valid as a generalization only for siliceous sands. In biogenic calcareous sands, which prevail in most warm water regions (coral reefs), the pores often become clogged with fine-ground sediment although the numerical overall particle size can often be rather large, determined by shell hash and coral rubble of greatly varying shape.

In addition to size, the roundness of grains also determines the sorting of the sediment. Angular-splintery particles are packed more tightly than spherical ones. The direct correlation between pore dimensions and body size of meiofauna animals could be experimentally proven (WILLIAMS 1972). In general, mesobenthic species prefer coarse sands, while endo- and epibenthic ones are mostly encountered in fine to silty sediments. These sediment differences affect the two major groups of meiobenthos, nematodes and harpacticoids, finer sediments being preferred by nematodes, coarser often by harpacticoids (COULL 1985). Within the nematodes, preference for a specific grain size was found to relate to certain ecological types (WIESER 1959). "Sliders" live in the wide voids of

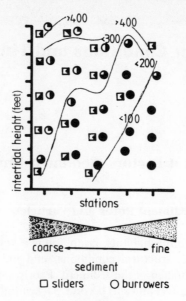

Fig. 1. Relation of grain size to ecological grouping in nematodes. (After WIESER 1959). *Size of black area in symbols* relates to number of species per sample belonging to the respective locomotor type

coarse sand. However, below a critical median grain size of about 200 μm, the interstices become too narrow, thus fine sand and mud will be populated by "burrowers" (Fig. 1). So, the shape of the particles, by indirect action via water content and by permeability (Chap. 2.1.1.2), is decisive for meiofauna colonization of the sediment.

In a more direct way is the colonization of sand with meiobenthos determined by the grain structure, the roughness of edges, and the shape of grain surfaces and cracks. These are important parameters structuring the microhabitats of different bacterial colonies (MEADOWS and ANDERSON 1966). Sand grains with diameters > 300 μm frequently have more plain surfaces than do smaller particles; they also have a different bacterial "aufwuchs." This diversification has been shown to attract different meiofauna species (MARCOTTE 1986). Likewise, in comparative experiments, columns of various sand grains have been found colonized by different meiobenthos, which emphasizes the capacity of meiofauna species to chose and "recognize" their "home sand." In the field, however, this structural impact of the sediment particles on meiofaunal preference is often masked by other factors and is difficult to document. The structure and dimensions of the pore system are directly correlated also to the anatomy of the inhabitants and function of their organs (Ax 1966; LOMBARDI and RUPPERT 1982).

The actual position of the animals in their microhabitat can be preserved and demonstrated only with considerable difficulties. Rather complicated methods have to be applied immediately after sampling to prevent the animals from changing their position. Micro-wave irradiation of the undisturbed (sub)sample has been tested with moderately good results (BERG and ADAMS 1984); however, it requires subsequent fixation (e.g. by hot formaldehyde vapours). Shock freezing of the core (even in liquid nitrogen) is not advisable for this purpose, since it has been proven to cause considerable distortions of the natural stratification (RUTLEDGE and FLEEGER 1988).

HAARLØV and WEIS-FOGH (1953) found that infiltrating a hot agar solution into a soil sample will, after cooling and hardening, leave the (soil) animals undisturbed in their original position. Furthermore, a step-wise dehydration of the core by ethanol makes it possible to produce serial sections with a razor blade. Slices 750 µm thick obtained from a sandy soil sample, may be stained, and then finally mounted in glycerol for inspection or photography.

Another method used in sedimentology involves embedding of the core with resin (WILLIAMS 1971). The sediment core is drained with a water soluble resin (exsiccator!) to which some adequate stain is added to make the animals more conspicuous. After hardening, the block must be prepared for microscopical inspection by trimming, sectioning and polishing. Such permanent sections reveal the natural structure of the pores with the animals included.

When examined under a microscope on a glass slide, all movements of inter-stitial animals seem clumsy and ineffective contrasted with their behaviour in their natural microhabitats. Only by microscopic inspection of active meiobenthic animals does their intricate anatomical and functional adaptation become revealed, resulting in swift and efficient movements. In order to understand the function of structures and their ecological role, "microecological" observations in appropriate microscopic observation chambers are much needed (see MARCOTTE 1984; GILMOUR 1989). Direct observation of meiobenthic animals in their undisturbed habitat is usually prevented by the optical inaccessibility of sediment samples. A newly developed micro-videocamera for insertion into (coarse) sand and gravel (DANIELOPOL and NIEDERREITER 1990) might be a promising tool for direct inspection. An obstacle for microscopic application is the uneven surface of each natural sediment surface. There have been several attempt to mimic the sedimentary structure of the meiobenthic habitat for an analysis. One method involves an artificial sand system cast in transparent resin (GIERE and WELBERTS 1985). The near-natural structure is possible through photography of the genuine sand and subsequent production of a plastic cliche'. Another cast technique has recently been used to analyze the void structure of sea ice (WEISSENBERGER et al. 1992) and might be an interesting approach also for casts of the pore system in sand.

2.1.2 Granulometric Characteristics

2.1.2.1 Grain Size Composition

Granulometric measurements are based on the fractionated sieving of a sufficiently large sample (100 g dry weight or more) through a stack of sieves operated by a sieving machine. Samples from freshwater habitats can be shaken right after drying (80 °C, 24 h) while salt-containing marine samples should be wet-sieved. Where it is important, the quicker and often preferably dry-sieving can be performed if the salt is carefully removed prior to drying by repeated rinsing in tap water. However, in samples with a large amount of fine sediment, it may be less accurate due to loss of material. Also, where gastropod faecal pellets may consolidate large amounts of fine sediment, wet-sieving provides a better representation of particle size values as, if dried, pellets tend to disintegrate into their constituent fine particles.

For meiofauna studies, the silt-clay fraction is usually not differentiated any further. After sieving, its portion can be calculated indirectly by loss of weight. It may be further graded by sedimentation analysis using the elaborate methods common in soil sciences. Mesh sizes of the sieve set follow usually a geometric series (WENTWORTH scale) with 1.0 or 0.5 φ (phi) intervals, φ being the $-\log$ dualis of mesh size in mm (KRUMBEIN 1939). Commonly, for meiofauna studies a series of sieves with 1.0/0.5/0.25/0.125/0.063/0.044 mm mesh size = 0.0/ $+1.0/+2.0/+3.0/+4.0/+4.5$ φ-units is used. For practical reasons (clogging, and excessive, relatively unproductive time consumption) the 0.044-mm sieve is sometimes omitted. On the other hand, very small meiofauna (e.g. some nematodes) would pass even this sieve and can only be quantitatively retained by using a 0.031 mm-sieve ($= +5.0\,\varphi$).

The simple process of sieving has some pitfalls which can render the procedure needlessly tedious or misleading:

a) If in marine samples, sediments passing through the 44 µm-sieve are relevant, and yet the cumbersome retention and sedimentation of this fraction should be avoided, it is important to weigh the core as soon as possible, in order to ensure correct determination of water content and salinity (see below). These values are necessary to recalculate the silt/clay fraction: total weight of sample minus weight of water content (calibrated for its salt portion according to specific density) minus weight of sieved fractions = loss through the finest sieve, i.e. the finest sediment fraction. If this treatment is not possible within a short time after sampling, care must be taken to keep the fresh core in a tight bag to minimize evaporation.

b) If water-unsaturated, massive shaking of the core during transport should be avoided because this can alter sediment structure and water saturation considerably.

c) If the samples contain only a few fine particles, a test run will show whether this percentage is negligible. Omitting the finest sieve always shortens time required for the sieving.

d) If the sediment core contains only a few coarse pebbles or shells in an otherwise relatively homogeneous and fine sediment, these should be removed. Since calculation of the character indices solely depends on weight, one or two massive particles can completely change the granulometric curve without having a relevant impact on the meiofauna. I feel this alteration of conditions is justified in biological studies, provided the manipulation is mentioned in the text.

Having calculated the dry weight percentages of each sediment fraction, a block diagram can be designed, but usually the fractions are computed as cumulative percentages starting with the coarsest fraction. These values are listed for further mathematical treatment or plotted as a cumulative frequency curve (Fig. 2). It becomes apparent that the use of the φ-notation (abscissa) has the advantage of giving relatively more detailed information on the important finer particles, and also renders equidistant intervals which are relevant for the assessment of the following important statistical indices.

Grain size composition of a sample is characterized by a few statistical indices which can be read directly from the diagram. These include the median particle diameter (Md), the first (Q_1), and the third (Q_3) quartile. The Md value corresponds to the 50%-point of the cumulative scale (φ 50), Q_1 to $\varphi25$ and Q_3 to $\varphi75$. These values indicate the average grain size and the spread of the grain size fractions towards both ends. The spread distance is conveniently expressed by the:

Quartile Deviation (QD_φ) or Sorting Coefficient: $QD_\varphi = \dfrac{\varphi75 - \varphi25}{2}$.

It measures the number of phi-units between the third and the first quartile. A homogeneous sediment with a small QD which between the quartiles encloses only a few phi-intervals is regarded as "well sorted" (Table 1). An ideally sorted sediment consisting of just one grain fraction would have the QD = 0.

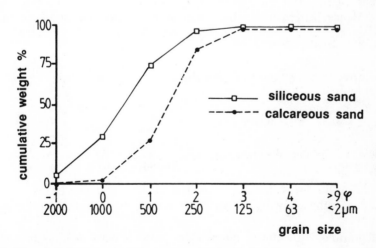

Fig. 2. Granulometric curves of two exposed Atlantic beaches. *Open squares* Portugal; *solid circles* Bermuda

Table 1. Sediment sorting classes. (GRAY 1981)

Sorting class	Classification of sediment
<0.35	Very well sorted
0.35–0.50	Well sorted
0.50–0.71	Moderately well sorted
0.71–1.00	Moderately sorted
1.00–2.00	Poorly sorted
2.00–4.00	Very poorly sorted
>4.00	Extremely poorly sorted

Only if sediment fractions tend to follow a normal distribution, will the frequency curve attain a sigmoid shape. However, it will become "skewed", i.e. it will have an asymmetrical slope when certain fractions are over- or underrepresented. The degree of curve symmetry is measured by the:

φ-Quartile Skewness: $Sk_\varphi = \dfrac{(\varphi Q_1 + \varphi Q_3)}{2} - \varphi Md.$

A strongly asymmetrical curve may make it difficult to obtain an exact reading of the above statistical values. Plotting the cumulative values on a probability paper results in straightening of the curve, and thus avoids problems of graphically ascertaining the relevant values (Fig. 3). Probability paper with a special logarithmic scale on its y-axis is available from major manufacturers of graphical/mathematical papers.

The above indices are based only on very few φ-values, they tend to neglect the "tails" of the curve. Using additional φ-values would render the indices more precise. Hence, better computations comprising a wider portion of the fraction, are for the Graphic Mean: $Md = \dfrac{(\varphi 16 + \varphi 50 + \varphi 84)}{3};$

the inclusive Quartile Deviation or inclusive Sorting Coefficient:

$QD_1 = \dfrac{\varphi 84 - \varphi 16}{4} + \dfrac{\varphi 95 - \varphi 5}{6.6};$

the inclusive Graphic Skewness: $Sk_I = \dfrac{\varphi 16 + \varphi 84 - 2\varphi 50}{2(\varphi 84 - \varphi 16)}.$

From the grain size composition analyzed above and plotted as curves in Figs. 2 and 3, the following characteristic granulometric values can be derived (Table 2):

All granulometric indices can be more precisely calculated mathematically by interpolation using the cumulative percentage values of grain size fractions (HARTWIG 1973b), although the graphical method is more convenient.

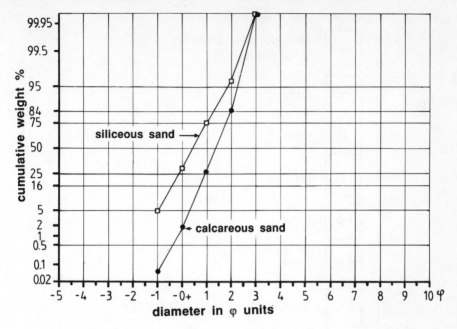

Fig. 3. Cumulative frequency curves of Fig. 2, plotted on probability paper

Table 2. Characteristic granulometric indices for the sediment samples plotted in Figs. 2 and 3

Granulometric index	Siliceous sand, Portugal	Calcareous sand, Bermuda
Median Md	$0.4 \varphi = 740 \, \mu m$	$1.3 \varphi = 410 \, \mu m$
Lower Quartile Q_1	$-0.2 \varphi = 1140 \, \mu m$	$0.9 \varphi = 528 \, \mu m$
Upper Quartile Q_3	$1.0 \varphi = 500 \, \mu m$	$1.8 \varphi = 285 \, \mu m$
Inclusive Sorting Coefficient QD_I	0.93 = moderately sorted	0.74 = moderately sorted
Inclusive Graphic Skewness Sk_I	0.003	-0.089

In scientific papers it is more illustrative to convert φ-values into mm and μm. The use of a conversion chart (PAGE 1955; Fig. 3.2 in BUCHANAN 1971) is often recommended, although calculation is just as easy, and more exact:

1. Calculation of φ from x [mm] (using the common $\log_{10}$ normally preset on a calculator):

Formula: $\varphi = -\dfrac{\log x}{\log 2}$

Sequence of keys to be used:

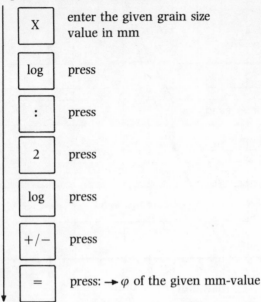

enter the given grain size value in mm

log   press

:   press

2   press

log   press

+/−   press

=   press: → φ of the given mm-value

2. Calculation of x [mm] from φ (using the common $\log_{10}$, normally preset on the calculator):
   Formula: $x\ [mm] = 2^{-\varphi}$

Sequence of keys to be used:

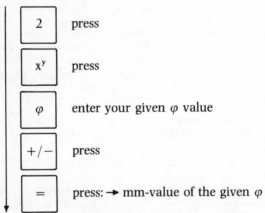

2   press

x^y press

φ   enter your given φ value

+/−   press

=   press: → mm-value of the given φ

Depending on local geological and physiological conditions, generalizations on the occurrence of certain sediments are seldom possible. In temperate and boreal regions, siliceous sand prevails, while in the warmer regions, biogenic calacareous sediments dominate. In volcanic areas, often black basalt and lava sand can be found. The deep sea floor is usually muddy, often consisting of foraminiferan (mostly calcareous) or radiolarian (mostly siliceous) skeletons. In shallow seas, offshore bottoms will usually consist of medium sand, while in nearshore flats attenuated currents will allow fine sand and mud to deposit. In areas where ripple marks indicate strong currents, crests contain coarser sediments than troughs. In troughs, fine sand and often a flocculent surface layer tend to accumulate, resulting in a higher content of organic material. This varying small-scale sediment pattern represents different microhabitats for meiobenthic animals (ECKMAN 1979; HOGUE and MILLER 1981; HICKS 1989). Near the shore line, the sediment structure may vary rapidly due to irregular water agitation, sedimentation and resuspension of shore vegetation and wrack material.

2.1.2.2 Exposure and Sediment Agitation

Exposure of a shore line is largely determined by the impact of waves and currents. Although of eminent importance for the movement of water through the sediment, exposure remains a more or less summative character. It depends on many single parameters and can hardly be objectively recorded. Direct measurement of exposure is mathematically and instrumentally too complicated to be of value to most biologists. HUMMON (1989) estimated exposure of a coast line from a fetch-energy index which he calculated using wave height and shoreline configuration. In most cases, however, exposure is estimated from the degree of sediment agitation and turbulence measuring the grain size composition (see above) as its indicator. In practice, a meiobenthologist is well advised to combine a good part of comparative experience to any measurements in order to denote the rate of exposure.

ELEFTHERIOU and NICHOLSON (1975), on the basis of granulometry, discriminated exposed beaches from sheltered plus semi-exposed ones by the critical median grain size of 232 µm. McLACHLAN (1980, 1989) attempted to create a general rating system for beaches which accounted for a compound set of parameters including the height of the incoming waves. MUUS (1968) and DOTY (1971) related exposure to the weight loss of plaster test blocks distributed in/on the sediment. The dissolution of calciumsulphate was considered proportional to the velocity of the surrounding water currents, and thus to reflect the exposure of the habitat. But in the field, the process of dissolution is a complex of various factors which makes this method rather inaccurate.

Current velocity largely determines the grain size composition of a beach because of the differential sedimentation of the particles. Due to the physical interaction between particle weight, specific resistance, and sediment packing, particles with a diameter of approximately 180 µm are most easily eroded by currents (SANDERS 1958). A threshold around 200 µm, earlier defined as a "critical grain size" for the occurrence of many animals (see Chap. 2.1.1.1), is of prime importance for the water content of sediments. The lower threshold for the existence of a mesopsammon is often reported to be 150 µm. In freshwater sediments, 250 µm has been considered the size limit for the circulation of interstitial water (RUTTNER-KOLISKO 1961).

Neither tightly packed silt nor permanently agitated coarse sand offer favourable living conditions for most meiofauna. Only specialized species will occur deep in the muds of sheltered flats or in the swash zone of exposed beaches.

Massive agitation of the sediment by storms can destroy the meiofauna populations. Most meiofauna react by attempting to escape by downward migration.
Avoidance reactions of meiofauna to increasing current and wave action, e.g.
the tidal wave front and vibration of the sediment have been documented also
in experiments (McLACHLAN et al. 1977; MEINEKE and WESTHEIDE 1979; FOY
and THISTLE 1991).

Erosion and shear strength of the sediment are influenced not only by
abiotic factors. Biogenic factors, such as reworking of the sediment by intensive
burrowing and pelletization as a result of defaecation, contribute considerably
to its physical and biological properties. Heavy bioturbation and deposition of
faecal pellets may reduce sediment shear strength and enhance its erodability
by water currents (RHOADS et al. 1977). In addition, protruding tubes and plant
culms may cause water turbulences and erosive forces with negative impact
on meiofauna (COULL and PALMER 1984; HICKS 1989). On the other hand,
particularly in fine-grained sediments, biotic factors like accretion by mucus
(bacterial and diatom colonies) and deposition of detritus, as well as compaction
in tube walls of small infauna, will solidify the texture and affect sediment
stability and resuspension (RHOADS et al. 1978; LUCKENBACH 1986; MEADOWS
and TAIT 1989; DECHO 1990; HEINZELMANN 1990). When diatom populations
become massively reduced by browsing so as to lower the mucus production,
shear strength and sediment cohesion will be decreased.

2.1.2.3 Permeability, Porosity and Pore Water Flow

The actual configuration of the individual pores, the structure, and the coherence
of the void system have considerable influence on the permeability of the sediment.
Permeability is a measurement of the rate of water flow through a given volume
of sediment. In freshwater biology it is also termed "hydraulic conductivity".
Since water flow via the exchange rates of interstitial and supernatant water
determines most chemical and physical factors in the sediment, it is responsible
for the supply of oxygen, dissolved and particulate nutrients, and, thus, controls
the living conditions for meiofauna in general. The hydrodynamic pattern within
the sandy interstitial habitat and its relevance have been documented by RIEDL
and MACHAN (1972). Permeability (in $cm\,s^{-1}$) is easily calculated with a
permeameter (see Fig. 5.4 in GIERE et al. 1988a).

In contrast to permeability, the total pore volume of a sediment core, the
porosity, is biologically of limited significance only, because it does not correspond
to the pore volume available to animals. It is, of course, of relevance for physico-
chemical fluxes in the sediment. For measurement of porosity see BUCHANAN
(1984). More recently, porosity profiles have been recorded measuring the
resistivity of the sediment matrix with electrodes (ARCHER et al. 1989).

The velocity of the pore water flow passing through the interstitial system
depends on the hydrodynamics of the overlying water, but also on the irrigational activity of tube-dwelling macrobenthos. Velocity of pore water flow
can be measured by microflowmeters. In these minute thermistor flow meters
(LABARBERA and VOGEL 1976; DAVEY et al. 1990), the cooling effect of the

currents on a heated wire corresponds to a voltage signal on a monitor which, after calibration, indicates microflows of water. Similarly, the platinum wire used to measure the oxygen diffusion rate in sediments (see Chap. 2.1.3.4), can also be calibrated for recording microflows of water. Recently, MALAN and McLACHLAN (1991) introduced a new method for measuring the pumping effect of waves on the oxygen milieu. They pointed out that most authors have underestimated wave-induced benthic oxygen fluxes due to inappropriate methods.

2.1.2.4 *Water Content and Water Saturation*

The water content or the rate of water saturation of the sediment is closely related to its permeability. Fine-grained sediments, saturated with water, have a higher water content than coarse sands. Mud cores can often contain > 50 weight % of water, while medium sand will only hold about 25%.

A fluctuating water content causes a repeated replacement of pore water and supplies the meiofauna with oxygen and nutrients. On the other hand, water-unsaturated surface layers with their reduced capillary forces cause steep gradients of many abiotic factors such as temperature and salinity (see Chap. 2.1.3.1 ff) which have a negative impact on the meiobenthos. Contrastingly, in the deeper, water-saturated layers, capillary forces dampen fluctuations of water content.

Water saturation and water flow play a dominant role in structuring meiofauna settlement. Above the water line, occurrence of meiofauna can become restricted because of reduced water content. Particularly in eulittoral shores at ebb-tide, the degree of moisture is often correlated with the distribution of meio-benthos. Lack of water in the surface horizons can force meiofauna into the deeper horizons. Many eulittoral meiofauna species adapt to the regular tidal alterations of water content with preference reactions and migrations (Fig. 4; see Chap. 7.1).

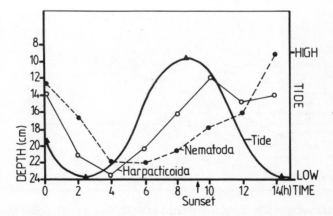

Fig. 4. Migration of beach meiofauna in relation to the tidal cycle. (McLACHLAN et al. 1977)

McLACHLAN (1980), basing on earlier work by DELAMARE DEBOUTTEVILLE (1960), SALVAT (1964) and others, suggested a stratification pattern of meiofauna distribution in intertidal beaches in relation to desiccation and water-saturation (Fig. 5).

a) An upper "dry sand stratum" is characterized by low water saturation and high fluctuations in temperature and salinity. Here, prevalence of (specialized) nematodes and oligochaetes is contrasted to scarcity of harpacticoids and turbellarians.

b) A partly underlying "moist sand stratum" ("zone of retention" in SALVAT 1964) offers a more favourable water supply and has less fluctuations in temperature and salinity. Meiofauna abundance and diversity increase, often with a particular increase in harpacticoids, under the always well-oxygenated conditions in this zone.

c) In the "water table stratum" around the (permanent) groundwater layer the sand is always water-saturated, but moderate oxygen tensions and often brackish salinities lead to a reduced meiofauna in both diversity and abundance.

d) In the "low oxygen stratum", oxygen deficiency can extend into considerable depth, and, in beaches with a high content in organic matter, can become a zone of reduced conditions harbouring a thiobios (see Chap. 8.5).

Variations of this general four-strata-pattern with respect to changes in tidal phase, grain size, slope of the beach profile, freshwater influx, high temperature and organic contents will modify the picture. In exposed beaches a swash zone ("zone of resurgence", SALVAT 1964), where the interstitial water intensively circulates and infiltrates, will be characterized by a very poor and specialized interstitial fauna ("otoplanid zone" of REMANE 1940). It will displace the low

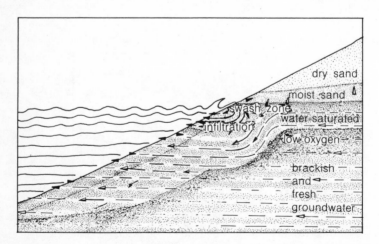

Fig. 5. Stratification of a beach profile related to water content. (Compiled from various authors)

oxygen stratum, if it occurs, to deeper layers. In general, however, the zonation derived from South African beaches, as suggested by McLachlan (1980), is mostly applicable.

In completely saturated sublittoral sediments, meiofauna communities may be structured by lack of pore water-flow and resulting poor supplies in oxygen and food particles.

Usually, moderately well-sorted medium sand harbours the most diverse meiofauna. In coarser sand, the species number remains relatively high but population density may decrease. Muddy sediments are often characterized by rich populations of a limited number of species restricted to the surfacer layer. In general, the correlation between sediment structure and meiofauna distribution is strong enough, particularly in littoral areas, to dominate all other factors. It often relates directly to dominance and diversity of meiofauna.

More detailed reading: Gray (1981, German edition 1984); Buchanan (1984); Giere et al. (1988a)

2.1.3 Physico-Chemical Characteristics

2.1.3.1 Temperature

Meiofauna is present in polar ice and tropical shores, around hot hydrothermal vents and under the rigid temperature fluctuations of supralittoral fringes. It seems that in most environments temperature does not prevent meiofauna existence. Nevertheless, temperature can have a structuring impact on meiofauna, particularly in exposed tidal flats with a steep vertical thermal gradient. In sublittoral bottoms, the influence of temperature on meiofauna distribution is negligible.

The steepness of the temperature gradient is primarily related to permeability (see Chap. 2.1.2.3). In little-permeable water-saturated mud flats, surface and depth layers can vary widely in temperature, particularly at ebb-tide. While summer temperatures can rise to $> 40\,°C$ at the surface, simultaneous measurements in the depth can record only $10–15\,°C$ due to a strong vertical dampening. In winter time, even under a thick ice cover, the frozen ground at the surface does not extend beyond the uppermost 5 cm and allows for tolerable conditions beyond this depth (Fig. 6).

This dampening effect, particularly evident after calculation of monthly ranges (Table 3) is important for the existence of meiofauna in climatically harsh biotopes and often causes vertical migrations of the more sensitive species if other conditions like oxygen supply are favorable.

Temperature can be conveniently and routinely measured by a variety of pointed semiconductor probes connected with electronic (field) instruments. Since only the narrow surface of the thin metal probes is temperature-sensitive, in situ measuring, even at considerable penetration depth, is possible without much compaction or displacement of the sediment.

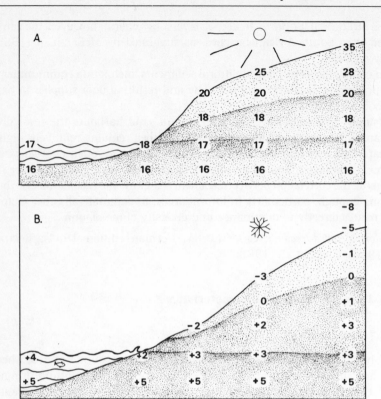

Fig. 6A, B. A typical temperature distribution in a boreal beach. **A** Summer aspect. **B** Winter aspect. (After Jansson 1966a)

Table 3. Monthly ranges of temperatures in 1964, measured at various sediment depths of a Scandinavian beach (minus sign: when minimum temperature was below zero and numerically higher than maximum temperature. Jansson 1967a)

Depth	March	April	May	July	October
Air	−9.0	19.9	19.7	18.8	14.5
Surface	−13.0	34.2	38.4	33.5	24.1
2 cm	−8.7	26.0	33.0	24.0	12.8
10 cm	−4.1	9.8	11.2	9.9	4.2
25 cm	−1.0	2.8	3.4	–	–
70 cm	−0.5	0.8	0.5	0.6	1.0

2.1.3.2 Salinity

As with temperature, meiofaunal organisms exist under all salinity regimes from freshwater to brackish shores, from oceanic bottoms to brine seep areas. Because many species are able to adapt to a wide range of salinities, there is often a diverse meiofauna even in those critical brackish water zones where

REMANE (1934), mainly for macrofauna, described a minimum in species number. In contrast, irregular salinity gradients can massively determine species composition and occurrence of meiofauna. In tidal shore lines, a steep vertical and horizontal salinity gradient is often developed depending, as with temperature, on water permeability of the sediment. In muds with their water-saturated fine sediments and much reduced vertical water exchange, surface salinity due to evaporation at ebb-tide can rise up to hypersaline conditions. After strong rain falls, it can suddenly drop to almost freshwater. ALONGI (1990b) reports a drastic decline in meiofauna abundance after tropical monsoonal rains. The effects of these fluctuations are greatest in the uppermost centimetre. Already at a depth of 2–5 cm, salinity (as well as temperature) fluctuations are much dampened and often remain amazingly constant. This stability offers the mobile meiofauna a favourable refuge zone. At the highwater line of a North Sea mud flat, salinities up to 40‰ have been measured on a warm summer day, while in 5-cm depth the usual 30‰ have not been exceeded. At depths approaching the level of groundwater, salinity often drops to brackish-water conditions (about 20‰). In contrast, on coarser, well-drained sandy beaches with a high permeability, precipitation can affect salinity to a depth of 30 cm (REID 1932–33), creating unfavourable conditions to meiofauna.

Consequently, in sandy beaches the drainage system is complicated by effects of both precipitation and groundwater currents. Even in fully marine shores without any direct freshwater influx at the surface, groundwater often has a markedly reduced salinity depending on local geological, climatic and geographical conditions. In sublittoral bottoms the salinity is less fluctuating and is mostly identical with that of the overlying water and, thereby, hardly a limiting factor for meiobenthic populations.

In many meiobenthic studies, differentiated salinity gradients are difficult to record because the amount of available water is often restricted. Two methods, microtitration and electrical conductivity measurements, are often used. In conductivity electrodes, an electric current, applied between two platinum foils, will be modulated in relation to the conductivity of the pore water. After calibration, this can be used to determine the salinity of the water. Since the conductivity depends on the distance between the Pt surfaces, each electrode has its particular geometrical pattern and must be calibrated against a "normal electrode" with the factor 1. In sediment with little moisture, conductivity measurements may be misleading since, for a correct reading, the Pt surface must be completely covered by a water film.

Salinity of sediment water can be also determined by an optical refractometer. Special salinity refractometers have a scale converted and calibrated for direct readings of salinity from only one drop of water extracted from the sediment. The precision of these instruments (± 1‰ S) is usually sufficient for ecological studies of meiobenthos. As with all methods for salinity measurement, the precision of both conductive and refractometrical methods suffers in brackish water because the altered ion composition will cause deviations from their constant relation which, in pure ocean water, classically defines salinity measurements.

2.1.3.3 pH Value

In the marine biome, pH plays only a minor role for meiobenthos, since the slightly alkaline seawater (pH 7.5–8.5) is well buffered against pH-fluctuations. In anoxic, hydrogen sulphide-containing sediments, the pH of pore water can fall below 7, but will seldom drop below 6.0. On the other hand, in the surface layers of tidal flats, intensive assimilation of the dense cover of microphytobenthos can cause a rise of pore water pH to > 9. In tropical tidal flats, a pH up to 10 has been measured. Here, high pH in combination with other stress factors such as extreme temperatures and salinity can be detrimental. Isolated from other factors, a noxious impact of extreme pH can be proven only experimentally. In the field, it is concealed by the synergistic effect of other environmental factors (GIERE 1977).

In natural freshwater biotopes, extremes of pH can occur in mires and limestone waters. Here, organical (eutrophication) and chemical pollution often cause drastic fluctuations of the pH level. However, because of the buffering capacity of the sediment, fluctuations in the pore water are often dampened compared with the overlying water. Thus, a negative impact of pH on freshwater meiofauna has rarely been demonstrated (PENNAK 1988). Measurement of pH in situ is unproblematical when using glass insertion electrodes with the reference electrode combined in the same shaft. Usually, the recording is done parallel to redox potential measurements with a mV-meter calibrated for pH. The construction of modern electrodes even prevents clogging of the diaphragm. Correct readings require that the temperature of the electrode's internal filling equals that of the ambient water or sediment to be measured, since equilibration of temperature persists for a considerable time.

2.1.3.4 An Interacting Complex: Redox Potential Oxygen and Hydrogen Sulphide

The overall reducing or oxidizing capacity of a sediment can be measured as its redox potential and mirrors, to a certain extent, the oxygen conditions in the pore water. The electrical potential between a small outer platinum ring or wire and the internal reference electrode recorded with a mV-meter after immersion into the water or moist sediment depends largely on the type of internal reference. This will differ in its composition (a silver/silverchloride system is the most common reference), and one needs correction factors to relate the actual measurement to the "Eh-value" of the "normal hydrogen electrode" which is conventionally quoted in literature.

In the field one can encounter values between $+ 550$ to $- 300\,mV$. In exposed sandy bottoms one will find positive values throughout. Frequently, there occurs a wide vertical transition zone between positive and negative values, indicating a broad redox potential discontinuity layer (RPD layer). In most sandy shores, this layer will be encountered underneath an oxidized surface zone of some depth (some centimetres), while in muds the RPD zone usually is sharply delimited closely underneath the surface (1–2 mm). Depending on the diurnal temperature curve, but also on light conditions influencing the assimilation activity of surficial diatoms, the RPD zone can periodically change its position.

In sediments rich in iron ions, there is a marked change in sediment colour caused by iron compounds. In the brownish to yellowish surface layers, fully oxidized ferric iron occurs, while the black sediment in the depth indicates the presence of iron sulphides formed under sulphidic and anoxic conditions. However, in many calcareous sediments, this optical indicator of changing oxidation status may be absent due to the low iron content in the sediment.

The vertical change in sediment colour has usually been related to changes in the redox potential of the sediment. It was generally accepted that the shift from positive to negative redox values, the redox discontinuity (RPD), would then coincide with the transition from the bright surface to the dark deeper layer. It was taken without detailed measurements to refer to the threshold between oxic and anoxic sediments. However, detailed Eh measurements have shown that the change from light (yellowish) to dark in fact mostly indicates the transformation of ferric iron to ferrous iron and does not refer to the RPD (the redox potential can still be +125 mV). Only in the deeply black layers, stained by iron sulphides, will the sediment be reducing. This means that the shift in sediment colour from bright to dark is not eo ipso an indicator for the change from oxic to reduced (and sulphidic) conditions (SIKORA and SIKORA 1982).

Only in those cases where the transition-layer is narrow and the change from bright to deeply black rather abrupt (in muds or in freshly stratified sediments after disturbances), may the colour change, by ecological standards, coincide with the redox discontinuity and allow for conclusions on the animals' supply with oxygen. In other words, in sediments with a diffuse redox gradient and a wider transition zone from bright over grey to black, the gradual shift from a bright to a dark coloration simply indicates the zone where free oxygen disappears, but oxidized compounds still prevail.

For a long time it was considered a general and practical rule for interpretation of Eh values that values $> +100$ mV indicate the presence of oxic pore water in the sediment, and below -100 mV its absence. Today, oxygen microelectrodes have proven that often free oxygen can be absent from sediments with redox potentials as high as $+300$ mV and more. A differentiation between oxic (i.e. with free dissolved oxygen available to animals) and oxidized sediments (with compounds in an oxidized chemical state) is of prime importance for correct ecological interpretations (SIKORA and SIKORA 1982; REVSBECH and JØRGENSEN 1986; JØRGENSEN 1988). Often the upper layers of a sediment are oxidized, but not oxic (Fig. 7).

Considering the earlier difficulties of measuring oxygen directly in the sediment (see below), redox values have been one of the most frequently recorded environmental parameters in meiobenthos studies. However, the results mentioned above emphasize that the measurement of redox potentials, used as a direct surrogate for the oxic situation, becomes deceptive: all the electron

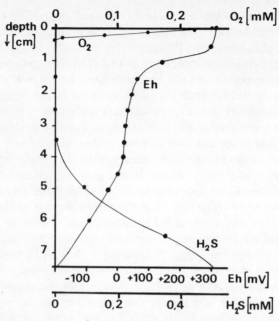

Fig. 7. Comparison of redox potential, oxygen and hydrogen sulphide concentration in a sediment profile. (After REVSBECH and JØRGENSEN 1986)

transport systems occurring in the sediment are integrated in the redox potential reading and all redox couples in addition to free oxygen will influence the measurement.

Eh measurements and their interpretation become even more complicated through the low reproducibility of replicate recordings from directly adjacent spots. This problem is caused mostly by microchambers of decaying organic matter, entrapped air bubbles or animal tubes encountered by the pin-pointed electrode. Moreover, the response of the electrode depends to a high degree on the properties of the platinum surface. Thus, in its empirical application, Eh should be taken as an operational parameter of integrative character only. A detailed description of the limitations and possible sources of significant error in Eh recordings is given by WHITFIELD (1974) and JØRGENSEN (1988). Even more than for other parameters recorded electrometrically, redox potential recordings require a basic understanding of electrometry and some information about flaws and errors inherent in electrodes in order to avoid bias and misinterpretation.

Oxygen is the predominant factor among the abiotic parameters determining the habitat conditions and presence of meiofauna. Most meiobenthic organisms have relatively large surface areas and high oxygen demands; only specialists will prefer hypoxic conditions (see Chap. 8.5). Thus, the distribution of meiofauna communities can be correlated to the oxygen supply of the pore water.

Recently, microelectrodes showed a much differentiated pattern of oxic conditions in the sediment. The microgradients of oxygen available to meiobenthic animals seem, in fact, extremely steep. Changes from fully oxic to anoxic conditions occur frequently in the mm range. If the oxygen supply via bioturbation/ irrigational fluxes ceases, a rapid oxygen depletion by bacterial consumption soon seems to cause a change to hypoxic and anoxic conditions. In sandy bottoms, a very narrow transition zone (in mm) from oxidized to anoxic layers exists. Here, microoxic conditions can overlap with hydrogen sulphide (MEYERS et al. 1988). In mud or algal mats, even steeper oxic-sulphidic gradients have been recorded (REVSBECH and JØRGENSEN 1986; VISSCHER et al. 1991). It is around this threshold zone that one must differentiate between oxic and oxidized sediments (see above).

The in situ assessment of oxygen stratification directly in the pore water system became possible in the last decade through very thin oxygen micro-electrodes introduced into ecology mainly by REVSBECH and his group (see REVSBECH and WARD 1983; REVSBECH and JØRGENSEN 1986). In a coastal sandy sediment they found that the oxic zone was only 2 mm thick, while an oxidized sediment layer with a positive redox potential extended to 3.5 cm (Fig. 8 in REVSBECH and JØRGENSEN 1986). In North Sea tidal flats, diffusion of oxygen from *Arenicola* burrows extended only about 1 mm into the surrounding fine sand (unpubl. data). This observation became supported by more measurements from other biotopes using microelectrometrical techniques (e.g. ARCHER and DEVOL 1992; FÖRSTER and GRAF 1992) and led to the principally new conception of oxygen conditions in the benthic environment pointed out above. REVSBECH et al. (1980) inferred already that "the meiofauna and bacteria that live below the upper few millimetres must obtain their energy from anaerobic metabolism." This was corroborated by similar results of SIKORA and SIKORA (1982).

On the other hand, it is also possible that in a microoxic environment above the RPD threshold, low oxygen concentrations are still present. These small amounts would be available to microbes and also to meiobenthos. These organisms are perfectly adapted to instaneous oxygen uptake and consume any residual oxygen intensively and immediately. So it would not be detected by the relatively insensitive electrodes (WATLING 1991).

In contrast to the usual polarographic oxygen electrodes, microelectrodes do not require any water flow and, if gold-coated, the presence of dissolved sulphide does not tarnish the minute measuring surface. Protected from abrasion by a layer of semipermeable silicon rubber and provided with a sturdier glass shaft, field versions have been recently designed for use in muds and sand (Fig. 8; REVSBECH and WARD 1983). By ensheathing the electrode in a thin stainless steel shaft (HELDER and BAKKER 1985; VISSCHER et al. 1991), the risk of breaking during field use is further reduced.

Measurements of pore water obtained with syringes or by centrifugation of the fractionated sediment core (see Chap. 3.3) can never yield a detailed picture, whether a micro-Winkler titration (BRYAN et al. 1976; PECK and UGLOW 1990) or a polarographic electrode is used. The drainage area of the pore water

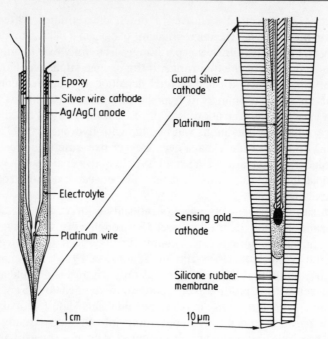

Fig. 8. Diagram of an oxygen microelectrode with a sturdy glas shaft. (After REVSBECH and WARD 1983)

is not clearly reproducible. Detailed studies with microelectrodes are the only way to resolve the little understood microscale pattern of oxygen in the sediment and its relation to anoxic and sulphidic (see below) conditions. Considering the ecological potential of this terrain for small benthos, microelectrometrical investigations are urgently needed.

A field method related to the polarographic recording of oxygen and much in use, mainly in the 1960s and early 1970s, in meiobenthological studies is the measurement of the oxygen diffusion rate (ODR). It is derived from soil biology and was introduced into meiobenthology by JANSSON (1966a). If a low voltage is applied to a Pt wire, it functions as a cathode reducing the oxygen molecules dissolved in the pore water. The resulting current (in µA) is directly proportional to the amount of oxygen molecules diffusing onto the surface of the wire. Rather than measuring the absolute concentration of oxygen, this signal was considered to reflect the availability of oxygen. The Pt wire would correspond to an oxygen-consuming meiobenthic animal whose available oxygen also depends on water permeability, entrapped air bubbles, temperature gradients etc. Because the same negative and positive factors which influence the RPD reading are relevant also for an animal's oxygen supply in the sediment, this method was held to be fairly "ecological". Regrettably, it is known that oxygen diffusion depends not only on temperature, but is highly variable with sediment water currents. This is the basis for a possible use of this method as a microcurrent meter (see Chap. 2.1.2.3). Hence, peaks in ODR recordings on beaches may indicate water currents caused by incoming waves as well as true pulses of available oxygen, thus leaving a reliable interpretation arbitrarily. Today, with the introduction of microelectrodes for direct oxygen measurements, the ODR recording has lost its former acceptance.

Beside oxygen, hydrogen sulphide is perhaps the most relevant environmental parameter in many benthic habitats. Blocking (reversibly) the cytochrome C oxidase of the intracellular respiratory chain, hydrogen sulphide is toxic for animals already in a nanomolar to micromolar concentration and can profoundly determine the distribution of the benthos. Especially in marine biotopes, H_2S can develop in high concentrations through microbial reduction of sulphate, which is abundant in seawater. Recent papers emphasize that even in the presence of (low) oxygen, sulphate reducers remain active, producing sulphide (JØRGENSEN 1988; FUKUI and TAKII 1990; JØRGENSEN and BAK 1991). Under normal marine pH conditions, hydrogen sulphide is predominantly dissolved as HS^- ion. Only in more reducing habitats with their slightly acidic pH, do the particularly toxic undissociated H_2S molecules prevail.

Because of the chemical balance between iron and sulphide ions, in temperate regions a considerable part of the dissolved sulphide will precipitate as iron sulphides, forming a black layer in the depth of the sediment. Particularly in the warm season, the deeper layers of muddy intertidal flats rich in organic matter develop high hydrogen sulphide concentrations sometimes exceeding 1 mmol l^{-1} (own unpubl. data; REY et al. 1992). In polluted sediments, dissolved hydrogen sulphide has an indirectly important impact in fixing the toxic heavy metals such as cadmium as solid precipitates. In calcareous sediments of warm-water regions with their low iron content this precipitation process is limited. Here, high concentrations of toxic dissolved hydrogen sulphide can develop without any blackening of the sediment.

In freshwater biotopes, periods of anoxia rarely give rise to comparably high concentrations of H_2S. Here, sulphate ions are rare and sulphide is mostly derived from the organically bound protein sulphur.

Despite its ecological relevance, hydrogen sulphide, as a decisive environmental parameter, has been grossly ignored both in field work and in experiments. Before the discovery of a thiobios (Chap. 8.5), hydrogen sulphide was not given appropriate attention in meiobenthic studies. In part, this may be caused by the widely reported prejudice that H_2S-smelling sediments were simply held to be azoic. Partly, the difficulty in measuring this labile substance quantitatively may have contributed to its neglectance.

For stating the presence of even low concentrations of hydrogen sulphide, smell is a good criterion because our nose senses concentrations of hydrogen sulphide down to $0.1 \mu m$ H_2S (DANDO et al. 1985). An illustrative semi-quantitative in situ method is provided by the "dipstick sampler for sulphide" designed by REEBURGH and ERICKSON (1982): a polyamide gel impregnated with lead acetate is mounted in a furrow of a measuring stick which is then pushed into the sediment. Any dissolved sulphide will precipitate as lead sulphide, blackening the gel in exactly those positions where hydrogen sulphide was present, thus indicating its distribution. After complete dissolution of the sulphide precipitate in the gel, the hydrogen sulphide concentration can be measured photometrically.

Today, hydrogen sulphide in sediment pore water is usually measured spectrophotometrically after addition of methylene blue (CLINE 1969; GILBOA-GARBER 1971). There are some modifications of this method which mostly concern the calibration and the range of concentration. Pore water is obtained by suction corers (see Chap. 3.3.1) and injected immediately after extraction into the prepared vials filled with alkaline lead acetate which is dissolved later in the the laboratory. Although somewhat cumbersome, this "pore water method" with subsequent photometrical analysis remains the most versatile and frequently used method to measure hydrogen sulphide.

As with oxygen, the stratification pattern of hydrogen sulphide in the sediment column is extremely complex and consists of steep gradients. In order to obtain a realistic picture, microelectrode measurements are required. Several companies offer a silver sulphide electrode combined with a reference system of calomel or Ag/AgJ for measuring hydrogen sulphide. The recorded voltage (mV) is directly proportional to the logarithm of S^{2-} partial pressure. Therefore, it is essential to convert all hydrogen sulphide in its various pH-depending stages of dissociation to S^{2-} or to calibrate the electrode in relation to the ambient pH. Particularly the microversions of this electrode, designed originally by REVSBECH and WARD (1983), are of interest for use as insertion electrodes in the field. The early constructions have suffered from technical problems (mechanical fragility and abrasion of the silver sulphide coating). Mounting sulphide micro-electrodes in a steel injection needle of a medical syringe, recent versions became more sturdy and sufficiently protected for use in sandy substrates even (VAN GEMERDEN et al. 1989). A particularly promising tool for simultaneous recordings of the closely interacting chemical couple oxygen and hydrogen sulphide is the combination of these two electrodes within one thin metal needle, as suggested by VISSCHER et al. (1991).

2.1.3.5 Pollutants

Today, the natural environment of meiofauna in wide parts of the world is altered by anthropogenic factors, especially by pollution, which rightly must be considered a relevant environmental factor for meiobenthos. The numerous chemical agents and the locally diverse situation of pollution render it impossible to consider details of their contamination pattern and relevance for the meiofauna. Instead, some features of general importance for meiofauna studies will be mentioned here.

Due to physical adsorption and chemical bonds, pollutants become highly enriched in sediments compared to the overlying water. Once sorbed in the sediment, noxious substances are only slowly released into the water column. Since lack of oxygen in the deeper layers of many bottoms leads to an extremely retarded oxidative decomposition (e.g. petroleum hydrocarbons, see Chap. 8.6.3), sediments often represent a sink for pollutants. This can cause long-lasting or chronic contamination of the benthos.

2.1.4 Conclusions

In the natural environment, all factors described in this chapter separately will interact with each other in many ways, often synergistically enhancing the effects. This interaction of intermingled factors creates a complex system that can be termed the abiotic environment of meiobenthic animals. It is governed by grain size composition as a key factor which, in turn, depends on exposure to currents and waves. Meiofauna certainly reacts in a variety of ways to the various abiotic parameters; however, the causative relationships to single factors can only be rarely demonstrated (mostly by experiments). Unfortunately, experimental results are easily misinterpreted, since a monofactorial experiment hardly reflects the normal response within a multifactorial network. In addition, in the field, the numerous abiotic factors are constantly interacting with the equally complex system of biotic factors (see following section and Fig. 12).

More detailed reading: REVSBECH and JØRGENSEN (1986); JØRGENSEN and BAK (1991); WATLING (1991)

2.2 Biotic Habitat Factors

While studies of the abiotic factor complex dominated the 1960s and early 1970s, meiobenthic studies focussing on biotic factors have increased since about 1975. Today, there is profound knowledge that biogenic factors and bioturbating microprocesses have a massive influence on meiofauna colonization and population structure via sediment-mediated interactions, whether of positive or negative influence on meiofauna (WOODIN and JACKSON 1979). It is realized that biotic factors are most relevant for the understanding not only of meiobenthic assemblages, but for the benthic habitat in general. WATLING (1991) even pointed out that the classical methods of sediment analysis, derived from geology, give a misleading, denaturated and biologically rather irrelevant picture of the sedimentary habitat. Only by consideration of the detrital input, bacterial colonization, and pore-water chemistry can we achieve a natural conception of the real world of benthic animals. This is not characterized by the mineral particles but by the delicate, flocculent organic matter which binds them into an interwoven fabric. Only in extreme biotopes such as the swash zone of exposed beaches will abiotic factors predominate (HOCKIN 1982b); but even in the wrack zone of tidal flats, trophic or predator-prey relationships are of significant importance (GIERE 1973; REISE 1987b). Biotic factors are problematical to quantify and, hence, difficult to measure and interpret. This probably accounts for their being much less investigated than abiotic parameters. The result of such investigations is a multitude of often incoherent data whose interpretation is often controversial and difficult to incorporate into meaningful generalizations. This has been the case in the better-studied macrobenthos (see review by WILSON 1991), and is more valid so for meiobenthos.

The array of biotic factors influencing the meiobenthic habitat extends from organic content and trophic aspects to biogenic structures, and includes interactive factors such as disturbance, competition, and predation (Figs. 11, 12). Many of these factors not only have a structuring impact on the habitat, they also affect the mutual interactions of inhabitants. In this section, they will be described only in terms of their biotopical influence and methods of measuring. All other aspects of this subject will be discussed in Chapter 9.

2.2.1 Dissolved Organic Matter (DOM)

Dissolved organics are often actively absorbed by meiofauna via transepidermal uptake. This trophic pathway is particularly relevant in soft-bodied, small animals with their relatively large surface (see reviews by JØRGENSEN 1976; SEPERS 1977; STEWART 1979; FERGUSON 1982; GOMME 1982; STEPHENS 1982; especially for meiobenthos: TEMPEL and WESTHEIDE 1980). Because of transepidermal uptake, sediments with high DOM concentrations are favoured by meiobenthos, primarily by the soft-bodied groups (e.g. ciliates, turbellarians and annelids).

Sugars (glucose, galactose, sucrose) and free amino acids (aspartic acid, glutamic acid, β-glutaric acid, alanine) are principal organic molecules which become highly enriched in the pore water. They originate from bacterial decomposition, leaching of decaying plant material or exsudation from viable bacteria and plants. Concentrations of these substances are often one to two orders of magnitude higher in pore water than in the overlying water, especially in neritic coastal regions more than in deeper lying oceanic sediments (Table 4).

DFAA are particularly enriched in the upper 0-2 cm and usually vanish in sediment strata below 10 cm. Because of the enrichment phenomenon, dissolved

Table 4. Concentrations of dissolved free amino acids (DFAA, JØRGENSEN et al. 1980) and total organic carbon (TOC, FARKE and RIEMANN 1980)

DFAA (μmol l^{-1})	Oceanic region	Neritic region
Pore water	12–50	50–225 (in mud)
		6–12 (in sand)
Open water	0.5–1.0	3.0
TOC (mg l^{-1})		
Pore water		20–100 (max. 390)
Open water	0.5–2.0	

organic matter (DOM) is released from the sediment into the water column. This usually slow diffusion process is accelerated by increased hydrodynamic forces (storms, currents and waves) but also by bioturbation (CULLEN 1973) and sediment disturbance by meio- and macrobenthic animals (ALLER and ALLER 1992). Extrapolations suggest that in tidal flats the upper 10 cm of the sediment will become completely bioturbated once within 3 years; even from 50-cm depth sediment particles will be transferred to the surface. Experiments by CULLEN (1973) underline the bioturbative impact of meiofauna: by the burrowing activity of meiofauna alone all the surface traces of macrofauna ("Lebensspuren") were eliminated within 14 days. Through the concomitant exchange processes, activities of benthic animals can enhance by a factor of 10, the physical diffusion of sediment-bound substances (GRAF pers. comm.), thus counteracting the adsorptive and accumulating sedimentary processes mentioned above (see also Hylleberg and Henriksen 1980; Aller and Yingst 1985; Kristensen et al. 1985). Recent studies have shown that also under anoxic conditions mineralization processes are considerable and may even exceed the aerobic degradation (Kristensen and Blackburn 1987; Hansen and Blackburn 1992).

Additionally, the presence of animal tubes and burrows has been shown to increase the pore water circulation and the flux of dissolved substances (ALLER 1980; JØRGENSEN et al. 1980; see Chap. 2.1.2.3).

Why is the concentration of DOM in the sediment so high compared to the open water despite these releasing processes? Animals of a larger biomass populate the bottom more densely than the pelagos. Deposition of degradable detritus is higher in the sediment and decomposition generally lower due to the frequent lack of oxygen. Hence, the sediment particles with their relatively large surface area and considerable adsorptive forces "retain" DOM in the pore water of the sediment. "New" sand grains with sharp edges have been found to adsorb more glucose than "old" ones with a "smoothed" surface (MEYER-REIL et al. 1978). Habitats with high DOM concentrations are often indirectly attractive for the meiobenthos, since DOM is intensively utilized by bacteria (MORIARTY 1980) which, in turn, are utilized as food by many species of meiobenthos (see Chap. 7).

More detailed reading: HYLLEBERG and HENRIKSEN (1980); STEPHENS (1982); HINES and JONES (1985); ALLER and ALLER (1992)

2.2.2 Particulate Organic Matter

Both dead organic matter (detritus) and living organic substances (bacteria, microalgae, other animals) represent important environmental factors structuring the habitat of meiobenthic animals. MARE (1942) stated already that the content of organic matter has a significant influence on the distribution of meiofauna, a fact which has been confirmed by many authors since (LEE 1980a; TENORE and RICE 1980; TIETJEN 1980; WARWICK 1989).

In silty muds, the dry weight of the organic particles can come up to 10% of a sample while in sandy shores this value is often < 1%. The ash-free dry weight (by combustion, at about 500 °C for 1–2 h) of the dried sample is the most generally used method for grossly measuring total organic matter. Care has to be taken that the combustion temperature does not exceed 580 °C or volatilization of sediment carbonates occurs and gives incorrect data. A more accurate discrimination between the different components of organic matter (e.g. organic carbon, proteins, lipids, carbohydrates) requires special methods (HOLME and MCINTYRE 1984; GREISER and FAUBEL 1988). One of these is the determination of organic carbon using titration to measure the reduction activity of organic matter. A problem inherent in the assessment of organic content and relevant for analysis of meiofauna distribution is the inclusion of the animals themselves in the bulk measurement of organic carbon. This is prone to cause a biased conclusion regarding the microbial and trophic potential of the sediment. A separation of the live fauna from the detrital parameters is possible by measurement of the live organism's ATP content. The extraction procedure of ATP from sediment samples by KARL and LA ROCK (1975) uses the sensitive reaction of the luciferine-test for oxidizing substances. However, the ATP content varies with life conditions and ontogenetic events of organisms. Consequently, the procedure must be carefully calibrated and the data suitably replicated. As an appropriate conversion of ATP content to weight in (nematode) biomass studies, GOERKE and ERNST (1975) measured an average ATP concentration of 1.35 mg ATP g^{-1} wet wt of meiobenthos. The difficulty involved in many of these methods is not so much the technical procedure or sensitivity of the measurement, but the correct interpretation of the data obtained.

2.2.2.1 Mucus and Exopolymer Secretions

Recently, it has been shown that bacterial exopolymer secretions attain an important role in the exchange of nutrients, fixation in biofilms, and transport of aggregates via biological flocculation. Mucus excreted by microorganisms and covering degrading detritus may gain considerable nutritional importance for meiofauna due to its rich content of amino acids and labile nitrogen compounds (BIDDANDA and RIEMANN 1992). For allogromiid foraminiferans from tidal flats, it could be shown experimentally that mucous biofilms represent a substantial trophic resource (BERNHARD and BOWSER 1992). The development of sophisticated methods revealed that these particulate substances, also excreted by diatoms, form "an extensive matrix of amorphous organic material which may provide the bulk of carbon sources for many benthic organisms" (DECHO 1990). These substances, when patchily distributed over a cm-scale, can significantly affect exchange of dissolved nutrients, and are not only of direct trophic relevance for meiofauna but also of microdistributional importance for meiofauna populations. Today, it is conceivable that the study of mucus secretions and their ecological role will gain considerable relevance in future (meio-) benthological research and should be much promoted despite the inherent technical problems.

2.2.2.2 Detritus and Bacteria

The content of organic debris and bacterial biomass are closely linked (JØRGENSEN et al. 1981). Detritus, especially through its rich coating of bacteria, can entail high concentrations of meiofauna. The oxic/anoxic interface of decomposing detrital particles harbours huge populations of "sulphur bacteria" and is of considerable importance to meiofauna (YINGST and RHOADS 1979). This probably refers also the rich colonization of bacteria in and on faecal pellets produced by meio- and macrobenthos. Faecal pellets represent reduced microchambers with a steep gradient system favourable for the growth of (sulphate oxidizing)bacteria. MEADOWS and TAIT (1985) found bacterial numbers in faecal pellets several orders of magnitude higher than in the surrounding deep-sea sediment.

With the ageing of plant debris, the bacterial colonization grows and the protein content of a sample increases. This, in turn, makes it more attractive to meiofauna (WARWICK 1989). In lenitic coastal sediments, the rich detritus/ bacteria complex is important for meiofauna, especially for nematodes and other components of the "detrital/bacterial-based food chain". In contrast, the "microalgal-based" harpacticoids are less dependent on these factors (MONTAGNA et al. 1989). The larger meiofauna populations in depressions and troughs of sand ripples correspond to an increased detrital content in these microhabitats (HOGUE and MILLER 1981; HICKS 1989). Rottening animal tissues can attract meiofauna (nematodes) in the sediment (GERLACH 1977a; OLAFSSON 1992).

Much of the detritus found in sediment samples is derived from plankton organisms. Phytoplankton blooms result in the deposit of considerable phyto-detritus on the sea floor, both in coastal and deep-sea bottoms. These organic deposits enhance the bacterial activity after relatively short time periods (a few weeks to months) and, subsequently, can cause a significant increase in meiofauna abundance and diversity (THIEL et al. 1988/89; FLEEGER et al. 1989; LAMBSHEAD and GOODAY 1990). In Kiel Bight, benthic metabolization of a phytoplankton bloom was completed after only 3 weeks (GRAF et al. 1982). Coastal sediments represent a huge filtering system (RIEDL 1971; MCLACHLAN 1989) trapping large amounts of plankton. In shallow, well-illuminated bottoms, also decaying benthic macroalgae and sea grass provide an ample source of detritus known to promote abundant meiofauna populations (NOVAK 1989; BLANCHARD 1991). In return, meiofauna has been shown to stimulate decomposition of plant litter (ALKEMADE et al. 1992).

Experimental observations indicate that detritus is not indiscriminately ingested by meiobenthic animals: brown algae are preferred over red algae (GIERE 1975; RIEPER-KIRCHNER 1989). The differing origin, the multitude of stimulating and inhibiting substances contained in detritus, and the diversity of degrading processes and bacterial assemblages on the different substrates led TENORE et al. (1982) to use the discriminative term "available" detritus.

In the field, the impact of detritus as a habitat factor can hardly be distinguished from that of its decomposing bacteria. In experiments it has been demonstrated that the "detritivorous" meiofauna mostly utilizes the bacterial films and not the detrital substrate (FENCHEL 1969, 1970; HARGRAVE 1972; MEYER-REIL and FAUBEL 1980).

A bacterial zonation established in laboratory sediment tanks was mirrored by meiofauna distribution (mainly nematodes, BOUCHER and CHAMROUX 1976). Discrimination between various bacteria and microfungi and selection of certain groups as food for meiofauna has been repeatedly shown experimentally, e.g. for ciliates (FENCHEL 1969), harpacticoids (CARMAN and THISTLE 1985), oligochaetes (CHUA and BRINKHURST 1973; DASH and CRAGG 1972), and polychaetes (GRAY 1966, 1971). In the field, this would lead to bacterial food niches preferred by meiofauna on the basis of varying structural, chemical or physiological details of the microbes (LEE 1980a; CARMAN and THISTLE 1985). In some cases, the functional correlation between the structure of the mouth parts or buccal armatures and shape of the bacteria could be demonstrated in more detail (WIESER 1959, 1960; JENSEN 1983, 1987a, for nematodes; MARCOTTE 1984, 1986, for harpacticoids).

Assessment of the community structure and abundance of bacterial microorganisms is problematical and not without serious methodological flaws. Estimates of bacterial number or biomass on the basis of the classical cultivation method yields severely biased results. Quantification using antibody methods is probably the most efficient modern method. Also direct counting of stained cells on the particle surface (fluorescence microscopy), although tedious, remains one of the more reliable procedures (DEFLAUN and MAYER 1983). MORIARTY (1980) recommends the determination of muramic acid as a good basis for calculations since this substance is a cell-wall component of almost all prokaryotes. Another indirect method to quantify sediment bacteria is calculation of their biomass by phospholipid analysis (FINDLAY et al. 1989). Experiments with fluorescent dyes (e.g. crystal violet) might help to differentiate between bacteria as food sources (HED 1977). Depending on evaluation methods, on the local microtopography and physiography, and on water quality and climate, the number of sediment bacteria can vary considerably (Table 5).

Usually, the bacterial density in sediments corresponds to the amount of organic matter and is reciprocally related to the degree of exposure. Muddy bottoms and sea grass beds are microbially richer than sand, just as the wrack zone of a beach is richer than its surf zone or its sublittoral bottom. MORIARTY (1980) found five times the amount of bacteria in sea grass beds than in the adjacent open sediment. A rich supply of detritus and oxygen, for most microbes, makes the surface layers of a bottom more attractive than groundwater layers or the anoxic depths. At the surface of sea grass beds, 18% of all organic substances measured were contributed by live bacteria (MORIARTY 1980), a significant nutritive amount even for larger animals. On the other hand, bacteria tend to concentrate in chemical gradients: around the redox-chemocline or in sediments exposed to irrigational fluxes of infauna (FENCHEL and RIEDL 1970; YINGST and ROADS 1979).

Table 5. Abundance of bacteria in various marine sediments

Habitat	Bacteria per g sed. (dry wt)	Reference	Bacteria per cm^3 sed.	Reference
Beach sand	9.8×10^6	WESTHEIDE (1968)	50×10^3 to 500×10^3	BOADEN (1964)
	140×10^6 to 1.1×10^9	ANDERSON and MEADOWS (1969)	14×10^6	WESTHEIDE (1968)
	4 to 28×10^8	MEYER-REIL et al. (1978)		
Mud to fine sand				
Eulittoral, tidal flats	20×10^9	MEYER-REIL (1987)	7.5×10^8	EPSTEIN and SHIARIS (1992)
Tidal flats	1×10^9	FINDLAY et al. (1989)		
Continental Shelf slope			10 to 45×10^6	VANREUSEL, pers. comm.
Deep sea			2–3×10^8	THIEL et al. (1988/89)
Deep sea			0.1 to 3×10^7	TIETJEN (1992)

A warmer climate or season favours bacterial growth. In temperate regions peak populations occur in summer. Extensive development of a bacterial mucus coating is a characteristic of both spring and summer conditions, rendering it likely that these substances often form the nutritive basis for deposit feeders rather than the bacteria themselves (see below).

But the world of meiobenthic animals is determined by a three-dimensional pattern of microniches and particle surfaces. Even in this microscopic scale the colonization with bacteria is highly differentiated. Both qualitative and quantita-

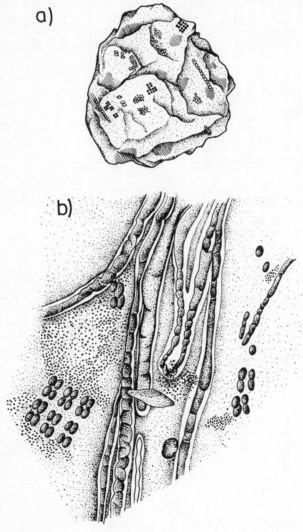

Fig. 9a, b. Colonization of a sand grain by microorganisms. **a** Overall aspect. **b** Detail. (After MEADOWS and ANDERSON 1966)

tive differences in bacterial colonization imply a difference in bacterivorous meiofauna composition. Colonization of sand grains seems to be proportional to surface area. One mm^2 of grain surface may be populated by up to 260×10^3 bacterial cells (ANDERSON and MEADOWS 1969)! It has been calculated that this enormous number of bacteria is concentrated on a very small portion of the huge overall surface (3 to 8 m^2) only (often $< 1\%$). The restriction may be caused by the limited diffusion of oxygen and nutrient molecules through the tortuous and long distances between densely packed particles (WATLING 1991). However, also mechanical forces seem to be responsible for the patchy colonization of bacteria. Microscopical inspection of sand grains revealed a hetero-geneous colonization of grain surfaces with rich microbial clusters in depressions and cracks and barren areas along the edges (Fig. 9; see MEADOWS and ANDERSON 1966, 1968; WEISE and RHEINHEIMER 1978; DeFLAUN and MAYER 1983). In the exposed edges, the destruction of bacteria through intense agitation of grains by waves or heavy rain falls is particularly massive, while they have more protection in smooth areas. Large grains ($> 300\,\mu m$ diameter) with fairly smooth surfaces are inhabited by a bacterial flora differing quantitatively and qualitatively from that on smaller particles with many crevices and depressions (MARCOTTE 1986). The relatively small surfaces of silt particles of $< 10\,\mu m$ diameter were found to be devoid of bacteria (DeFLAUN and MAYER 1983). Because of this phenomenon, the majority of meiofauna inhabiting silty bottoms are unlikely to be bacterivorous. Exposure and particle size contribute to the relative scarcity of bacteria in the exposed upper eulittoral of beaches compared to sheltered flats.

Details of the numerous relationships between bacteria and meiofauna, reviewed by TIETJEN (1980) and MONTAGNA (1989), will be reported later dealing with the position of meiofauna in the benthic food web (Chap. 9.3). Since microbes have a predominant and often direct impact on the habitats of meiofauna organisms, cooperation between meiobenthologists and sediment microbiologists is indispensable for a better ecological understanding.

More detailed reading: ALLER (1980); TENORE and RICE (1980); TIETJEN (1980); YINGST and RHOADS (1980); WARWICK (1987); DECHO (1990); KEMP (1994)

2.2.3 Plants

Unicellular algae, as in the case of bacteria, mainly structure the habitat of meiofauna because of their trophic value. Aside from bacteria, diatoms are the most important live food source. Many ciliates, harpacticoids, nematodes and oligochaetes have specialized mouth structures enabling utilization of, and even differentiation between, various algal cells and cell shapes. They often feed exclusively on diatoms, which apparently requires a particular trophic/morpho-logical specialization because of the hard siliceous diatom tests (BROWN and SIBERT 1977). Diatoms occur preferably in the upper, light-exposed few millimetres of lenitic beaches where they can grow in considerable density (20 000 cells cm^{-3} sand). Although the penetration of light is only 1–2 mm in mud and 2 cm in coarse sand (longwave red light), living diatoms occur regularly down to 5 cm; they have even been recorded at horizons as deep as 20 cm (TAYLOR and GEBELEIN 1966; STEELE and BAIRD 1968; FENCHEL and STRAARUP 1971; WASMUND 1989). Correspondingly, 20–50% of the surficial chlorophyll concentrations have still been measured in a depth of 5 cm. This unusual distribution pattern for photosynthesis-dependent cells is interpreted as a result of sediment turbation. For the diatoms, this suggests a phase of passive resting rather than heterotrophic metabolism. Vertical migrations of diatoms caused by the diurnal light rhythm are known to occur in the upper few millimetres (TAYLOR 1964; PALMER and ROUND 1967). However, regarding the much higher mobility range of vagile animals, the millimetre migrations of the microalgae will hardly relate to the diurnal rhythmicity of meiofauna occurrence, as has been postulated by BOADEN and PLATT (1971) and WIESER (1975).

The occurrence of diatoms determines the spatial and seasonal microscale distribution of some meiobenthic food specialists (e.g. peaks of naidid oligochaetes in tidal flats coincide with spring blooms of diatoms, GIERE and PFANNKUCHE 1982), and diatom patches were found to correlate directly with patterns of meiofauna distribution (MONTAGNA 1983; BLANCHARD 1990). This is reasonable considering the utilization rate of microphytobenthos by meiofauna: ESCARAVAGE et al. (1989) calculated that 27% of all benthic microalgae are devoured by meiofauna in a shallow lagoon, 38% in oyster beds, and as much as 89% in sea grass beds.

Sessile macroalgae and sea grasses mainly have a physically structuring effect on meiofauna biotopes. The reduction of sediment agitation and enhancement of particle suspension under a plant canopy favours meiofaunal abundance. Culms, thalli and holdfasts provide numerous niches and an important protection for small animals. In addition, these plant structures expand the available living space for meiobenthos from the benthal to the phytal. Delicately branched algae are more densely inhabited by meiofauna than the smooth surfaces of sea grasses. Thus, structural complexity is often correlated with meiofauna abundance and diversity (see Chap. 8.4; REMANE 1933; HICKS 1985; HALL and BELL 1988). Plant roots have a similar structural effect.

In addition, there are also chemical and nutritional effects by which plants can influence the habitat conditions of meiofaunal biotopes. The mucus production of diatoms has an indirect trophic effect. Mucus can form a dense coating on the grains which, in turn, promotes bacterial growth. In addition, it increases the critical shear strength of the sediment thereby influencing its structuring properties considerably (DEFLAUN and MAYER 1983). An enhanced bacterial growth at the frequently damaged and leaching frond ends of plants indirectly promotes the trophic basis of meiobenthos. Live plant roots have been found to metabolically create a favourable micro-oxic gradient system in their surrounding sediment (TEAL and KANWISHER 1961) with enhanced density and heterogeneity of (nematode) meiofauna (OSENGA and COULL 1983).

A discussion of other effects of plants relevant for the habitat of meiofauna will be presented in Chapter 8.4 dealing with the phytal.

2.2.4 Animals (Meio- and Macrofauna)

Only those animal-mediated factors will be discussed here which have a direct structural effect on the meiobenthic habitat. Among meiofauna species, structuring effects account mainly for some kind of competition which can ultimately lead to amensalism or mutual exclusion. Competition is often difficult to separate from the other negative factors discussed above. Among meiobenthic animals, it is mostly caused by specialization on common nutritional resources. Trophic competition can cause spatial niche segregation and result in mutual exclusion (e.g. FENCHEL 1968a for ciliates; JOINT et al. 1982 for nematodes). This exclusion has been documented to exist not only as a within-group effect among meiobenthic species (nematodes, OTT 1972; ALONGI and TIETJEN 1980; harpacticoids, CHANDLER and FLEEGER 1987), but also from taxonomically distant meiofauna groups (foraminiferans and harpacticoids CHANDLER 1989; oligochaetes and turbellarians, DÖRJES 1968; capitellid polychaetes and nematodes, ALONGI and TENORE 1985). Competition can also result in a differentiation of life history characteristics (HEIP 1980a, MARCOTTE 1983). A good example is the mutual exclusion of two species of the ciliate genus *Condylostoma, C. arenarium* and *C. remanei*, which have contrasting population dynamics, with maximum numbers occurring in June and November respectively (HARTWIG 1973b).

The specific reasons for an effect summarized as "competition" and accounting for the absence of certain animal taxa in meiofauna samples are mostly difficult to ascertain. Is it a result of a direct, animal-mediated factor or merely an unknown, less direct factor? For instance, the mutual exclusion of enchytraeid oligochaetes and turbellarians in the upper sandy beach, reported by DÖRJES (1968), and that of the naidid *Amphichaeta sannio* and the nematode *Tobrilus* in the freshwater flats of the river Elbe (SCHMIDT 1989) remain unexplained. The causative factors for the inverse relationship between the harpacticoid *Tisbe furcata* and nematodes (WARWICK 1987) are as unresolved as are the contrasting population fluctuations in experiments with the gastrotrich *Turbanella hyalina* and the archiannelid *Protodrilus symbioticus* (BOADEN and ERWIN 1971). For the negative interaction of the foraminiferan *Ammonia beccari* and the harpacticoid *Amphiascoides limicola* in muds from tidal flats (CHANDLER 1989), a conclusive explanation could not be found.

Concerning macrobenthos, the impact on meiobenthos refers mainly to negative interactions by (a) mechanical disturbance, (b) alterations in the chemical milieu and (c) reduction by sediment ingestion or predation.

a) Disturbance through disruption and reworking of the sediment: digging of horse shoe crabs, rays, crabs, pipetting of *Tellina* (Bivalvia) siphons, reworking by anthozoans, polychaetes, echinoids, fish and birds (CREED and COULL 1984; DYE and LASIAK 1986; REISE 1987b; WARWICK et al. 1990; HALL et al.

1991). These negative effects gradually merge into predator-prey relationships or disturbance, particularly when meiofauna and macrofauna interact (meiofauna vs. burrowing polychaetes, crustaceans or fish (JENSEN 1986; COULL and BELL 1979; NELSON and COULL 1989). Disturbance will affect mainly the meiofauna of the epibenthic interface and upper sediment layers (BELL 1980). Some groups avoid the disturbance by downward migration, which creates surface layers with a reduced meiofaunal abundance. But the main reduction will be caused by passive suspension and subsequent drift of meiobenthic animals out of the habitat (see Chap. 7.1.2). The general impact varies depending on the meiobenthic group and the various habitats. The populations of more passive, strictly sediment-bound meiofauna groups (nematodes, annelids) will be less impoverished by disturbance. In contrast, the more epibenthic and temporarily suspended harpacticoids will show massive short-term reductions. Especially in tidal flats, however, these losses will soon be compensated (SHERMAN and COULL 1980; 'OLAFSSON and MOORE 1990; WARWICK et al. 1990; HALL et al. 1991). In the shelter of seagrass beds, the disturbing effects on sediment-dwelling meiofauna are only small (WEBB and PARSONS 1991). Compared to macrofauna, meiofauna seems less sensitive to disturbance and will be less persistently affected (AUSTEN et al. 1989; HALL et al. 1991). In a study performed by WARWICK et al. (1990), this cognition even enabled a discrimination between a merely mechanical disturbance which reduced the stability of the sediment and pollutional stress.

b) Accumulation of thick layers of organic debris and faecal pellets with subsequent oxygen deficiency: beneath mussel and oyster beds (DITTMANN 1987; DINET et al. 1990),

c) Sediment ingestion: the feeding galleries of *Arenicola* contain lower numbers of nematodes (JENSEN 1987a); however, a corresponding reduction was not noted in habitats of other sediment feeders. Direct predation on meiofauna by small crabs, polychaetes, small fishes and others (see reviews by REISE 1979; GEE 1989; COULL 1990; WILSON 1991).

Of particular relevance and wide-spread occurrence is the mutual regulation between permanent and temporary meiofauna (ELMGREN 1978; WARWICK 1989). It is probably a combined series of effects changing from predation to sediment reworking. Permanent meiofauna, through their more effective and intensive feeding activity, initially attenuates population development of newly settled polychaetes, bivalves and very young amphipods. Subsequently, the growing macrofauna species, through their intensive sediment reworking and their direct predatory impact, have an aggravated negative influence on permanent meiofauna (BELL and COULL 1980; WATZIN 1986).

Examples of positive mutual effects between species are mostly more indirect and their structuring impact on the habitat of meiobenthos is often not well understood. This will be reported in another context (Chaps. 2.2.5 and 9).

▓   *More detailed reading*: see citations in the text of this chapter

2.2.5 Animal Biogenic Structures (Tubes)

Directly linked to the effects of animals reported in the previous section are biogenic structures such as tubes of benthic animals. Biogenic structures can have a massive impact on the habitat conditions of meiofauna. Mainly macrobenthic annelids and crustaceans, but also some meiofauna (harpacticoids: CHANDLER and FLEEGER 1984; WILLIAMS-HOWZE and FLEEGER 1987; nematodes: CULLEN 1973; RIEMANN and SCHRAGE 1978; PLATT and WARWICK 1980; NEHRING et al. 1990) produce a network of more or less permanent tubes, preferably in silty sediments. The relevance of the burrowing activity has been recently assessed by FORSTER and GRAF (1992) who studied the decapod *Callianassa subterranea*: under each square metre of sediment in the southern North Sea, there is a burrow surface of $0.7\,m^2$ (only from this species) subjected to oxygenation at least once a day. Biogenic structures stimulate bacterial growth usually in two ways, and thus exert a positive effect on meiofauna populations.

The first way involves secretion of exopolymers, mucus, and the enrichment of the sediment with fine organic particles which solidify the texture and enhance the bacterial content (ECKMAN et al. 1981; MEADOWS and TAIT 1989; MEADOWS et al. 1990; DECHO 1990; NEHRING et al. 1990).

The second stimulus on bacterial growth is a consequence of the flushing effects of tubes which profoundly influence the pore water circulation and the geochemistry of the bottom. They verticalize the RPD layer and its concomitant gradient system, and thus stimulate geochemical fluxes, often creating better growth conditions for microorganisms (ALLER and YINGST 1978; KRISTENSEN 1984; REICHARDT 1989; DAVEY et al. 1990; GROSSMANN and REICHARDT 1991). As has recently been shown, these fluxes of solutes become modified and often greatly enhanced by wave action in the overlying water (WEBSTER 1992).

Tubes protruding above the surface of the sea floor, irrespective of being inhabited or not, are thought to have a positive effect on meiofauna, protecting it from predatory macrofauna (BELL and COEN 1992). Moreover, an enhanced structural complexity in/at the bottom will increase meiofauna diversity.

Considering this array of factors beneficial to meiofauna, it is understandable that the positive effects of biogenic structures have been suggested to contribute by 10 to 50% to the colonization of tidal flats with meiofauna (Fig. 10), and with even higher figures for harpacticoids and gnathostomulids (REISE 1981a).

Negative meiofaunal effects of biogenic structures have been reported in only a few cases. Through alteration of the geochemical system along animal tubes, particularly through the import of oxygenated water by irrigational fluxes into anoxic layers, will precipitates of heavy metals (sulphides), buried in the depth, become dissolved and released into the surficial, oxygenated layers and the sediment/water interface. Bioturbation can evoke a considerable increase in phosphates released from the oxygenated sediment and may concurrently have undesired eutrophicating effects. Brominated compounds, excreted into the tube-wall lining of some echiurids and enteropneusts, have been suggested to exert a toxic impact on bacteria and meiobenthos (KING 1986; JENSEN et al.

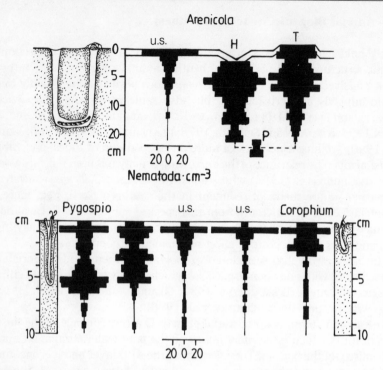

Fig. 10. Impact of biogenic structures on meiofauna density. Nematodes around tubes of tidal flat macrofauna in comparison with unstructured sediment (*u.s.*). (After REISE 1981a)

1992), causing a zone of impoverished meiofauna around the tube system. In deep-sea bottoms, mangenese diagenesis may be affected by meiofaunal oxygen consumption (SHIRAYAMA and SWINBANKS 1986). The impact of this biogenic mobilization of enriched chemical pollutants on meiofauna has not yet been properly addressed.

An aggravating impact on meiobenthos by mechanical disturbance has been shown by HICKS (1989), whose field experiments indicated that (artificial) sea grass not only promotes meiofauna populations, but can also disturb mainly epibenthic meiofauna assemblages, probably through the sweeping action of the blades and alteration of the microtopography (depressions, ripples). In areas of turbulent water currents, irregularities in the flux system caused by protruding tubes and culms, can destabilize the sediment and thus negatively affect meiofauna (ECKMAN 1979).

2.2.6 Conclusions

The intricate interaction of biotic and biogenic factors and the complexity of their impact renders interpretation of effects difficult, particularly since most of them as yet can hardly be measured quantitatively. The interactive and multiple nature of numerous determinants in the field, illustrated in Fig. 11 and, schematically, in Fig. 12, has been emphasized in a detailed study of WARWICK et al. (1986b), who investigated the impact of the macrobenthic, tube-dwelling polychaete *Streblosoma* on the meiofauna assemblage. Surrounding the tubes which extend slightly above the surface, there is a rich meiofauna, probably because of improved flux conditions influenced by the tube and also because of the mucus secreted by the worm. Slightly more distant, in the grazing area of the polychaete, meiofauna was impoverished through mechanical disturbance and perhaps trophic reduction by *Streblosoma.*

Biogenic sedimentary microstructures such as animal burrows and tubes, mucus tracks, and faecal pellets are considered "sediment-mediated facilitative interactions" for meiofauna. Mainly through the activation of geochemical fluxes and microbiological activity they create an inhomogeneous small-scale topography (Fig. 11), support aggregative tendencies in meiofauna and patchy distribution of meiofauna even in a superficially uniform sediments (SUN and FLEEGER 1991). The mucus film secreted by many burrowing animals like nematodes and annelids, and the metabolic waste products of such organisms, attract bacteria and promote a rich meiofauna (FENCHEL 1969; ALLER and YINGST 1978; YINGST and RHOADS 1980; REISE 1981a; MONTAGNA et al. 1983; ALONGI 1985). The spatially and temporally variable interactions from animals, plants, bacteria and detritus (TENORE and RICE 1980; MEADOWS 1986) become further complicated by the release of substances which exert specific stimulatory or inhibitory effects on all components of the system, including bacteria. Other,

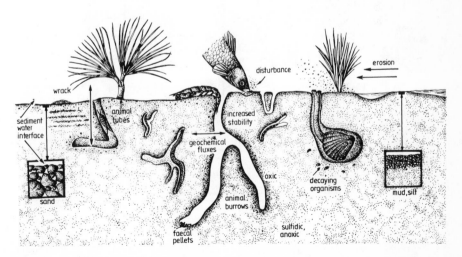

Fig. 11. An illustration of biotic factors structuring the occurrence of meiofauna in a tidal flat sediment. (Compiled from MEADOWS 1986; ANDERSON and MEADOWS 1978, and others)

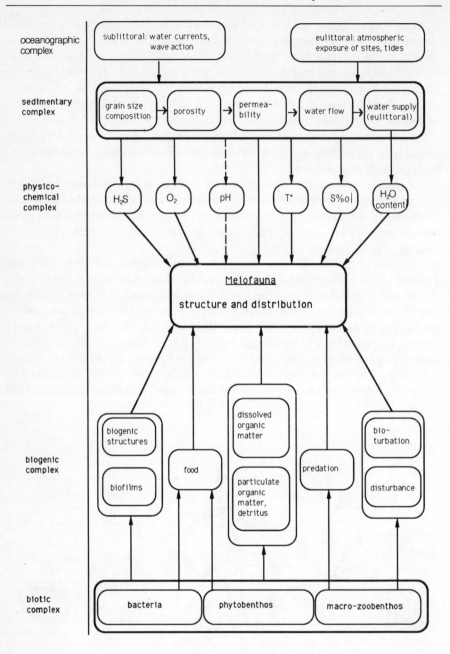

Fig. 12. A schematic factorial web structuring the habitat of meiobenthos

mainly mechanically adverse or even inhibitory structural effects on meiofauna result mainly from the activity of macrobenthos.

The entire web of biotic ecofactors which influence the occurrence of meiofauna is intricately combined and influenced by the multitude of abiotic parameters described in Chapter 2.1. Figure 12 is an attempt to schematically illustrate the complexity of the "meiobenthic habitat".

In addition to the more regular factorial system, depicted in Fig. 12, the meiobenthic ecosystem is also subject to stochastic or "accidental" factors, such as local disturbances (e.g. storms) and benefits (e.g. food input through setting of larval forms) which are irregular and temporary. These frequent local and temporal changes in the habitat of meiobenthos cause continuous "oscillations". Already small-scale variations may influence the system unpredictably. Together they certainly contribute to the notoriously patchy distribution pattern of meiofauna as well as support their high diversity, two characteristics of meiobenthic communities which make generalizations very difficult. A high diversity is also maintained by the well-developed nutritional selectivity in meiofauna which exceeds that of macrofauna. The resulting much differentiated resource partitioning of the available food stock gives biotic (trophic) factors more relevance compared to the physiographic parameters.

More detailed reading: COULL (1973, 1986); COULL and BELL (1979); RHOADS et al. (1977); ECKMANN (1985); WARWICK (1989); DECHO (1990)

CHAPTER 3

Meiofauna Sampling and Processing

Since methodology is the dominating aspect of the *Introduction to the Study of Meiofauna* (HIGGINS and THIEL 1988), this chapter should be read in close conjuction with that reference. Only the most important sampling methods will be presented here. In some cases I will include some supplementary hints not given in the previous book. Additional information, particularly more details on sampling devices and on the problems of sampling in soft-bottom sediments, may be found in the review by HOLME and McINTYRE (1984), the methodological recommendations edited by ELMGREN and RADZIEJEWSKA (1989), and the recent compilation by BLOMQVIST (1991). Since sampling design and strategies and statistical evaluation vary with the scientific problem addressed, it is almost impossible to describe here universally applicable methods. Valuable hints are given in McINTYRE et al. (1984).

In addition to the usual textbooks for (biological) statistics, the critical review on calculation of diversity by HURLBERT (1984) is highly recommended. The problem of control sites, important for each evaluation of disturbances, is competently covered for meiofauna in ESKIN and COULL (1984).

3.1 Sampling

Prior to each investigation, it is very important to carefully consider both the area to be sampled and the equipment to be used. Among other factors, these logistical aspects depend greatly on sediment structure and animal group to be studied. The abundance of animals in the sediment will determine the appropriate sample size and sampling strategy.

3.1.1 Number of Replicates, Sample Size

It is indispensable for reliable quantitative meiofauna studies to take parallel samples, since patchiness is a characteristic of meiofauna (SUN and FLEEGER 1991). In practical work, the mathematically optimal number is usually unrealistic because of limited time and manpower. As a general rule, variation between replicates should be less than between sampling stations. Instead of taking one large sample, it is more advisable to take several small subsamples and evaluate them separately. In this context, notice should be taken that the replicates obtained by multiple corers do not represent statistically independent and random samples, but rather nested two-stage pseudoreplicates (HURLBERT 1984). They are not equivalent to repeated samples done with a single corer. For any reliable assessment of population development, the strong temporal fluctuations in many meiofauna groups require to repeat the sampling within narrow time intervals (e.g. each week, see ARMONIES 1990).

The classical procedure to ascertain the optimal sample size (which causes least effort to evaluate and still yields reliable results), is to initially count a larger unit and then compare it with data from subsamples of decreasing size units. The expected data should not deviate $> 10\%$ from the calculated one. However, in view of the extreme inhomogeneity in meiofauna distribution pattern, it remains questionable whether this method is always applicable. To a priori follow the rule that the surface sampled by the corer should exceed patch size does not help very much, since the latter is usually unknown.

Sample size and, accordingly, sampling gear have to be related to the wide size range of meiofauna groups, have to consider their abundance and patchiness. If just one size class is considered (e.g. small, but numerous ciliates or the fairly large, but infrequent oligochaetes), sampling can be optimized by using specialized corers and evaluating different sample volumes by different extraction methods. For studies on the entire assemblage of meiofauna within one sample site, either varying samplers of different size have to be used in parallel or one must compromise in selecting the sampling gear. In any case, pre-sampling and, if possible, examination of live meiofauna, gives valuable information about qualitative composition and local distribution of the meiofauna.

The limits of exactness and sources of sampling error have been well investigated for the most frequently used sampler of meiofauna, the perspex tube corer (see Chap. 3.1.2.2). WELLS (1971) recommended as the minimal sample area $10\,cm^2$ ($= 3.6\,cm$ diameter). This small tube is applicable mostly for ciliates and other very abundant small meiofauna. It is often constructed from large medical syringes whose lower end has been cut.

Sampling problems with all tube corers are the shortening of the core by sediment compaction, and, under water, the disturbance effect of shock waves which tend to flush away the sediment surface (see JENSEN 1983).

These shortcomings of sampling are particularly serious in soft, flocculent bottoms and can be minimized only by using corers with a fairly wide diameter (8 cm more, MCINTYRE and WARWICK 1984) which, for most purposes, yields an inacceptably large sediment quantity. RUTLEDGE and FLEEGER (1988) tested corers between 2.6 and 10.5 cm diameter without any major error in the results when compared with controls, provided the corer is gently and slowly pushed into the sediment. In shallow water this can be done best by SCUBA divers. HOLOPAINEN and SARVALA (1975) found that 4.4 cm ϕ-cores, obtained by hand in this manner, contained 20% more fauna than cores of identical size obtained mechanically. This underestimation of numbers by mechanical sampling was not compensated for until the corer diameter was increased to 8.3 cm! In a detailed investigation on the reliability of core sampling, BLOMQVIST (1985) evaluated the effects of factors such as the shortening of the core, sediment resistance, lateral extrusion (bow wave effect) and penetration of the corer.

3.1.2 Sampling Devices

Depending on sampling locality, research objectives and the prevailing animal group, meiofauna studies are based on a variety of special constructed sampling gear. Only the most commonly used and generally available sampling gear will be described and commented upon here. For a valuable recent review on benthic sampling gear see BLOMQVIST (1991).

3.1.2.1 Qualitative Sampling

Straining Groundwater. The simplest, most extensively ("classical") method used since REMANE's studies of eulittoral shores is to dig a funnel-shaped excavation into the sediment down to the groundwater depth, allow water to fill the bottom of the pit and then strain the inflowing water rapidly with a small hand net (mesh size 45 or 63 µm) with swift, circulating movements. Meiofauna from the surrounding sand layers are washed into the pit and collected by the net. Repeating this procedure several times will result in a representative qualitative overview of the meiobenthic inhabitants. If this in situ procedure is impracticable, a sample of the sand (usually 5–10 l taken at or near the bottom of the excavation) may be brought into the laboratory, mixed with filtered seawater, agitated to suspend the meiofauna and then decanted through a fine mesh sieve. This will render similar, if not better results.

Climate Deterioration. A convenient method to concentrate meiofauna for qualitative inspection from a large sample. Leaving a jar or bucked with the moist sediment for about a day under normal room temperature will force most meiofauna to aggregate near the surface (mainly due to oxygen deficiency). From the surface layers, enriched subsamples can then be taken for analysis. However, it should be cautioned that this method is not representative: sensitive taxa deteriorate quickly, "stationary taxa" remain where they are and do not migrate to the surface (e.g. oligochaetes, nemerteans).

Dredging. From sublittoral grounds, a fine-meshed dredge (e.g. OCKELMANN 1964; HIGGINS 1964), sometimes combined with a pushing plate (MUUS 1964) will scoop the uppermost centimetre off the sediment. The MUUS sampler is sometimes considered to be semi-quantitative and may even catch some of the more mobile meiofauna and small macrofauna (e.g. isopods).

Suction Corer. This is very rough method for collecting large quantities of soft bottom fauna, irrespective of spatial distribution and abundance. Also termed "air lift", the hose of the suction corer is connected to a diver tank. The gas pressure sucks water plus sediment up into a sieve or an adequate net. Since in this method the strong sucking pressure damages many soft-bodied animals, it is rarely used by serious meiofauna investigators.

3.1.2.2 Quantitative Sampling

For meiofauna, tube corers are the most versatile instruments and superior to most box corers and bucket grabs. According to BLOMQVIST (1991), there are four major sources of error in coring: (1) core shortening, (2) resuspension and loss of enclosed sediment, (3) loss of surface sediment and (4) repeated penetration due to wave action. The latter three refer to sampling in deep water (sublittoral).

The classical meiofauna corer is a perspex tube with externally bevelled lower edge and an internal surface between 10 and $20\,cm^2$ ($=3.6–5\,cm$ diameter, for size restriction see Chap. 3.1.1). Holes drilled into the perspex wall allow insertion of thin electrodes for measurement of abiotic parameters at various depths (RIEDL and OTT 1970). A suction device consisting of a rubber stopper connected with a hose helps to reduce friction when inserting the corer by lowering the pressure above the core (Fig. 13), but serves also to press the sediment column out of the tube for fractionated subsampling.

In directly accessible (eulittoral) muddy shores, the open tube can be pushed into the bottom and, after closing its upper end, the core can be removed with ease. In sandy areas with a potentially more compact sediment, insertion of the corer is more difficult (greater resistance) so that sometimes more force must be applied, such as with a (rubber) hammer. In non-consolidated sediments, a closing device is required for a precise, quantitative removal of the core (see piston stopper in BLOMQVIST and ABRAHAMSSON 1985), unless the tube cannot be dug out carefully and closed from below by hand. One device which reduces friction in sandy bottoms and allows deep sampling consists of a square stainless steel tube with one side fitted for an interlocking sliding blade (Fig. 13).

Extrusion of the core is accomplished by use of a piston (see below) or by air pressure. Subdivision is usually done by fractioning the core with a knife or a thin blade (beryllium bronze is considered best). A convenient fractioning device has been described by DANIELOPOL and NIEDERREITER (1990); probably more useful for muddy than for sandy cores is another construction designed by BLOMQVIST and ABRAHAMSSON (1987) for serial sectioning.

In corers to be opened from one side, tightly fitting blades can separate the subsamples horizontally. In any case, subsampling of the core should be done immediately after retrieval of the sample in order to avoid bias through faunal migration.

The length of the corer (usually 30 cm) is determined by the depth to which it can be driven into the bottom without major disturbances, compression or otherwise altering the depth distribution of meiofauna. Core sampling in exposed beaches where such taxa as nematodes and tardigrades occur in greater depths (up to 180 cm deep, KRISTENSEN and HIGGINS 1984) must be incremental, e.g.

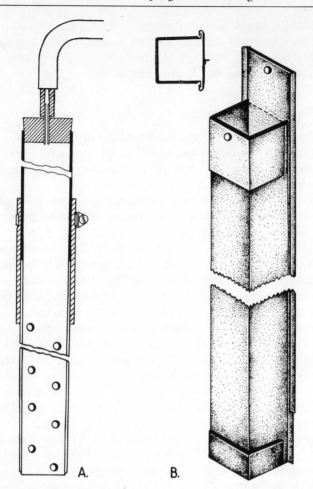

Fig. 13A, B. Quantitative meiofauna corers. **A** Perspex corer with rubber stopper and hose. (After RIEDL and OTT 1970). **B** Square steel corer for sampling deep sand layers. *Upper left* cross-section. (After McLACHLAN et al. 1979)

short core samples must be taken repetitively at increasing horizon depths. The unit by McLACHLAN et al. (1979) described above is designed to allow sampling down to the depth of 1 m. Exact vertical subsampling of the core is facilitated by ring-marks (e.g. every 2 or 5 cm) on the outer surface of the tube. JANSSON (1967b) constructed a metal tube which enclosed a thin perspex corer cut into several rings which served for the subdivision of the sediment core. The sharp-edged metal mantle reduced compaction of the sediment and shortening of the core. If the study object requires a finer subdivision, the corer designed by JOINT et al. (1982) can serve as a valuable tool.

For small coring tubes (e.g. for sampling ciliates), medical syringes are usually cut at their lower end and used as piston corers. For studies on bigger or rare meiofauna animals, larger samples must be obtained. Sampling marine oligochaetes, the author used a square 10 cm × 10 cm metal box corer with sharp edges at its open lower end and one removable side wall (Fig. 14). Having obtained the sample filling this 20-cm-long box (2000 cm³), the core could be subdivided inserting tightly fitting thin metal sheets from its opened side. For convenience and exact subdivision these blades were guided into position by a perspex top piece conveniently attached to the side of the metal box corer.

Sampling and subdividing of submersed sediment cores is difficult because, once removed above the surface, the whole core or at least the last sediment fractions in the corer tend to become washed away by the overlying water in the tube. If the core is taken from a submersed locality, the overlying water covering the core plus the flocculent interface layer should be carefully syphoned off into a fine sieve prior to subdivision of the sediment in order to avoid distortion by percolating water flow. In shallow water, tightly closing the tube by a stopper immediately after withdrawal from the sediment, while the corer is still under

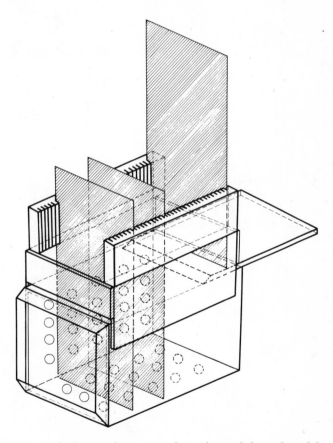

Fig. 14. A box corer for large sediment samples with metal sheets for subdivision. (After GIERE 1973)

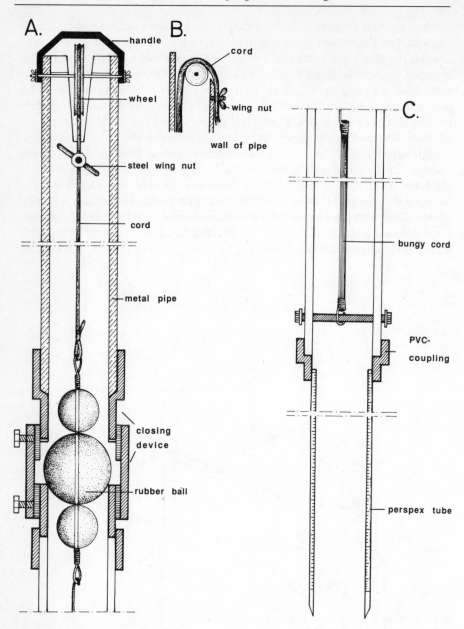

Fig. 15A–C. A pole corer. (After ALI 1984). **A** Upper part with closing device of the tube. **B** Detail of upper locking and releasing mechanism for the cord. **C** Lower end of corer with perspex tube inserted in coupling

water, will secure the core. In order to properly obtain also the upper centimetre-fractions of the core, the use of a metal piston with a sturdy handle is recommended. Pressing this handle against the ground or on another solid support, the sediment core can be gently pushed up to the required height and the slice exceeding the upper edge of the core easily be taken. This is done without any loss with the blade of a broad knife or spatula. A rubber plug of exactly the internal diameter of the coring tube, positioned between the piston and the sediment, will prevent the piston from becoming wedged in the tube due to jammed sand grains.

In shallow sublittoral soft bottoms, the pole corer of FRITHSEN et al. (1983) combines simple handling with reliable closing (lid from above triggered by a bottom plate on contact with the sediment). A similar device has been designed by ALI (1984) for work in shallow lakes. Here, tight closure of the coring tube by a stopper enables also sampling of the flocculent surface layer (Fig. 15).

For coring in deeper sublittoral sediments, remotely operated corers will be needed. Another option is the use of sampling devices by means of the hydraulic arms of research submersibles. Since the often unconsolidated surface layer is particularly rich in meiofauna, a tight closing device which retains the flocculent fraction is imperative. In order to achieve a reliable quantitative sample without a bow wave effect, the coring tube should be mounted in a supporting stand which, after touching the bottom, allows the tube to slide slowly and vertically into the sediment. Connected with a pinger, the coring apparatus can electronically record the bottom surface and be lowered gently during the last metres of descent.

Reliable gravity corers for sublittoral sampling are the Kajak- or the Haps-corer (KANNEWORFF and NICOLAISEN 1973) with its several modifications (JENSEN 1983; BLOMQVIST and ABRAHAMSSON 1985; CHANDLER et al. 1988), some of which allow multiple sampling with three or four parallel tubes. In all versions, the tube is tightly closed from above by a lid released when the corer comes in contact with the sediment (Fig. 16). Since in the corers of the Kajak type a closing valve from below is lacking, their function is best in fine sediments.

Another corer frequently used in many types of sediment has been developed by CRAIB (1965). When pulled out of the sediment, the tube is closed from below with a ball valve (Fig. 17).

A remarkable addition to the various modifications of the CRAIB corer (e.g. FENCHEL 1967; JACKSON 1986) has been recently given by DANIELOPOL and NIEDERREITER (1990), who combined in their sampler the lid device of the Kajak corer with the ball valve of the CRAIB corer. Its multiple coring frame (six tubes) enhances its value for reliable quantitative work.

When working at greater depths from a larger research vessel, the multiple SMBA corer described by BARNETT et al. (1984) seems to be the best choice. With their solid supporting frame and their elaborate technique (hydraulically dampened penetration of the tubes, electronic release system, optional visual inspection by video camera surveillance) multiple corers can provide reliable quantitative samples with replicate cores even from the deep sea. They have

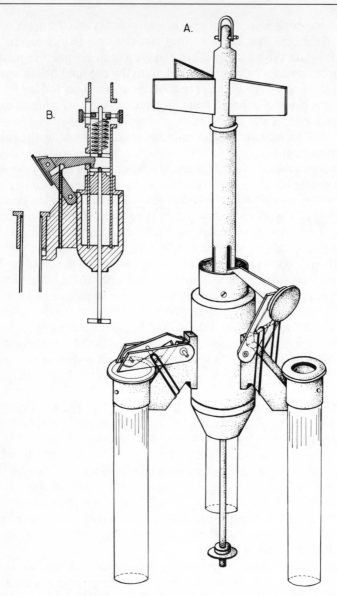

Fig. 16A, B. A modified Kajak corer. (After HAKALA 1971). **A** Overall aspect. **B** Schematic detail of closing mechanism

been used successfully for meiofauna sampling on board of many research vessels (see Chap. 8.3). Most of the deeper-water corers are equipped with some weights and guidance wings for swift and untilted descent to the bottom.

When both macro- and meiobenthos has to be studied simultaneously, e.g. in many ship-board situations, collection of meiobenthos is mostly accomplished

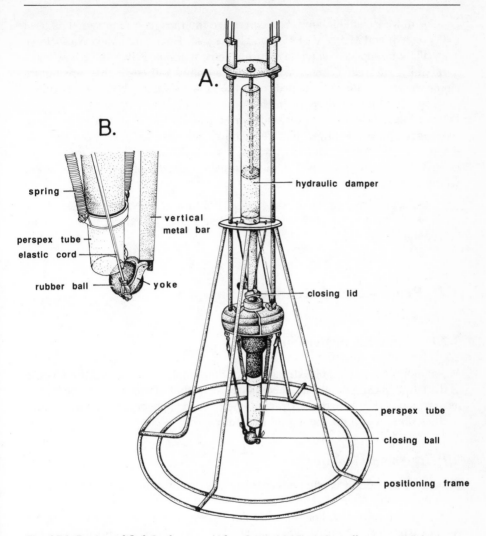

Fig. 17A, B. A modified Craib corer. (After CRAIB 1965). **A** Overall aspect. **B** Schematic detail of closing mechanism

by subcoring the sediment taken with larger bucket grabs or box corers which are needed for obtaining the macrobenthos (see HOLME-McINTYRE 1984). It should be noticed, however, that the popular van VEEN grab and EKMAN grab in their traditional design are notoriously non-quantitative samplers (ANKAR 1977; BLOMQVIST 1985, 1990) and their results should be used with caution. Although their construction can be improved (HÅKANSON 1973; BLOMQVIST 1990), even better grabs do not penetrate deeper than 5 cm at their edges into the sediments (RIDDLE 1989).

For more exact sampling, box corers are the most common samplers used today (SMITH and McINTYRE 1954, and others, see HOLME and McINTYRE 1984). They allow unimpeded water flow during descent and largely avoid a bow wave if carefully lowered. When the hoist line becomes tautened, they are tightly closed with a "closing arm" pulled below the square sampling box from the side.

For the study of deep-sea meiofauna, THIEL (1966) designed a narrow Perspex box whose open lower end can be closed after insertion into a large sediment sample pushing a flexible metal or plastics "tongue" from the side down to the bottom. Even the improved version (KÖLMEL 1974) of this device, called the meiostecher, is useful only in muddy sediments since each sand grain will impair its sliding mechanism.

More detailed reading: HURLBERT (1984); HOLME and McINTYRE (1984); RIDDLE (1989); BLOMQVIST (1985, 1991); ANDREW and MAPSTONE (1987); BURD et al. (1990)

3.2   Processing of Meiofauna Samples

3.2.1   Extraction of Meiofauna

Meiofauna must always be extracted from the sample because the total volume of the fauna is exceedingly small compared with that of the sample. The choice as to which of the numerous extraction methods to apply depends on the nature of the sediment and on the aim of the study.

3.2.1.1   Sample Staining

Prior to extraction, it is often advisable to selectively stain the fauna for easier recognition and discrimination between dead shells and freshly dead animals (e.g. in foraminiferans). For unfixed samples, classical vital stains like neutral red or carmine can be used, fixed samples are typically bulk-stained with Rose Bengal, added to the formalin fixative (for details of concentrations see PFANNKUCHE and THIEL 1988). Rose Bengal is absorbed differently by the various meiofauna groups: after 10 min, annelids and turbellarians stain dark red while nematodes may not even become pink; kinorhynchs may even require > 48 h. As a general rule, staining for 1 h is sufficient for most meiofauna. Overstaining is difficult to remove (acid alcohol) and impairs structural analysis under the microscope. The bright fluorescence of Rhodamine B is another convenient method to facilitate sorting (HAMILTON 1969). Counterstaining is often helpful to differentiate between dead cells (detritus) and live animal tissue. Rose Bengal can be counterstained with Chlorazole Black or Phloxine B (WILLIAMS and WILLIAMS 1974; MASON and YEVICH 1967). According to HAARLØV and WEIS-FOGH (1953), living or freshly dead cells are selectively stained by Violamin, which contrasts them against the detritus and sediment particles. The different fluorescence of living (active) and dead cells is exploited by a method suggested by WILLIAMSON and PALFRAMAN (1989) originally for microbiology, but certainly applicable also for meiobenthology. A corresponding method using Crystal violet has been used for differentiation of bacteria by HED (1977).

3.2.1.2 Qualitative Extraction

Sandy Substrates. The classical method of decantation by hand using a 45-μm or 63-μm sieve and a bucket is still the most convenient and most widely applicable method for quickly obtaining abundant numbers of meiofauna from large samples (fixed or unfixed). It will yield sufficiently representative results if preceded by anesthetization of the often adhesive meiofauna by isosmotic $MgCl_2$ (in full-strength seawater: 7% solution). Another anaesthetizing substance, relaxing the animals very gently, is a saturated solution of Mephenesin (3-o-Toloxy-1,2-propandiol). A method to obtain even adhesive meiofauna is to rinse the sample briefly in freshwater, decant into a sieve and immediately reintroduce the decantate into seawater. The freshwater shock will cause the animals to detach from the substrate. Soft, fragile taxa may be destroyed, but much of the meiofauna remains in suitable condition. Reliable decantation is best accomplished with medium or coarse sand and cannot be used for finer sediments. From fine sand and mud substrates, meiofauna must either be extracted by tedious hand sorting or by using a flotation method (see below: quantitative methods). Special procedures for specified animal groups (e.g. bubbling method) will be dealt with in context with the respective systematic overview (Chap. 5).

3.2.1.3 Quantitative Extraction

1. Medium and coarse sand

– Decantation: see above
  If seawater decantation is repeated twice, ensued by freshwater extraction and done in small portions, enumeration can be considered quantitative (ANKAR and ELMGREN 1976), particularly after staining with Rose Bengal. This has been successfully applied even for "soft meiofauna" (NOLDT and WEHRENBERG 1984).
– Elutr(i)ation. Developed by BOISSEAU (1957), this method has been modified by McINTYRE. His "Aberdeen elutriation apparatus" proved to work both reliably and quickly (UHLIG et al. 1973). There exist several more complicated versions, but all are based on the same principle: to use a water jet for separation of fauna and sediment particles by specific gravity. When elutriating samples with a higher content of fine or detrital material, it is advisable to pore the water through two sieves mounted in series to minimize the risk of overflow due to mesh clogging, and thus losing animals. In fresh samples pre-treatment with $MgCl_2$ is needed for quantitative evaluation, in fixed samples staining

with Rose Bengal is recommended. Thorough inspection of the sediment residue helps to estimate necessary extraction time and number of runs. Careful elutriation is usually more exact and convenient than decantation, although, for soft meiofauna, elutriation is a rather rough method and may cause damage. The restriction of elutriation to sandy samples is evident from its working principle.

– The "seawater ice method" (UHLIG 1964), although applicable to a wide spectrum of sediments, is not quantitative for all meiofauna taxa. On freezing full-strength seawater, initial concentrations of the perfusing thawing water may rise to 80‰ S and more, while the last molten water is oligohaline (below 5‰ S). This gradient causes a massive salinity shock for the fauna and a subsequent downward migration of the animals. The avoidance reaction of (many) meiofauna to extreme temperature gradients caused by the melting seawater ice supplements the effect of this method.

In its various modifications which essentially enhance the throughput of samples per time unit and facilitate the handling, the seawater ice method is simple enough to be used in each lab. As much as the method is suitable for soft meiofauna (e.g. ciliates, turbellarians), it is not reliable for quantitative extraction of nematodes, halacarids, tardigrades, annelids, and some harpacticoids. Care has to be taken that the thawing process continues long enough (about 3 h), and that the bottom gauze is exactly in flush contact with the seawater in the collecting dish (there must be no air being trapped); but most important is the restriction of the method to really fresh samples with active animals. Only animals from freshly taken samples will react adequately by avoidance migration. A suction corer specifically designed for subsequent use of the seawater ice method has been designed by RUPPERT (1972) by adjusting the lateral drill holes for insertion of special tubes that could later directly be used as thawing tubes.

2. Fine sand and mud

– Flotation: For quantitative work the tedious hand sorting of samples has been largely replaced by flotations methods. The fauna, suspended after thorough mixing in a medium of a definite specific density, is separated from the sediment by short centrifugation. Introduced into soil biology in the 1950's, the earlier flotation medium was mostly a sugar solution. In modern meiobenthology, this has been replaced by Ludox AM (Du Pont), a colloidal silica sol with a specific gravity of 1.21. JONGE and BOUWMAN (1977), when introducing Ludox, had used it in the TM specification (specific gravity 1.39); however, for use in seawater, the AM version has proven superior since it avoids gel formation (NICHOLS 1979).

Ludox is meant for processing fixed samples, although most animals extracted from live samples will still remain moving and only slightly malformed if they are quickly reintroduced into seawater. Flotation without any physiological harm to the animals can be performed by Percoll or a Percoll-Sorbitol mixture

(SCHWINGHAMER 1981b), but the high cost of these substances makes it too expensive for routine processing. For reliable results it is advisable to repeat the Ludox treatment several times. The number of repetitions for quantitative evaluation has to be assessed by test runs (HEIP et al. 1974), but even in soft muds, two repetitions and thorough centrifugation usually yield 95% retrieval of fauna. With some routine, this can be done in about 30 min per sample. The size of each subsample in the centrifuge depends on the size of the centrifugation beakers; however, in each case, for sufficient accuracy the addition of at least four times the amount of Ludox (or a Ludox-seawater mixture, see below) to the sediment sample is required. It is advantageous to remove pebbles and other heavy particles prior to the addition of Ludox. In clayey sediments which tend to form stable clumps, addition of a water-softening agent like Calgon (BARNETT 1980), accompanied by prolonged stirring has been proven helpful in the extraction of fauna without excessive damage (add one part to five parts of sample volume). Great care has to be taken to quickly flush the animals from the sieves into a seawater dish and to rinse the sieves immediately in order not to destroy them by clogging. If the sediment can be subjected to Ludox flotation only once without any repetition, HEIP et al. (1974) suggested empirically assessing a correction factor for better reliability of the data.

Since Ludox AM is available only in expensive 260-kg drums, one should also test the effectivity of Ludox LS or Ludox HS-30%. These products have the same specific gravity but are available in smaller containers. In order to minimize the amount of Ludox used, FURSTENBERG and WET (1982) mixed with very good results Ludox HS-40% with water (specific gravity 1.28) for flotation. However, our own tests showed that dilution of Ludox AM was not successful since sometimes the silica suddenly became gelatinous, thus destroying the sample. On the other hand, Ludox HS-30% is dilutable with seawater to a concentration of almost equal parts.

– Decantation: gentle sieving and decanting of suspended fauna after sonification of the sample for 10 s proved an effective method even in "difficult" muddy and clayey sediments (MURRELL and FLEEGER 1989). Several repetitions of the procedure are recommended.

3.2.2 Fixation and Preservation

The usual fixative for (marine) meiofauna is formalin, which should be added in amounts to make its end concentration approximately a 4% formaldehyde solution (1/10 of saturated formalin). For dilution of saturated formalin, membrane-filtered seawater is preferable, since it has simultaneously an excellent buffer capacity and enhances the osmolarity of the fixative with favourable results for preservation. An excess of $CaCO_3$ (chips of natural chalk, limestone or marble) in the stock solution will help to increase alkalinity. For longer preservation without damage to calcified structures, the necessary careful buffering of the acidic formalin can be done with any alkaline buffer, most commonly hexamethylene tetramine, Borax, or a mixture of sodium hydrogen phosphates.

In samples retrieved from the deep sea, the distribution pattern of the meiofauna in the core may change markedly during the considerable time needed for hawling the sampling device. This should be avoided by using devices which allow injection of the fixative directly after sampling while the instrument is still at the bottom (e.g. the GUIDI-corer).

A better fixative than formalin, especially for the problematical soft-bodied fauna, is glutaraldehyde. However, if the animals are not specifically prepared for ultrastructural studies, the high price and risks of handling do not make glutaraldehyde a routine fixative. In all those cases where an ultrastructural investigation of the animals additional to the routine treatment is desirable, the use of Trump's fixative (McDOWELL 1978), a modified DDD's fixative, is recommended. The advantage of this fixative is its durability and versatility, so that specimens or samples can be stored for years at room temperature well fixed for both light- and electron microscopical inspection. Trump's can be directly used (even in the field) without any subsequent change of buffer and further handling.

Recipe
For 100 ml solution

– dissolve 1.16 g $NaH_2PO_4 \cdot H_2O$ and 0.27 g NaOH in 88 ml distilled H_2O
– add 10 ml saturated formaline ($\approx$ 37%, high grade)
– add 2 ml 50% glutaraldehyde

to make a mixture of final pH = 7.2.

We did not observe any growth of bacterial colonies on the sample, even after long storage at room temperature. Any risk of bacterial deterioration is avoided if the highly toxic sodium cacodylate buffer is used instead of the phosphate buffers. Sodium cacodylate buffer is necessary in cases of bulk fixation of a sample containing much seawater because it has been observed that phosphate buffer in seawater causes precipitation.

Microwave fixation is a method described by BERG and ADAMS (1984) for histological purposes. It has been tested, modified for bulk fixation of whole meiofauna samples. However, for durable preservation of the animals, microwave treatment has to be ensued by formalin or some other fixative/preservative. The suitability of this technique remains doubtful, since a superiority to one of the traditional fixation and preservation methods cannot be seen.

The appropriate method for longer-term storage of selected specimens depends very much on the meiofauna group studied and can hardly be generalized (for information see Part 3 in HIGGINS and THIEL 1988).

There is one interesting resin technique (after fixation in glutaraldehyde or DDD's fixative) that allows fixation and permanent storage of larger animal numbers (RIEGER and RUPPERT 1978). The animals are embedded in resin which is cast in the form of microscopic slides, optionally together with some sediment particles. The resulting resin slide is applicable for taxonomic as well as morphological and ecological purposes and allows storage of significant parts of the samples as "ecotype material". Depending on the taxonomic group, a disadvantage is the definite enclosure without leaving an option to later change the animals' position if so required during inspection.

In cases where a sample contains too many animals for easy enumeration of all specimens, the extracted meiofauna assemblage should be subdivided with a sample splitter to facilitate quantitative evaluation. A precision-made and statistically tested instrument is required to warrant subdivision in exactly equal parts. Today, the most reliable type, especially made for meiofauna, is the JENSEN splitter (JENSEN 1982). In our laboratory, some modifications have made this instrument indispensable for meiofauna work: quantitative drainage of the chambers could be improved, giving the bottom of the chambers a slight slope towards the stopper holes. If mounted on somewhat longer supporting stands than suggested by JENSEN, rinsing of the subsamples into dishes becomes easier.

3.2.3 Selected Instruments for Processing Meiofauna Organisms

A few instruments, needed for the various procedures involved in handling and studying meiofauna organisms, shall be mentioned here. They have been proven very helpful in the daily routine work with meiofauna, but are usually omitted in descriptions.

– Quantitative sorting of large numbers of organisms is much easier and more reliably done in specially made rectangular perspex trays, called Bogorov trays, named after a Russian plankton specialist. The inner surface of these trays is subdivided by a system of levelled bars, leaving a meandering system of stripes for counting the extracted organisms. The width of these stripes should be slightly less than the diameter of the eye field at normal magnification of the dissecting microscope used for routine scanning. The advantage of this device for quantitative enumeration over any kind of petri dish is the square shape and the bars. In case of an occasional push, these largely prevent swashing and dislocation of the specimens, and thus enable continuation of precise counting. Straight lines, 1 cm apart, scratched across the bottom of the tray support orientation. The overall size of the dish should be conveniently adjusted to the size of the microscopic stage. Slats at the undersurface of the dish prevent undesirable adhesion, if the microscopic stage surface should become wet.

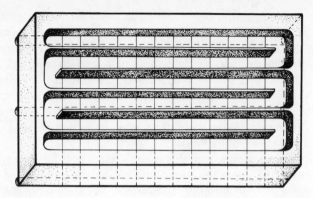

Fig. 18. Sorting tray for meiofauna (a modified Bogorov tray), size: 8 × 14 cm, height: 1.5 cm

A recent development, derived from the typical Bogorov tray, is a sorting tray where the meandering trough system has been milled by a cutter out of a solid Perspex plate (Fig. 18). Its precise manufacturing avoids the inevitable crevices and imperfections resulting from glueing of Perspex strips. For quantitative evaluation of the complete meiofauna assemblage in a sample, the extracted content should be inspected once with 10–15x magnification and another time at 30x to 50x.

– For removal of single specimens, the use of a "mouth pipette" allows far quicker working than normal pipetting. As mouth pieces for the pipettes, we use the rimmed plastic protective tubes of syringe needles cut to convenient length. They allow a good and yet relaxed hold between the teeth and are easily and inexpensively exchangeable. Each pipetting carries the risk, however, of losing specimens in the glass tip and sucking up other unwanted particles. Coating with silicone can partly solve this problem. In live samples, animals clinging to the walls and bottom of the dish are difficult to remove with a (mouth) pipette.

– By far "cleaner" and more selective, yet more tedious, is the "fishing out" of the specimens with thin needles or loops. Stainless steel needle holders, heavy enough to effectively reduce the natural slight vibrations (the resting tone) of the hands, can be purchased from medical instrument suppliers. Their grip should firmly hold even the minute stainless steel pins used in entomology for small insects. For removing more spherical organisms, the use of loops is recommended. The loops can be easily manufactured from very thin copper wire available at suppliers of electronical equipment. Then, the wire is tightly twisted from both ends around a thin needle to produce a loop in any desired width which can be slightly flattened by careful hammering. The twisted "handle" of the loop is then inserted into the grip of a needle holder.

– For detailed scrutinization of very small meiobenthic forms (e.g. tardigrades, gastrotrichs etc.), a dissecting microscope is desirable with magnifications beyond the usual 50x. Recently, leading optical companies have produced high-power stereomicroscopes of a previously unknown resolution. They allow optically "meaningful" magnifications up to about 120x, and yet maintain the large working distance necessary for inspection of animals on a sorting tray. In combination with a glass fibre spot light source or a (dark-field) transmittant light from below, these instruments often reveal details which otherwise become recognizable in the compound microscope only.

– Satisfying meiofauna microphotographs are extremely rare. For a professional quality standard with sufficient depth of focus special and expensive photo-microscopes (often better: photostereomicroscopes) are required, if possible combined with microflash to enable live photography. The problem of main-taining the animal in a desired position without the risks of smashing or loss, is elegantly, though expensively, solved by the "compression chamber" devised by UHLIG and HEIMBERG (1981).

Use of video cameras adapted to microscopes has considerably broadened the possibilities of optical documentation of (live) meiofauna. Moreover, optometrical studies and image analysis, if necessary also by computer-aided software programms, can now conveniently be performed on the monitor screen (FARRIS and O'LEARY 1985). On the basis of conversion factors selected from literature, THOMSEN (1991) calculated number, size, volume, and biomass of bacteria and of meiofauna using a semi-automatic image analysis system.

3.3 Extraction of Pore Water

The importance of pore water for meiobenthic animals requires detailed analysis with methods which are mostly based on the extraction of free pore water. The inherent problem in this frequently used analytical procedure is the destruction of fine-scaled physical and chemical gradients which often characterize the meiobenthic habitat.

3.3.1 Suction Sampling

In accessible intertidal areas, the easiest way to extract pore water is to gently push a disposable glass pipette or a hypodermic syringe directly into the sediment and to obtain the pore water by suction. From a sediment core, water in the different horizons can be collected through lateral openings in the corer which are initially sealed (Figs. 13, 14). Pressing the core by a tightly fitting piston facilitates extraction of the pore water (JAHNKE 1988). A piece of fine nylon gauze (20 μm mesh size) or a commercially available copper grid for transmission electron microscopy, sealed on the tip of the pipette, prevents clogging of the fine opening by sediment particles.

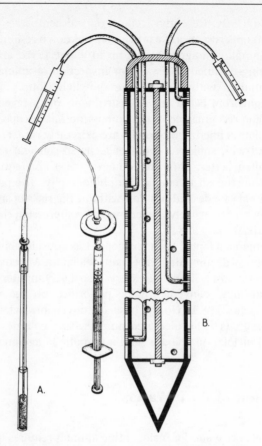

Fig. 19A, B. Pore water samplers. **A** Capillary syringe. (After Howes and Wakeham 1985); **B** Pore water lance. (After Giere et al. 1988a)

An effective and easily built device to obtain pore water is a modified glass capillary (Fig. 19A). It has a tightly fitting syringe cap into which many holes have been drilled with a microdrill. A thin gas chromatography capillary tubing (note for determination of oxygen and hydrogen sulphide in pore water: Teflon tubing is not oxygen-tight!) is connected via two sealed syringe stubs (Luer stubs) to the capillary tube at one end and to a small syringe (1 ml) at the other. Pore water from muddy bottoms which may be undesirably turbid can be filtered by interconnection of a disposable microfilter in the tubing. Drawing the piston of the syringe with some care will allow for obtaining pore water without air bubbles, which tend to penetrate into the tubes at their connecting ends.

A pore water sampler particularly designed for flushing with nitrogen, and thus well suitable for measuring oxygen or sulphide content has been constructed by Zimmermann et al. (1978). Its considerable size renders it more suitable for permanent installment in the sediment to be investigated.

DYE (1978) constructed a pointed metal tube with a sample port covered with fine gauze that, during insertion into the sediment, is tightly capped by a PVC sleeve. When arrived in the desired depth, the sleeve is slid up and pore water can enter into the inner chamber, from where it is obtained by a connected syringe.

From sublittoral sediments, a diver-operated pore water lance allows extraction of pore water from a series of horizons. A modified version of the pore water lance, originally constructed by BALZER (Kiel), is described in detail in GIERE et al. (1988a). Through the series of samples extracted with this instrument, differentiated gradient profiles of pore water can be obtained for subsequent measurements. The versatility and sturdiness make the pore water lance a reliable and effective tool in depths accessible to diving (Fig. 19B). Essentially similar designs are now in use by several working groups.

For soft bottoms in deep water, more complicated harpoon samplers with a triggering device have been constructed. As a spring loaded piston moves up in the sampling chamber, pore water is sampled and filtered (BARNES 1973).

Another suction corer for use in deep water has been designed by SAYLES et al. (1976) for geochemical studies, but it can be modified for the smaller dimensions of meiobenthic purposes. A dampened piston, moving in a cylinder, is mounted on top of a pointed metal tube with a series of ports capped with some filtering device. Through the action or the triggered piston, pore water is sucked through the ports into a series of chambers mounted in the inner lumen of the tube. Hauled in on deck, the water can be extracted from the chambers by syringes.

3.3.2 Squeeze Sampling

REEBURGH (1967) used a plastic (Delrin) squeezer equipped with filter screens on top of a drain tube. Mounted on a glass bottle, the squeezer is subjected to gas pressure and the filtered pore water is efficiently squeezed out of the sediment.

3.3.3 Centrifuge Sampling

In coarse sand, particularly in eulittoral sediments of low porosity, extraction of pore water by suction or squeezing becomes problematical and inefficient. Here, centrifuging yields much better results. After only a short centrifugation time and without further contamination, pore water can be extracted in quantities larger than obtained by squeezing methods. SAAGER et al. (1990) developed especially manufactured centrifuge tubes with a built-in filter unit for filtering the samples without any cotamination. An further step to maintain an undisturbed pore water sample is to use special corers adapted to function directly as centrifuge tubes.

CHAPTER 4

Biological Characteristics of Meiofauna

4.1 Adaptations to the Biotope

The heterogeneity of meiofauna biotopes outlined above explains why there are only few morphological adaptations of general validity for the diverse meiofauna. It is in the interstitial system of medium and coarse sands that the mesopsammic fauna displays the most clearly adaptive features related to this peculiar environment. The formative constraints to which interstitial meiofauna are exposed, the premises for entering the world of a narrow void system, will be detailed in the following sections.

4.1.1 Adaptations to Narrow Spaces: Miniaturization, Elongation, Flexibility

It is first of all necessary to be small, at least in one dimension (e.g., body width). This becomes particularly evident comparing body size within an animal group that contains more than just meiobenthic forms (Tables 6, 7, p. 65).

Miniaturization of body size becomes particularly evident in interstitial animals belonging to mostly macrobenthic groups that prevail in other habitats (Table 8). It is less striking in those numerous groups in which most, if not all, representatives belong to the meiobenthos, such as ciliates, tardigrades, (free-living) nematodes, plathelminths, halacarids and harpacticoids. Diminution of body size is believed to have intrinsic lower limits for the various animal groups: 0.5 to 1 mm in many taxa (SWEDMARK 1964), 0.3 mm in copepods (SERBAN 1960). It often entails reduction of cell number since the average cell size is fairly constant. However, the high number of small cells in the minute Loricifera (Chap. 5.10) shows that this rule has its exeptions. Dwarfism often leads to a simplification of body organization or to loss of organs (number of gonads, loss of eyes) and, along the lines of regressive evolution, it can phylogenetically lead to new taxa restricted to interstital refuges (see Chap. 6).

Adaptation to the interstitial habitat is often achieved by reduction in width only, while the slender bodies can remain surprisingly long (e.g. in ciliates the Trachelocercidae, in turbellarians the Coelogynoporidae, in nematodes the Stilbonematinae, in polychaetes the "archiannelid" *Polygordius*, in oligochaetes the gutless phallodriline genera; in Acari the oribatid Nematalycidae).

Table 6. A comparison of copepod size in various habitats (Kunz 1935)

Habitat	Size (mm)
Pelagic environments (plankton)	1.4
Mud	0.8
Phytal	0.7
Sand (interstitial)	0.5

Table 7. The portion of small copepods (< 0.5 mm length) in various habitats. (Lang 1948)

Habitat	Percentage
Phytal	13.6
Mud (mostly burrowers)	14.0
Coarse sand	35.7
Medium sand	72.7

Table 8. Examples of dwarfism in interstitial animals belonging to larger-sized taxa (After Remane 1933, extended)

Animal group (genus)	Body size (mm)
Cnidaria (*Halammohydra*)	0.4
Polychaeta (*Nerillidium, Diurodrilus*)	0.4
Gastropoda (*Microhedyle*)	1.0
Priapulida (*Meiopriapulus*)	1.5
Echinodermata, Holothuria (*Labidoplax*)	3.0
Ascidiacea (*Psammostyela*)	3–4

A filiform body shape, sometimes with a corresponding reorganization of internal organs, also offers advantages related only indirectly to the small habitat dimensions: the locomotory active surface (by ciliation or body musculature) becomes enlarged, effects of adhesion and anchoring become improved, the favourable relation of body surface to diameter facilitates effective transepidermal uptake and diffusion of dissolved organic substances and gases and enables a complex pattern of tactile stimuli.

While the length-width ratio is normally from 3:1 to 10:1, it can attain in typical interstitial animals 100:1, e.g. in interstitial nemerteans, some polychaetes and oligochaetes (compare the "width index" of Remane 1933); but also specialized nematodes, crustaceans and other groups attain similar size relations (Jensen 1986, 1987b). This convergent adaptive advantage of body elongation becomes particularly apparent in representatives that belong to groups which normally have a more oval to roundish shape (Fig. 20).

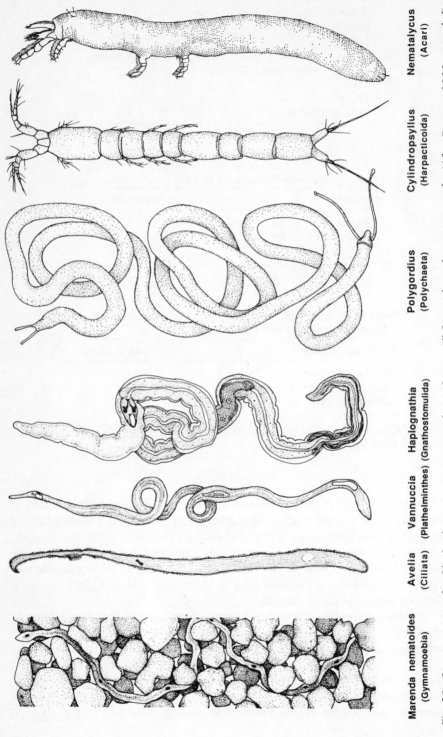

Fig. 20. Convergent trends of body elongation in interstitial meiofauna. All figures brought to same scale. (After REMANE 1933, extended)

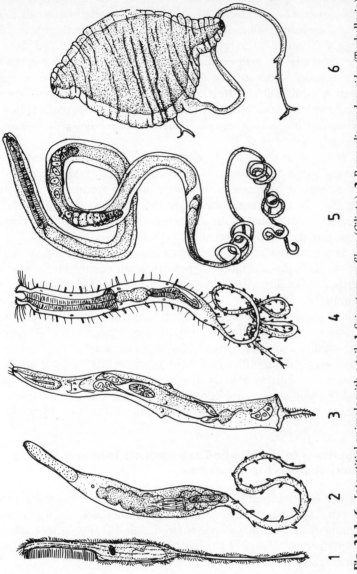

Fig. 21 1–6. Interstitial animals with a tail. **1** *Spirostomum filum* (Ciliata). **2** *Boreocelis urodasyoides* (Turbellaria). **3** *Thylacorhynchus caudatus* (Turbellaria). **4** *Urodasys viviparus* (Gastrotricha). **5** *Trefusia longicauda* (Nematodes). **6** *Heterostigma fagei* (Ascidiacea). All figures brought to same scale. (After Ax 1963, extended)

Extreme flexibility is an important adaptation for a life in the labyrinth of the interstitial system of sand grains. In soft-bodied, vermiform animals this is rather easily accomplished, but in animal groups with a chitinous cuticle, often forming compact "tagmata", the body must become more articulated, forming small, uniform segments (e.g. in interstitial harpacticoids, isopods and amphipods). Through this modified structure, the body atains a vermiform overall shape flexible enough to easily allow U-turns. In those interstitial worms where the number of segments is not constantly fixed, their number often is greatly enhanced (some oligochaetes with > 150 segments, the polychaete *Polygordius* with up to 185 segments).

Perhaps linked with the trend to elongate the body and to cling to the substratum is the frequent development of a tapering body end forming a tail, convergently present in numerous interstitial animal groups (Fig. 21). It is doubtful that there is a common function in this structure. While the tail in *Urodasys* suggests functioning as an anchoring device against displacement, in others it has rather been interpreted as a tactile organ.

Another frequent adaptation to the interstitial environment is flattening of the body. This enables the animals to squeeze through narrow crevices and to increase their body flexibility. Also, it enhances the forces of friction against the substratum important for an effective movement of wriggling. As in body elongation, there are also physiological and behavioral advantages for becoming flat (oxygen diffusion, transepidermal nutrient uptake, enlargement of the contact zone with the sand grains, development of a "creeping sole").

In groups with a roundish diameter the dorsoventral flattening is an exceptional phenomenon occurring mainly in interstitial forms (e.g. in the ciliates *Remanella* and *Loxophyllum*, in some chromadorid nematodes, in the oligochaete *Olavius planus*, even in some crustaceans like mystacocarids and ostracodes.

4.1.2 Adaptations to the Mobile Environment: Adhesion, Modes of Locomotion, Reinforcing Structures

Flattening of the body surface often enhances the chances to adhere to the surrounding sand. For many meiobenthic animals, adhesion is an important reaction adapting them to life in a mobile environment. Development of adhesive or "haptic" structures are therefore particularly frequent in the interstitial fauna of mobile sands (Table 9).

Table 9. Portion of "haptic" animals in meiofauna. (REMANE 1933)

Substrate	Assemblage	Percentage
Mud	Pelos	0–18
Plants	Phyton	24
Sand	Psammon	54

Adhering to small, mobile particles like sand grains is achieved through a wide variety of structural and behavioral adaptations (Fig. 22).

Functionally, these organs are often based on the "dual gland system" (TYLER 1977; RIEGER and TYLER 1979), where one gland contains the adhesive mucus, the neighbouring gland produces the releasing substance.

- Secretion of viscid mucus on the whole body surface: many nematodes, turbellarians, gastropods, annelids;
- excretion of sticky coelomic fluid through caudal coelomopores in interstitial oligochaetes (*Marionina*);
- adhesive glands aggregated in particular regions forming special structures: an adhesive "girdle" in the turbellarian *Rhinepera* and the polychaete *Dinophilus*, posterior haptic "toes" in many rotifers and in the polychaetes *Diurodrilus*, *Saccocirrus*, *Hesionides* and *Protodrilus*;
- haptic setae: in ciliates (cirri), gastrotrichs, many nematodes (Draconematidae, Epsilonematidae), kinorhynchs and in many ostracodes.

Beside the glandular adhesion, there are examples of mechanical adhesion (the discs on the toes of the tardigrade *Batillipes*, the sucker-like mouth parts in the harpacticoid *Porcellidium*.

Parallel to the structural and morphological features mentioned above, an effective adhesion process requires behavioral adaptations. The animals must react immediately with adhesion to increased water currents, sediment displacement or other sudden disturbances. After a while, the contact by glandular excretions will be chemically released in animals with a dual gland system (e.g. many nematodes, rotifers) or is physically ruptured by movements of the animal (*Marionina*).

In the phytal, where larger and complex substrates are present for clinging, typical adaptations of the meiofauna are clinging legs, often equipped with long bristles and spines (Halacarida, Tardigrada, Harpacticoida, Ostracoda; see Chap. 8.4).

Many meiobenthic animals have developed specialized locomotive organs and characteristic modes of locomotion, often in conjunction with adhesion or elongation of the body.

- Ciliation: The mesopsammon includes an unusually large number of ciliated animals (Table 10).
  Gliding by means of ciliation is typical not only of ciliates, turbellarians and gastrotrichs, where it is a group characteristic, but also in many meiobenthic polychaetes (archiannelids), molluscs, hydroids (*Halammohydra*) and, in a modified way, also in rotifers which move with their wheel organ. Many ciliated animals move very quickly, particularly members of the proseriate plathelminth family Otoplanidae. Some gastrotrichs seem to "swim" through the coarse sand although, in fact, as animals with positive thigmotactic behaviour they remain in contact with the substratum.
- Wriggling, i.e. an undulatory propulsion by alternation of pushing and bending, is typically developed in nematodes which possess only longitudinal muscula-

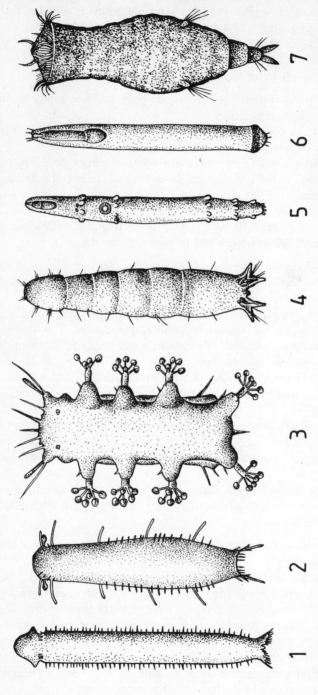

Fig. 22 1–7. Interstitial animals with characteristic adhesive structures. **1** *Turbanella* (Gastrotricha). **2** *Thaumastoderma* (Gastrotricha). **3** *Batillipes* (Tardigrada). **4** *Diurodrilus* (Polychaeta). **5** *Cicerina* (Turbellaria). **6** *Rhinepera* (Turbellaria). **7** *Lindia* (Rotifera). All figures brought to same scale. (After REMANE 1933, extended)

Table 10. Portion of ciliated meiobenthos among total fauna. (REMANE 1933)

Substrate	Percentage
Mud fauna (pelos)	7
Plant fauna (phyton)	15
Sand fauna (psammon)	54

ture, but occurs also in specialized oligochaetes like the enchytraeid *Grania*, which has a rather solid cuticle and thick longitudinal musculature, but only a thin circular muscle layer. Also some gastrotrichs with a thick cuticle and some turbellarians have writhing and kicking movements. Even some psammobiotic harpacticoids move in a similar way. Their vermiform body does not move by the strokes of their legs but by a general wriggling of the body which pushes the animal through the sand.

– Crawling on the sand grains occurs relatively seldomly in some slowly moving groups. It is typical for halacarids, ostracodes and tardigrades.

– Burrowing is common only in the meiobenthos of soft muds, be it by peristaltic contractions of the body musculature as in some annelids, or by the typical protrusion-eversion of an introvert (priapulids) or the eversion of the introvert of kinorhynchs which derive their name from this kind of motion. In crustaceans inhabiting soft muds, digging is exerted by powerful leg movements (some harpacticoids, ostracodes and cladocerans).

– Climbing through the thicket of sand grains or plants by extending long, elastic and often sticky appendages and drawing their body behind is a typical movement of some exceptional interstitial animals whose "normal" relatives are pelagic swimmers or sessile benthic forms: the mesopsammic genera of cnidarians (*Psammohydra, Otohydra*), the meiobenthic anthozoan *Sphenotrochus*, the slow-moving brachiopod *Gwynia*, the mobile bryozoon *Monobryozoon*, the small holothurian *Leptosynapta*, and the sand-living ascidian *Psammostyela*.

– Somewhat related to this movement is the iterative contraction-elongation of the whole body with alternating fixation of the body ends (looping). Characteristically known from inch worms (larvae of geometrid Lepidoptera) and hirudinids, this type of movement occurs fairly frequently also in meiobenthic animals, mostly in conjunction with rhythmic body elongation and subsequent adhesion to the substratum (Fig. 23). It characterizes the movement of very different meiobenthos such as the hydrozoans *Psammohydra nanna* and *Protohydra leuckarti*, epsilonematid and draconematid nematodes, the gastrotrich *Macrodasys*, the tiny gastropod *Unela*, rotifers adhering with their "foot" and their wheel organ such as *Philodina*, and the annelid *Rheomorpha*.

Members of meiobenthic groups living in the interstitial of sand are frequently "armoured" with structures interpreted as reinforcement and protection from mechanical stress (pressure, agitation of sediment; Fig. 24).

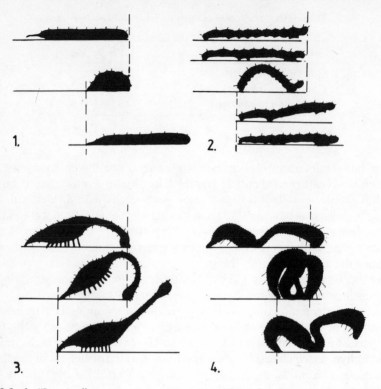

Fig. 23 1–4. "Looping" movements in various taxa of meiobenthic animals. **1** *Macrodasys* (Gastrotricha). **2** *Rheomorpha* (Annelida). **3** *Draconema* (Nematoda). **4** *Epsilonema* (Nematoda). All figures brought to same scale. (Compiled from various authors)

However, in those cases where the animals live in little exposed soft mud, the mechanically protective character of these structures is doubtful.

Solid external cuticular plates, shells and rings characterize some interstitial halacarid, ostracod and harpacticoid species as well as epsilonematid and desmoscolecid nematodes. Combined with scales, spicules and thorns, these structures are frequent also in gastrotrichs, in kinorhynchs, loriciferans, priapulids (particularly the priapulid larva), and in aplacophoran molluscs. Internal crystals, fibrils and spicules occur repeatedly in intersitial animals. Some ciliates aggregate conspicuous fibrils (e.g. *Diophrys, Aspidisca*) and spicules (*Remanella*) in their cytoplasm. The turbellarian *Acanthomacrostomum spiculiferum* derives its species name from its armature of internal spicules. Other meiobenthic forms belonging to typically soft-bodied groups like molluscs and holothurians contain a lattice of characteristic spicules: the snails *Rhodope* and *Hedylopsis* as well as the minute sea cucumber *Leptosynapta* (Fig. 24; see RIEGER and STERRER 1975).

Another means of stiffening the body is vascularization and turgescence of cells in tissues or intercellular spaces. Cushion-like structures occur in some

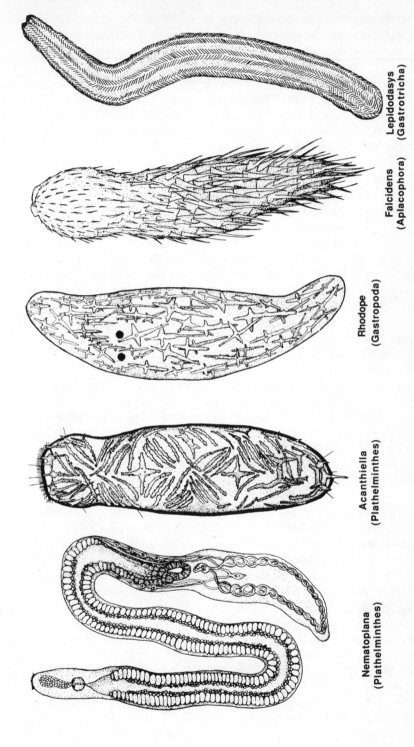

Fig. 24. Structures reinforcing the body of meiobenthic animals. All figures brought to same scale. (Compiled from various authors)

Lepidodasys
(Gastrotricha)

Falcidens
(Aplacophora)

Rhodope
(Gastropoda)

Acanthiella
(Plathelminthes)

Nematoplana
(Plathelminthes)

polychaetes, many gastrotrichs (RIEGER et al. 1974; TEUCHERT 1978), some monhysterid nematodes (VAN DE VELDE and COOMANS 1989) and turbellarians. In the turbellarians *Nematoplana* and *Coelogynopora*, intestinal cells form a stiffening "chordoid" tissue extending through the longitudinal axis of the body. Also the characteristic accumulation of turgescent cells in the prostomium of the interstitial oligochaete genus *Aktedrilus* (Tubificidae) probably has a stiffening function adapting the worm to burrowing through the sand. In this context, also the layer of vacuolized chambers in the egg capsules of some turbellarians from exposed sands should to be mentioned.

4.1.3   Adaptations Related to the Three-Dimensional, Dark Environment: Static Organs, Reduction of Pigment and Eyes

The repeated presence of static organs in the meiobenthos is considered an answer to the needs of orientation in a wide three-dimensional sediment system, comparable to the corresponding phenomenon in the members of the pelagos. Static organs (organelles) are common in members of the ciliates (*Remanella*) and hydroids (*Halammohydra, Pinushydra*), turbellarians (Acoela and otoplanid Proseriata), some enoplid Nemertinea, the isopod group Anthuridea and the meiobenthic synaptid holothurians. They are often idioadaptations, developed not only in the interstitial forms, but also among their larger relatives (e.g. holothurians).

Living in an environment without access to light, in many meiobenthic animals the body pigmentation is reduced; commonly they appear whitish-opaque or transparent. Favoured by the small body dimensions of meiobenthos, this makes scrutinization of internal organs for identification possible without dissection. The lack of body pigments and prevalence of transparent tissues is particularly striking in comparison with related larger, epibenthic species, and makes meiofauna studies aesthethically very satisfying; but there are also exceptions of the general pigment reduction in meiobenthos. Particularly in the calcareous sands of warm water regions, a common white pigmentation of meiofauna is striking. Symbiosis with white sulphur bacteria makes numerous gutless tubificids and stilbonematid nematodes shiny white. But also some aposymbiotic harpacticoids, turbellarians, ciliates and other meiofauna are brightly white in colour. Sometimes, even an orange to pink or yellowish colour (*Meiopriapulus fijiensis*, polychaetes, ostracodes) blends in well with the pink to orange shine of some coralline sands. Whether this coloration really represents a camouflaging adaptation to the environment remains dubious, at least for motion-oriented or non-selective macrobenthic predators. In any case, it makes them sometimes difficult to discern for the researcher.

Reduction of eyes is particularly striking in animal groups where eyes usually are present, e.g. in polychaetes. Endobenthic meiofauna are mostly blind, only a few epibenthic forms such as naidid oligochaetes, ostracodes and harpacticoids often have retained their eyes.

4.1.4 Adaptations Related to Reproduction and Development

The minute size of meiobenthos does not allow for shedding of abundant gametes, but in many cases also particular biotopical features (negligible water movements) require specialized ways of sperm transfer, fertilization and development.

Due to their reduced size many meiobenthic animals have only a single ovary. Most of them produce just one or only a few eggs (Table 11).

Table 11. Reduced egg number in some typical meiobenthos. (From various authors)

Taxon	Egg number
Halammohydra (cnidarians)	1–4
Kalyptorhynchia (turbellarians)	1–3
Trilobodrilus (polychaetes)	2–5
Microdriline Oligochaeta	1–4 (at one time)
Microparasellidae (isopods)	1–4
Angeliera (isopods)	1–2
Bogidiella (amphipods)	1–2
Uncinotarsus (amphipods)	1–2
Leptosynapta (holothurians)	2–3

Formation of loose sperm bundles (spermatozeugmata) or more complicated spermatophores considerably enhances the chances of fertilization and reduces the risk of sperm loss in a world where spermatozoa, released into the pore water, would not be carried by water currents to any greater distance. Sometimes sperm bundles are deposited on the substratum, e.g. in Halacaroida and in the polychaete *Nerilla*, where they are subsequently taken up by the partner. More frequently they are transferred directly onto the skin of the partner but the following process of transfer to the eggs often remains unknown (histolysis?): some tardigrades, hesionid polychaetes, tubificid oligochaetes (*Aktedrilus monospermathecus*) and kinorhynchs. Hypodermal injection occurs in the polychaete *Trilobodrilus*, the snail *Hedylopsis*, in some gastrotrichs, and, particularly well studied, in many rotifers where the initially produced sterile sperm "clears" histolytically a route towards the eggs for the subsequently released fertile sperm.

The most direct way of sperm transfer ensuring fertilization is copulation, which frequently occurs in meiobenthic animals (e.g. in polychaetes, see Fig. 46). It not only entails "construction" of organs directly participating in sperm transmittance (cirrus, penis) and often sperm storage (vesicula seminalis, spermatheca), but also of sense organs to find the partner from a distance and of behavioral cooperation to allow the correct positioning needed to complete copulation successfully.

In groups such as turbellarians or oligochaetes, the whole taxon is characterized by this most evolved way of reproduction; but also in taxa which normally
simply shed their spermatozoa into the ambient water like polychaetes have
meiobenthic representatives developed that possess penis-like organs, complicated
modes of copulation and an elaborated sexual behaviour. The sexual foreplays
developed in some turbellarians, gastrotrichs and polychaetes are astonishingly
complex for these "lower invertebrates."

An important way to further enhance the chances for successful reproduction
in a world with a limited distributional range for the individuum is hermaphroditism. Most macrodasyoid gastrotrichs (e.g. *Turbanella*) and some meiobenthic
polychaetes are hermaphrodites (*Ophryotrocha, Microphthalmus*), but also some
tardigrades (some Echiniscidae; many Eutradigrada), hydroids (*Halammohydra,
Otohydra*) and even some meiobenthic crustaceans (some Tanaidacea) and
holothurians (*Leptosynapta*). Groups such as gnathostomulids, turbellarians and
oligochaetes with hermaphroditism as a general feature of the taxon are in this
respect preadapted to a meiobenthic life.

Despite their low number of embryos and reduced vagility, the survival of
most meiobenthic species is favoured by direct development without any free-
swimming larvae, by relatively short generation times and frequently continuous
reproductive periods. Like the sperm, the fertilized eggs are not freely floating in
the pore water. Instead, with their sticky surface they adhere to sediment particles
or they are ensheathed in a capsule (turbellarians) or cocoon (oligochaetes,
some polychaetes) which becomes glued to the surface of sand grains.

Loss of offspring is further reduced by hatching of well developed and self-
sufficient juveniles. Viviparity is not uncommon (e.g. the gastrotrich *Urodasys
viviparus*, the rotifer *Rotaria*, the hydroids *Armorhydra* and *Otohydra vagans*).
Parental brood protection of the embryos occurs frequently in special capsules
or pouches of/on the body. Nerillid polychaetes attach their offspring to the
posterior part of the body, ovisacs (harpacticoids) or "marsupia" (peracarid
crustaceans) render protection to the developing embryos. Rapid multiplication
by asexual multiplication (budding or fission) occurs in some turbellarians
(Macrostomida), some polychaetes (Syllidae), in most Naididae (Oligochaeta)
and in Aeolosomatida. Parthenogenetic reproduction is regular in rotifers,
chaetonotoid gastrotrichs and in cladocerans).

In animal groups of meiobenthic size, the trend to abbreviate ontogeny
often led to neotenic or progenetic development (WESTHEIDE 1984, 1987a).
Progenesis characterizes many "archiannelids" (e.g. *Ophryotrocha, Dinophilus,
Apodotrocha, Trilobodrilus*), which have retained larval traits such as ciliar tufts
and bands, lack of coelomic compartimentation. But progenesis occurs also in
meiobenthic opisthobranch molluscs, in many gastrotrichs and interstitial
Cnidaria, and was recently found also in some loriciferans. In some (freshwater)
crustacean groups, progenesis led to phylogenetically interesting new groups (see
Chap. 6).

Encystment often helps to survive periods of hostile life conditions in marine
tardigrades, in many turbellarians from salt marshes, but it occurs also in

Dinophilus taeniatus (Polychaeta) and in the marine harpacticoid *Heteropsyllus nunni* (COULL and GRANT 1981; WILLIAMS-HOWZE and COULL 1992). Turbellarians, cladocerans and rotifers have stages of "dormancy". The "cryptobiotic" stages in freshwater tardigrades and some nematodes are perfectly adapted to survive under extreme conditions. Some antarctic harpacticoids seem to have a free diapause phase during their copepodite phase (DAHMS 1991). Larval stages of some deep-sea species of *Rugiloricus* (Loricifera) can retard their hatching by encysting (KRISTENSEN 1991b).

Despite the heterogeneity of the numerous animal groups represented in the meiobenthos, despite their different organization, complexity in structure, taxonomic rank and probably also phylogenetic age, they all have been subject to "integrating adaptations" (REMANE 1952a) by the constraints and dynamics of the habitat. These adaptations often formed "analogous specializations" with often surprisingly uniform convergent traits in groups of different systematic position and habitat (e.g. mesopsammal, edaphal, partly also phytal; REMANE 1952a). Their presence indicates the ecological links connecting all (mesopsammic) meiofauna.

More detailed reading: REMANE (1933, 1952a); AX (1966, 1969); SWEDMARK (1964); SCHWOERBEL (1967)

CHAPTER 5

Meiofauna Taxa – a Systematic Account

Many of the zoological groups belonging to the meiobenthos are either omitted in zoology textbooks or commented upon as "small and isolated groups" in a few lines in petit. However, for the advanced student, they often represent anatomically fascinating and phylogenetically important taxa. Nevertheless, the overview over these meiobenthic groups given here with its prevalence on ecological and gross diagnostic aspects and its omission of many anatomical details, cannot substitute textbook reading. The frame of this book does not allow for completely covering all groups that possibly have some isolated representatives of meiobenthic size. Its figures, which depict only some selected forms, cannot serve for any more detailed identification. Each chapter begins with taxonomical data, continues with biological, ecological and distributional comments and ends with some aspects of relationship and phylogeny (where appropriate, these topics are set off by separate subheadings).

5.1 Protista

5.1.1 Foraminifera (Sarcomastigophora, Rhizopoda)

Most foraminiferans belong to the size spectrum of meiobenthos, although there exist numerous species with larger, often concrete shells that can attain up to 12 cm and more. There have been 34000 species described altogether, whereof about 4000 are recent (LEE 1980b). The multitude of shell shapes is bewildering (Fig. 25). From small symmetrical or snail-shaped forms to plait- or ribbon-shaped ones, from disc-like to completely asymmetrical forms which are hardly recognizable as foraminiferans, and thus have often been overlooked. The original organic substance of the shell secreted by the cytoplasm will mostly be more or less completely substituted by calcareous material (exception: the Allogromiidae with their soft shell of organic material).

Although by far the most species are epi- and endobenthic forms of soft sediments and not planktonic drifters, the biology and ecology of benthic foraminiferans is poorly known. The large species often are heterokaryotic like ciliophorans, i.e. they have a macro- and micronucleus. Many of the disc-shaped forms live symbiotically with unicellular algae and cover the bottom of shallow waters where they can utilize the sunlight. Foraminiferans cling to the substratum by

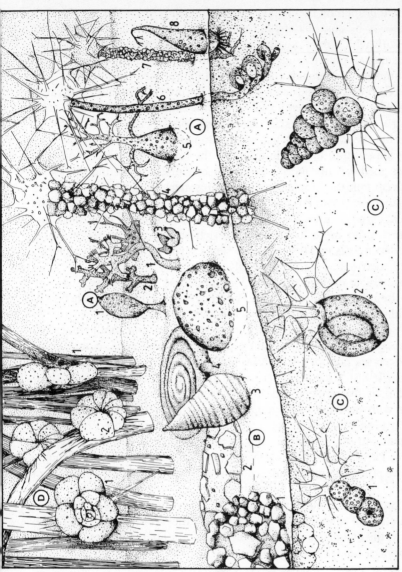

Fig. 25. Four morphogroups of Foraminifera in their typical habitat (based on JONES and CHARNOCK 1985). **Morphogroup A** (hyperbenthic suspension feeders): 1 *Haliphysema*; 2 *Dendrophyra*; 3, 5, 8 *Pelosina*; 4, 7 *Jaculella*; 6 *Bathysiphon*. **Morphogroup B** (epibenthic, surface-dwelling herbivores, detritivores, omnivores) 1, 2 *Psammosphaera*; 3 *Hippocrepina*; 4 *Ammodiscus*; 5 *Saccamina*. **Morphogroup C.** (endobenthic, sediment-dwelling herbivores and detritivores) 1 *Ammobaculites*; 2 *Miliammina* and *Quinqueloculina*; 3 *Textularia*. **Morphogroup D** (phytal herbivores) 1 *Trochemmina inflata*; 2 *T. labiosa*

their long, often branching pseudopodia (reticulopodia, rhizopodia, see Fig. 25). They move slowly by contractions of their amazingly large net of cytoplasmic appendages with which they prey on bacterial biofilms, pick detritus, but also bacteria and small animals for ingestion. In some cases, even animals of 2–3 cm in size, like cumaceans, caprellids and newly metamorphosed echinoderms, have been found in their cytoplasmic net. Apparently, the prey is immobilized (toxins?) before it is rapidly digested (even for large prey digestion lasts only 1 day). This uptake of nutrition seems to be a selective process. Some species are herbivorous, preferring certain benthic microalgae (e.g. diatoms) as food.

Relating life position and feeding habits, the non-parasitic foraminiferans have been assigned to four "morphogroups" (see Fig. 25; JONES and CHARNOCK 1985). These contain the tubular and branching suspension feeders which are semi-sessile, anchored in an erect position preferably in more exposed sediments in the sediment (Type A), the deposit feeding or (passively) herbivorous epibenthic Type B occurring on more lenitic, detritus-rich sediments, the more vagile, endobenthic detritivores and herbivores in sands (Type C) and the phytal forms living on phytodetritus or plants (Type D).

The considerable ecological role of foraminiferans was not assessed until recently. In the tidal flats of the North Sea, they usually occur in densities of about 700 individuals per 10 cm^2 (*Ammonia beccari*, CHANDLER 1989) or > 1000 ind. per cm^3 sediment (ELLISON 1984). Sandy bottoms on the continental shelf harboured more than 400 individuals per cm^3 (GROSS, pers. comm), and even in the deep-sea bottom 150 to 200 ind. per cm^3 sediment have still been found (GOODAY 1986). GABEL (1971) counted in the northern North Sea > 1000 ind. per g sediment (dry wt.) with a decreasing trend towards the south. Although Foraminifera with their net of pseudopodia can densely cover the sea bottom, their share in productivity and benthic turnover has hardly been quantified. They can dominate the meiofauna, but in most studies on meiobenthos they are neglected. Particularly in the deep-sea (see Chap. 8.3), the overall trophic and productive processes have to be reconsidered, having found that 50–80% of the meiofauna species and 30% of their biomass are made up of foraminiferans (see Fig. 83; SHIRAYAMA and HORIKOSHI 1989). They may even play a key role in the production of manganese nodules where they contribute 10% of the outer layer. In the cytoplasm of Foraminifera, granules of manganese and iron are quite common (MULLINEAUX 1987).

More detailed reading: tidal flats – HOFKER (1977); North Sea – RHUMBLER (1938); GABEL (1971); deep sea – GOODAY (1986); review – LEE (1980b)

5.1.2   Testacea (Rhizopoda)

There are about 60 interstitial species known to occur mostly in the moist and often brackish shore line (the "hygropsammal"). Marine species usually have a shell of organic secretions with only a few particles embedded in it, while in limnetic forms the shell essentially consists of foreign particles (e.g. *Centropyxis*).

As an anchoring device, the sac-shaped shell has a wide circular opening from which the pseudopodia extend. In some species spiny processes on their shell serve additionally against suspension when the sediment becomes agitated.

▓   *More detailed reading*: GOLEMANSKY (1978)

5.1.3   Ciliophora

These heterokaryotic protists (the genome is separated into micro- and macro-nucleus) are classically divided in Holotrichia, Heterotrichia, Hypotrichia etc. This artificial grouping becomes gradually substituted by phylogenetically more natural taxa; but many references still refer to the "old" classification.

Identification of ciliates is – in a first step – based on inspection of live animals. The diagnostic features of many forms, characteristic enough to allow relatively easy identification at the genus level (body shape, ciliation, position of mouth opening), will immediately disappear after fixation. However, species determination usually requires scrutinizing the nuclear apparatus and infraciliation by staining procedures.

Benthic habitats harbour more ciliate species than the open water. About 1000 of the 8000 species described belong to the meiobenthos. In coastal sediments, a 10-cm-long core of $1 \, cm^2$ surface area may contain between 30 and 100 species (FENCHEL 1992). Not only is the pellicle mostly thicker in benthic species than in others, but there are often clear structural correlations between general body shape and size of pore volume (FAURÉ-FREMIET 1950).

While "euporal" ciliates are apparently rather independent of a particular grain size and occur in both fine and coarse sediments, "microporal" ciliates populate the interstitial system of fine to medium sands (250 to 400 µm). Their rather thin pellicle renders them a high flexibility, the slender, often thread-like or flattened body allows them to quickly glide forward despite sometimes a considerable body length ($> 3000 \, µm$). The microporal species avoid sediment agitation. The most primitive ciliates, the Karyorelictida (the 2n-macronucleus is not capable of division), are characteristic members of these microporal ciliates with the common family Trachelocercidae (Fig. 26). The genera *Trachelonema, Tracheloraphis, Geleia* and *Remanella* occur frequently in the fine to medium sands along the coasts, but convergent adaptations to an interstitial life (body elongation, flattening, head-like anterior end, formation of a tail shaft, see Chap. 4.1) are also developed in the non-karyorelictid genera *Condylostoma, Spirostomum, Blepharisma* (Heterotrichia).

"Mesoporal" species live in coarse sands of 400 to 1800 µm pore width, where a thick and sometimes armoured pellicle (plates, spines) helps to survive the strong wave exposure. The wide void system allows the small to medium-sized, mostly oval to slightly flattened forms (often Hypotrichia) to move in the pore water with sudden jerks of their ventral cirri. They are also found to cling to the grains by adhesive cirri. These mostly hypotrichous ciliates are frequently

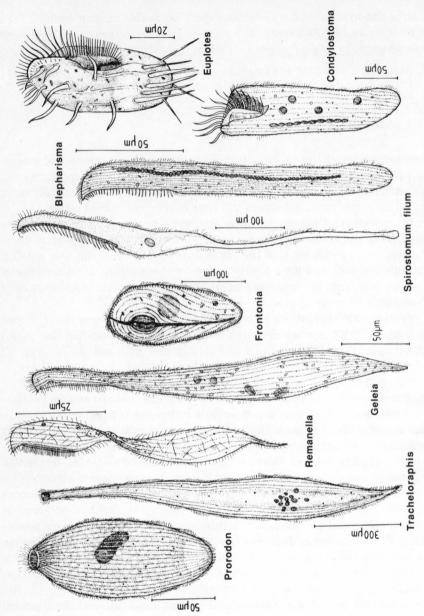

Fig. 26. Some benthic Ciliata. (Various authors)

represented by the genera *Aspidisca, Frontonia, Diophrys, Prorodon, Euplotes, Coleps* etc. (Fig. 26).

Most ciliates occur in medium to fine sands where they can numerically exceed all other animals: 10–100 ind. with 1–20 µg (fresh weight) per ml sand or 20 ind. per g sediment (dry weight). FENCHEL (1969, 1978) found up to 50×10^6 ind. m^{-2} corresponding to 2 g fresh weight. In favourable medium to fine sand ciliates can dominate all other meiofauna even in biomass. Contrastingly, in exposed coarse sands and in sediments of < 100 µm particle size ciliates decrease rapidly in abundance. Fine sediments whose voids become clogged by organic matter are devoid of ciliates. Highest abundance was found in sediments with < 1 wt% organic substance (SICH 1990).

Among ciliates all trophic types are represented. Although most of them are bacterivores, there are also detritivores, often even selective algivorous grazers (e.g. *Remanella* spp., feeding on diatoms of selected size, see FENCHEL 1968a), histophagous scavengers and carnivorous predators. SICH (1990) grouped the ciliate fauna from the brackish Kiel Bight (Baltic Sea) in bacterivores (50%), herbivores (30%), carnivores (1%) and omnivores (15%). He found experimentally that only those 10–15% of bacteria are ingested by ciliates which swim freely in the pore water, while those bacteria remaining on the sand grains are not utilized. EPSTEIN et al. (1992) concluded after inspection of the digestive vacuoles (gastrioles) that benthic ciliates from a temperate tidal beach at the American east coast were largely herbivorous, consuming mainly flagellates and to a lesser extent diatoms. Heterotrophic flagellates made up 17% of their food. In general, the herbivorous and omnivorous ciliates live in more superficial horizons while the bacterivores are encountered deeper down.

Most ciliates will be found in the oxidized, upper 2 cm of the sediment. However, many bacterivorous species tolerate periodically hypoxic conditions and aggregate around the RPD layer (*Geleia nigriceps, Paraspathidium, Remanella* sp.) where they feed on the rich sulphur bacteria (Fig. 27). In winter, many ciliates are found deeper down in the sediment than in summer (e.g. *Coleps, Condylostoma*).

The capacity to live even in permanently anoxic conditions mostly coincides with the presence of "hydrogenosomes" instead of mitochondria (FENCHEL and FINLAY 1991). Many of the more specialized ciliates like *Uronychia transfuga, Kentrophoros* sp. (FENCHEL and FINLAY 1989) harbour bacterial endosymbionts (methanogenes) which enable them to live in sulphidic and methanic sediment (see Chap. 8.5).

The annual population development of ciliates in spring closely follows the rising temperature and increasing food supply. In fall, food availability seems to determine population sizes. In boreal latitudes minimal populations are found in February and March. Generation time (sexual reproduction by conjugation) is variable, usually about 10 days, but ciliates can multiply asexually by division already after a few hours (SICH 1990).

Annual P/B values (see Chap. 9.2) of about 250 are not unusual for ciliates (compare 13 in metazoan meiobenthos and 2–3 in macrofauna!). It is this rapid

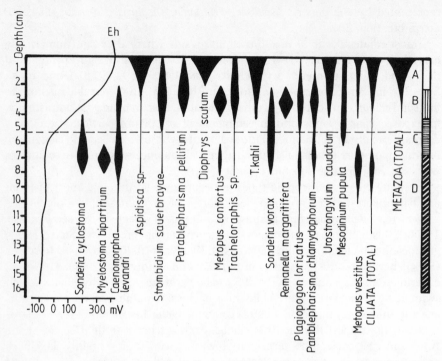

Fig. 27. Vertical distribution of ciliates in a Scandinavian beach in relation to the redox·
potential. *Broken line* indicates position of redox potential discontinuity. Sediment layers
A–D stand for: *A* light, oxic surface sand; *B* greyish sand; *C* decomposing *Zostera*; *D*
blackish anoxic sand. (After FENCHEL 1969)

turnover which underlines the productive relevance of ciliates while their
standing stock can stay relatively small, particularly so, since many carnivorous
ciliates feed intensively on their bacterivorous relatives.

The trophic relevance of bacterivorous ciliates has been documented in
experiments on the degradation of algal debris from wrack (RIEPER-KIRCHNER
1989). Addition of ciliates to the experiments caused a considerably higher
bacterial activity resulting in a quicker degradation of plant remnants (see also
ALKEMADE et al. 1992). Through ciliate grazing, the microbial populations
apparently remain in their exponential growth phase. But detritivorous ciliates
contribute also directly to the mineralization of organic matter through regeneration of nutrients. For pore water bacteria, SICH (1990) found that ciliates reduce
the annual bacterial biomass production by about 50%. A species-specific interaction between various ciliates and meiofauna has been experimentally tested by
EPSTEIN and GALLAGHER (1992) in a sandy tidal flat. These data characterize
the ecological role that ciliates play in the food web as a link mediating between
bacteria as their food and Metazoa as their predators. This is in contrast to the
results of KEMP (1988), who felt that the ecological role of ciliates has often
been overestimated.

More detailed reading: ecology – FAURE-FREMIET (1950); FENCHEL (1968a, b, 1969); HARTWIG (1973b); interstitial ciliates – DRAGESCO (1960); HARTWIG (1973a); systematic and phylogenetic aspects – CORLISS (1974, 1975, 1979); illustrated key – CAREY (1992); reviews on marine ciliates – FENCHEL (1978); PATTERSON et al. (1988)

5.2 Cnidaria

Representatives from most cnidarian classes have been found to belong to the meiobenthos, although their general organization and size seems not particularly preadapted to interstitial life. During their polyp phase, many cnidarians remain

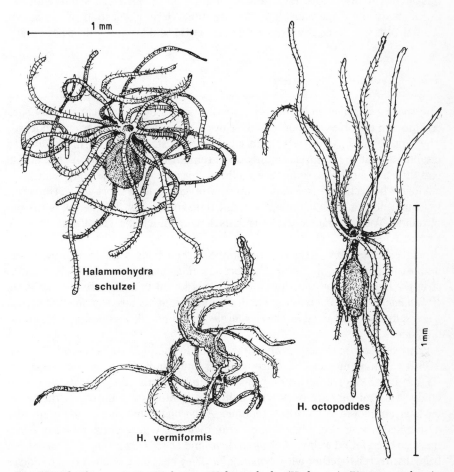

Fig. 28. The famous interstitial genus *Halammohydra* (Hydrozoa). (Various authors)

in meiobenthic size (e.g. cubozoans), hence, only some better-known polyp forms are included here as typical representatives. Since almost all polyps of cnidarians can detach from their underground and slowly change their position, it is often arbitrary whether to regard small polyps as true meiobenthos or just as temporary meiofauna, or even to omit them in a treatise on meiobenthos due to their prevalent sessility.

The first report on a "classical" meiobenthic medusa described the tiny (only 300–1200 µm) hydromedusa *Halammohydra* from the littoral sands of the North Sea (REMANE 1927). At that time it was a small zoological sensation (Fig. 28), and its lasting reputation is mirrored by the emblem of the International Association of Meiobenthologists. Today, the genus is known to occur worldwide in about ten species, and in favourable habitats *Halammohydra* can be quite common.

Due to the often extreme adaptation and reduction of diagnostic features, the metagenetic characterization and systematic position of meiobenthic medusae has often been disputed and varying among specialists. Progenesis and reduction of the medusoid stage often blur the situation. Here, a selection of the better-known, plus some recently discovered meiobenthic cnidarians, will be presented.

5.2.1 Hydroida (Medusae)

The species of *Halammohydra* have a more or less ovoid to slender body sack reminiscent of polyps with one to two rows of tentacles bearing large cnidocysts. The umbrella is completely reduced; however statocysts, normally located peripherally at the umbrella of medusae, remain well developed. At the aboral end there is a groove serving as an adhesive organ. The whole body is strongly ciliated, which enables *Halammohydra* to swim gently for a short while. However, its usual movement is climbing between the sand grains. The genus is hermaphroditic. Today, *Halammohydra* is considered a much-derived, if not neotenic hydromedusa (Fig. 28).

The species vary slightly in the shape of their body sack, their movements and their preferred biotopes. The famous *Halammohydra schulzei* is clumsy, strongly hapto-sessile and slow moving. It is found in sublittoral coarse sand. *Halammohydra octopodides* from sublittoral fine sand is more slender and active, a trend further developed in *H. vermiformis* which populates the upper eulittoral of beaches.

Armorhydra janowiczi, found at Roscoff, France (SWEDMARK and TEISSIER 1958) has more medusoid features with its small umbrella and a minute velum (Fig. 29). It lacks statocysts and has two kinds of tentacles, filiform ones with cnidocysts and capitate ones with adhesive desmonemes. Based on the set of various cnidocyst types, the cnidome, *Armorhydra* has been taxonomically designated a representative of the Limnohydrina, whose corresponding polyp has also been found today. *Otohydra vagans* and *O. tremulans* (Fig. 29) have both been discovered together with *Armorhydra*. They bear only one whirl of tentacles

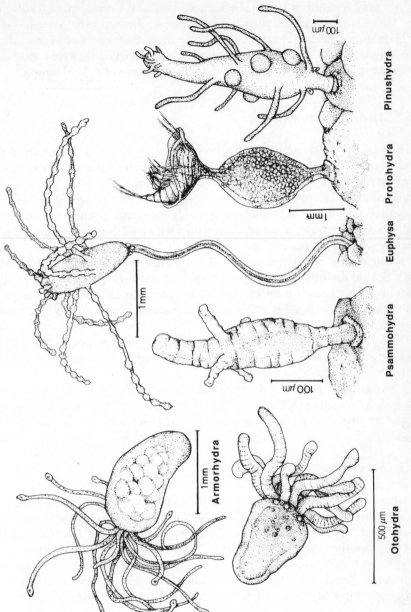

Fig. 29. Some meiobenthic hydroids. (Various authors)

and are only 350 μm in length. Lacking an adhesive groove, the animals are often found swimming with their body ciliation in the interstitial system. They are hermaphroditic and viviparous, brooding their Actinula-like larvae in their wide body sack. Although their appearance is rather polypoid, they are considered to be apomorphic medusae with a systematic position yet to be discussed.

5.2.2   Hydroida (Polyps)

The best-known representative of these hemisessile or hapto-sessile meiobenthic forms of only a few mm length is *Protohydra leuckarti*: it has no tentacles, its cnidocysts cover the whole body (Fig. 29). Gonads are rarely developed, asexual multiplication is by transverse division. Discovered already in 1868 in Belgian brackish waters, it lives in muddy, rarely sandy sediments. Occurring worldwide mostly in brackish water, it is a predator with a wide food spectrum (oligochaetes, insect larvae, nematodes and harpacticoids).

Psammohydra nanna was described from sandy bottom in the Kiel Bight (Baltic Sea) as an interstitial polyp carrying four short tentacles (SCHULZ 1950). Today it is also known from the North Sea, the Atlantic and Mediterranean Ocean. Gonads have never been observed, only asexual reproduction by division is reported.

Euphysa ruthae (NORENBURG and MORSE 1983) is an exceptional interstitial polyp through its mode of reproduction by budding with subsequent polarity reversal of the buds.

Pinushydra chiquitita (BOUILLON and GROHMANN 1990) is a newly discovered hydropolyp from interstitial sands of Brazilian beaches. It is remarkable through its basal ring of statocysts, sessile gonophores and an adhesive disc at its base.

5.2.3   Scyphozoa

Stylocoronella riedli, described by SALVINI-PLAWEN (1966) from Rovinj (Adriatic Sea), was probably seen first in 1962 by MONNIOT in the western Mediterranean (Banyuls-sur-Mer). It is a tiny interstitial scyphopolyp which has changed its body organization very little (Fig. 30). The mouth cone is expanded into a proboscis and surrounded by 24 tentacles with ocelli at their base. Sexual reproduction is unknown, the species multiplies by budding of planuloids which directly develop into young polyps. Strobilation has not been observed.

5.2.4   Anthozoa

The strange stony coral *Sphenotrochus* (ROSSI 1961) (Madreporaria, Caryophyllidae) occurs in coarse sand of the Mediterranean and Atlantic Ocean where it moves around by its numerous (up to 24) tentacles (Fig. 30). Its maximal size (about

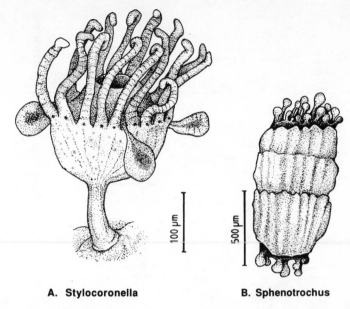

A. Stylocoronella   **B. Sphenotrochus**

Fig. 30A, B. Interstitial representatives of Scyphomedusae (**A**) and Anthozoa (**B**) (A SALVINI-PLAWEN 1966; B ROSSI 1961)

10 mm) certainly exceeds the range of meiobenthos, however, already at a length of 1–2 mm it starts to divide asexually. During fission it develops, unlike its sessile relatives, bipolar whirls of tentacles (as well as of the calcareous septa) before separation of the resulting individuals.

▓   *More detailed reading*: see citations in the text of this section.

5.3   Free-Living Plathelminthes: Turbellarians

Although recognized as an artificial group not based on the principles of phylo-genetic systematics (EHLERS 1985), the term turbellarians, well introduced in scientific literature and text books, will be maintained here for the sake of convenience. Almost the entire speciose group of flatworms is of small size, but only the large species are flat; most meiobenthic turbellarians are oval to round in cross-section. Their vermiform body, extreme flexibility, ciliation, hermaphro-ditic genital organs and internal fertilization by copulation, render them a classical meiobenthic group, preadapted for an interstitial life. As in ciliates, identification of the soft-bodied animals mostly bases on live specimens. For ascertaining further details, adequate preservation, sectioning and staining has to follow. In many studies this tedious procedure has contributed to a certain

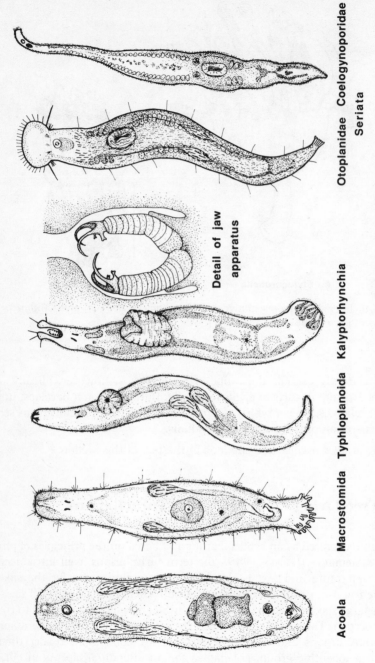

Fig. 31. Representatives of various meiobenthic turbellarian groups. (Various authors)

neglect of the group, although in meiofauna samples turbellarians usually rank third in abundance after nematodes and harpacticoids. At least identification of orders and some characteristic families is possible on the basis of well-recognizable features such as shape of the pharynx, the gonads, intestine, presence of statocysts etc.

There are probably more than 3000 species of turbellarians described both from marine and freshwater habitats, most of them from the marine littoral. In a relatively small area at the island of Sylt, about 450 species have been found. Based mostly on the thorough studies from the North Sea coasts, only some major groups can be characterized here, mentioning features recognizable already by the beginner and giving some additional ecological remarks (Fig. 31).

5.3.1 Acoela

This order contains about 250 species of small, mostly roundish size. They are characterized by an often inconspicuous mouth opening and the absence of a true intestine. Instead, the intestinal tissue is substituted by a solid "digestive parenchyma". Usually a statocyst is visible near the anterior end. In this group, often considered primitive, the ovary is not differentiated in discrete ovarian and vitelline tissue, and the diagnostically important male gonads are in posterior position.

Acoel turbellarians occur preferably in marine (fine) sand rich in detritus. Many feed on diatoms, some species are specialized to live in the sulfide biome. Significant genera: *Haplogonaria, Pseudaphanostoma, Oligofilomorpha, Convoluta* (this genus depends on the photosynthesis of numerous microalgal endosymbionts).

5.3.2 Macrostomida

There occur both marine and limnetic species. They have a *pharynx simplex*, a simple, unspecialized pharynx, in the anterior part of their body. As in the Acoela, the ovaries are uniform and not differentiated in various organs. Some species are characterized by numerous internal spiculae (Fig. 24), a few have "clepto-cnidocysts", functional cnidocysts (originating from interstitial cnidarians?) used as defense weapons in their dermis. Numerous Macrostomida have eyes, but there are no statocysts developed. Within the mostly limnetic family Microstomidae, asexual reproduction by fission is frequent. The resulting chains of zooids eventually separate in single individuals.

5.3.3 Rhabdocoela (= Neorhabdocoela)

In this order, the female gonads are differentiated in a germarium and a vitellarium.

Sub-Order Dalyelloida: these are mainly freshwater forms, characterized by a barrel-shaped pharynx (*pharynx doliiformis*) with an anterior opening. Many of them have eyes and are associated with symbiotic zoochlorellae in their tissue. Dalyelloids often feed on diatoms.

Sub-Order Typhloplanida: the marine and freshwater species (e.g. *Mesostoma*) belonging to this suborder are characterized by a *pharynx rosulatus*. Their male gonads are equipped with posterior "stylets"; the vitellarium and germarium form longitudinal tubes.

Sub-Order Kalyptorhynchia: the more than 100 marine species are mostly equipped with an anterior statocyst. In analogy to the Nemertinea (see Chap. 5.5), they have at their anterior end, independent of the mouth and pharynx, an eversible proboscis for catching prey. This proboscis is often armed with adhesive or toxic glands, chitinous teeth or elaborate grasping hooks or "jaws". In contrast to Gnathostomulida (see Chap. 5.4), these jaws always lie at the front end of the animals. Numerous kalyptorhynchs feed on harpacticoids.

5.3.4 Seriata

The germarium and vitellarium are of follicular shape; the pharynx is tubular and much folded (*pharynx plicatus*), its opening is directed posteriorly.

Sub-Order Proseriata: the numerous species of this marine group are mostly adapted to the interstitial system. Some of them are rather long, and their body is very slender (Otoplanidae often 10 mm long!). The pharynx opening is located in the posterior part of the body. The gonads are formed as longitudinal, diffuse follicles. There are various families of proseriates, and only some conspicuous ones can be mentioned here.

Otoplanidae: the "head" of these very typical interstitial turbellarians is slightly set off and equipped with long sensory cilia. It contains a refractory statocyst; the brain and posterior adhesive cells are conspicuous. The pharynx is short and lies horizontally. A very effective ventral ciliation lets the animals "rush" though the sand as if swimming. Characteristic genera are *Bulbotoplana, Parotoplana, Otoplanella*.

Coelogynoporidae: also these interstitial turbellarians have a very long and slender form with a marked "head", but, in contrast to the otoplanids, the body ciliation is uniform, not just ventral. Posterior to the statocyst the brain is visible in a conspicuous capsule. The pharynx is long and directed vertically.

Nematoplanidae: as indicated by the scientific name, these turbellarians are long and flexible. As an adaptation to their rigid interstitial habitat, they often have a vacuolized and turgescent intestinal tissue or parenchyma (see Chap. 4.1.2; Fig. 24). Nematoplanid turbellarians have no statocyst.

Sub-Order Tricladida: mostly terrestrial and limnetic genera with a clearly tripartite intestine. Some genera (e.g. *Atrioplanaria*) are stygobiotic: they live exclusively in subterranean habitats.

The well-developed diversification of turbellarian groups allows the allocation of the major groups to certain biotopes, resulting in a typical general distribution pattern: Acoela (particularly the smaller forms), Macrostomida and most Neorhabdocoela prefer fine sand and mud. Kalyptorhynchia dominate in the sheltered tidal flats; if occurring higher up the beach, they are restricted to the groundwater zone. Proseriata prevail in the lower slope of sandy beaches; however, the Otoplanidae are to be found subtidally and in the exposed swash zone with agitated coarse sand.

Highest diversity of turbellarians occurs in fine sands and tidal flats, where it is even comparable to the entire macrofauna of a coral reef. In tidal flats their species richness was reported as high or even higher than in nematodes and exceeded twice that of harpacticoids. In one North Sea tidal flat, REISE (1988) found 83 species with sometimes 20 congeneric species under $100 \, m^2$! Relatively small structures like *Arenicola* tubes seem to present "centres of population development" for turbellarians with no less than five different spatial niches and numerous co-occurring species (REISE 1987a). In Australian mangrove muds they made up 60% of all meiofauna and represented the most important taxon (ALONGI 1987). In more extreme zones like exposed sands and muds, the species number decreases, but even here it remains relatively high (REISE 1988). The same holds true for the sublittoral, so the usual steep decrease in species number from the more stable sublittoral up to the harsh supralittoral is not to be recorded in turbellarians (Table 12).

However, turbellarians become rare in bottoms beyond 400 m depth and are only scarcely recorded from the deep sea. In part, the few records from deep bottoms will relate to the delicate body structure of turbellarians and inadequate sampling/preserving methods. Not only is the species number rich in turbellarians, but their abundance is similarly high. From sandy beaches in Iceland 'OLAFSSON (1991) reported Turbellaria dominating all meiofauna with up to 700 individuals per $10 \, cm^2$. On the Belgian coast, between 100 and 500 individuals per $10 \, cm^2$ resulted in a biomass which locally exceeded nematodes (MARTENS and SCHOCKAERT 1986). In some rather extreme biotopes, e.g. in (European) salt marshes, and

Table 12. Species number of turbellarians in various habitats of the littoral zone around the island of Sylt (North Sea). (Based on data from REISE 1988)

Salt marshes		Beaches		Tidal flats			
High	102	Exposed	71	Sand	159	Subtidal sands	187
Low	104	Semi-exposed	136	Silty fine sand	142	Total no. of	
Creeks	116	Sheltered	154	Mud	63	species	435

in the exposed swash zone of beaches, turbellarians can become the dominating meiofauna group.

Among the abiotic factors it is the water content which greatly determines the occurrence of the soft-bodied turbellarians. This explains their rich number in the permanently moist fine sand of flats. In the much dryer upper beach, they prefer the moist deeper layers near the groundwater horizon. On the other hand, most turbellarians need an oxygenated environment which in less aerated habitats such as tidal flats restricts them to the superficial horizons and to the narrow oxic zone around macrofauna holes. Exceptions are, however, the acoel Solenofilomorphidae and some other turbellarian groups which contain typical thiobiotic species (see Chap. 8.5.2). Exposure to water currents and wave action, as well as general physical instability, does not seem to adversely affect turbellarian settlement (FEGLEY 1987; HELLWIG-ARMONIES 1988), which would relate to their frequent emergence and hyberbenthic occurrence (see below).

ARMONIES (1988c) showed that salt marsh species of turbellarians preferred distinct ranges of salinity and endured unfavourable conditions in protective cysts or in dormancy. This is exceptional, because salinity seems a subordinate factor for most turbellarians. Almost all species tolerate brackish conditions well. All the species occurring in the brackish Baltic Sea are also represented in the more haline North Sea (ARMONIES 1988d). The brackish water species are widespread and most of them are amphi-North Atlantic in distribution, many occurring even along the Alaskan coasts. This would corroborate the hypothesis of a boreal-subarctic brackish water community (Ax and ARMONIES 1990).

A rich food supply is the other determinant for the occurrence of voracious turbellarians. In a North Sea tidal flat, REISE (1988) found almost 65% of all species (mostly the larger ones) to be predators on other meiofauna, e.g. nematodes (OTT 1972). Particularly in sandy habitats, they can exert a considerable predatory pressure also on the smaller macrofauna (WATZIN 1985, 1986). Hence, high abundance of small fauna is a prerequisite for the presence of turbellarians. Selective predation has been observed in many turbellarian species (Fig. 32). Preference of otoplanids and coelogynoporids for oligochaetes has been documented. Their absence on beaches in New Zealand was related to the scarcity of oligochaetes (RISER 1984). Also MARTENS and SCHOCKAERT (1986) think the overwhelming majority is predacious, leaving hardly any bacterivorous turbellarian species. Other authors contend that 27% of turbellarians are diatom-eaters and still about 10% prefer small flagellates and bacteria.

As much as turbellarians are directly linked to a rich supply of suitable prey, they themselves seem to be utilized as food to a small extent only, despite their abundance. It is unclear whether the presence of defensive rhabdites in the dermis of turbellarians contributes to this avoidance. Small syllid polychaetes are one of the few predators that feed on them (WATZIN 1986).

Their differentiated distribution pattern becomes even more complex through the marked ontogenetic population dynamics with strong numerical fluctuations. About four to five annual generations, a restricted reproductive period, and a

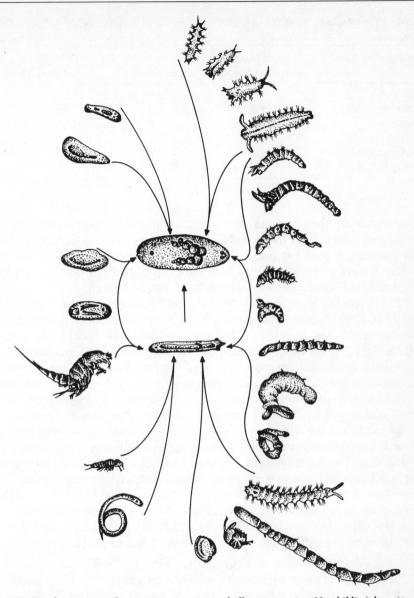

Fig. 32. Food spectrum of two representative turbellarian species, *Neochildia* (above), and *Archiloa* (below). (WATZIN 1985)

developmental time of 20–30 days are average values. However, many Acoela are "polyvoltine", reproducing throughout the year with ten annual generations. Other turbellarians are bivoltine or just univoltine.

The population dynamic is further greatly modified by seasonal and irregular temperature variations. Maximal abundance is usually in spring and fall (Acoela,

Table 13. Composition (in %) of turbellarian fauna in a sandy beach (island of Sylt, Germany)

Turbellarian taxa	FAUBEL 1976a Sandy beach	SOPOTT 1973 Sandy beach	REISE 1984 Sandy tidal flat
Acoela	61.8	26.3	.21.7
Macrostomida	5.8	10.4	8.3
Proseriata	22.0	41.0	22.3
Kalyptorhynchia	5.6	} 22.3	26.1
Typhloplanoida and Dalyelloida	4.6		19.5

Macrostomida), the extremes of summer and winter are apparently less favourable, although for the turbellarian inhabitants of supralittoral shores and salt marshes, population peaks have been recorded particularly in the cold season (FAUBEL 1976b). Many turbellarians, especially from salt marshes, can adapt to hostile conditions encysted or by phases of dormancy.

It has recently been found both in the field and in experiments (ARMONIES 1989a, b, 1990) that a considerable amount of turbellarian species, particularly the pigmented ones, regularly enters the overlying water column and, thus, after distribution via water currents, can colonize new and more distant areas in a short while (sometimes a matter of hours only). This "hyperbenthic" behaviour of many turbellarians is preferably nocturnal and may contribute to the wide geographic range of turbellarian species (Ax and ARMONIES 1990).

Considering the randomizing effect of a regular emergence and transport, it is amazing to still find differentiated patterns of turbellarian occurrence with the biotopical specializations described above resulting in many regional divergences evident from various studies (REISE 1988; JOUK et al. 1988). In most biotopes, the Kalyptorhynchia represent most species, followed by the Seriata. The Proseriata often dominate in abundance of individuals. The variability in turbellarian composition, evident in different studies from comparable biotopes or even from the same area, may reflect the ties which link turbellarian species to an intricate combination of local trophic and abiotic factors (Table 13).

> *More detailed reading*: ecology – REISE (1984, 1988); MARTENS and SCHOCKAERT (1986); phylogenetic aspects – EHLERS (1985); ultrastructure – RIEGER (1981); illustrated key to relevant genera – CANNON (1986)

5.4   Gnathostomulida

It took 40 years to realize that gnathostomulids represent a separate, well-defined phylum and not just an order of turbellarians. Discovered in 1928 by REMANE in Kiel Bight (Baltic Sea), it was REMANE's student Ax (1956) who, after finding additional specimens from the North Sea and the Mediterranean, wrote an account on *Gnathostomula paradoxa*, considering it to belong to a primitive turbellarian group. RIEDL (1969) finally classified the group as a separate phylum on the basis of characters which had no parallels to structures known from turbellarians (see below). Today there are about 20 genera and 100 species known, six of them with circum-mundane distribution, but many more will probably be found with more investigations, particularly in thiobiotic habitats (see Chap. 8.5).

All gnathostomulids are slender, vermiform, more or less cylindrical and soft-bodied animals of meiobenthic size (500–1000 μm in length, Fig. 33). Through their ciliation and their gliding movement they are similar to turbellarians. However, they are characterized mainly by two unique features:

a) Their epidermal cells are uniciliated, i.e. each cell has just one, relatively long, slowly beating cilium which propels the animal in a slow, gliding movement (some species can even move backwards).
b) They have conspicuous cuticularized jaws protruding into the pharynx, inserted in a bilateral-symmetrical musculature. A ventral mouth plate is flanked by two large jaws with a relatively complicated pattern of teeth and grooves (important diagnostic features) and a complex muscle apparatus (ethymological derivation of the group's name: "animals with a small mouth armed with jaws").

The head, mostly studded by some stiff sensory setae, and the hind end are often somewhat set off. As in turbellarians, the intestine has a blind internal ending, so there is no anus. The genital apparatus is hermaphroditic, but pure functional males appear beside females and functional hermaphrodites. During their life span, gonads and accompanying copulatory organs (stylet for sperm injection; bursa, vagina) become repeatedly fully developed and, after the reproductive cycle, subsequently degenerate, a feature not yet understood in its biological significance. In some species, the spermatozoa can be very large and conspicuous. In *Gnathostomula*, sperm, bundled in spermatophores, is injected into a subcuticular "bursa" of the partner (observed in cultures of *Gnathostomula jenneri*). Cleavage of the egg is spiral. Although a parenchyma is poorly or not developed, there is practically no body cavity since the internal organs fill the body completely. Fragmentation of the hind end occurs frequently, but asexual multiplication has not been observed.

Gnathostomulids were discovered relatively late, probably since their preferred habitats, hypoxic and sulphidic fine sands, have not been investigated in much detail and since they are not generally common animals. In fact, gnathostomulids populate the narrowest system of interstices known among metazoa (mean grain size 150 μm). Only in favourable habitats (see below), scrutinized with the right methods, one can find > 100 specimens per litre sediment. RIEDL (1969) reports > 6000 specimens of *G. jenneri* per litre! Gnathostomulids are invariably associated with detritus-rich, hypoxic or slightly sulphidic fine sand. Here they feed on bacteria, fungal hyphae and perhaps diatoms which they rasp off with their jaws. They reach maximal occurrence in the subsurface horizon

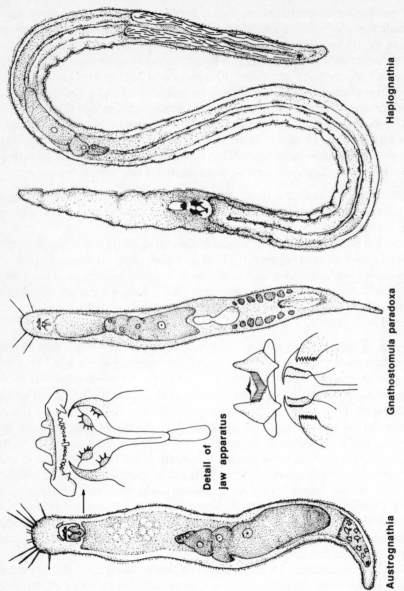

Haplognathia

Gnathostomula paradoxa

Detail of jaw apparatus

Austrognathia

Fig. 33. Some representative Gnathostomulida, with details of jaw apparatus. (Various authors)

near the oxic/sulphidic interface or along the tubes of burrowing macrofauna. However, they have also be encountered in the anoxic layers at greater depth. In the permanently sulphidic bottom underneath brine seeps in the Gulf of Mexico, they even represented the dominant animal group (POWELL and BRIGHT 1981). They are also frequent and speciose inhabitants of the reduced sand core in pebble-shaped nodules of cyanobacteria from Bermuda (WESTPHALEN, in press). As RIEDL (1969) puts it, in their main environment, fine sediments smelling of hydrogen sulphide, gnathostomulids can "dominate all the other groups of the biotope, even the nematodes."

A remarkable feature in gnathostomulids is the frequent co-occurrence of several species in virtually the same patch of sediment. REISE (1981b) found 3 species along a gradient of a few mm around *Arenicola* tubes, STERRER (1971) counted 13 species in one small sample. This high diversity is contrasted by the amphiatlantic distribution of some common species.

The systematic relations of Gnathostomulida are still arbitrary. While the jaw apparatus might superficially remind of some kalyptorhynch turbellarians, the monociliated epidermal cells and the structure of the pharynx musculature clearly set them aside any Plathelminthes. Scarcity of parenchyma and the presence of a jaw apparatus might suggest some affinities with the "Aschelminthes" (gastrotrichs, rotifers); however, lack of an anus and other features exclude any close relation. According to Ax (1985), they should be ranked as a sister group to the Plathelminthes into the taxon Plathelminthomorpha. In any case, occurrence of cosmopolitan species, despite the lack of any effective mechanisms of dispersal, combined with a high degree of sympatry and preference for very "exotic" biotope conditions, all point to a group with a long, phylogenetically isolated history.

More detailed reading: RIEDL (1969); STERRER (1972); STERRER et al. (1985); REISE (1981); Ax (1985)

5.5 Nemertinea

Nemertines are an isolated animal group without close relatives. Except for a size reduction, the whole phylum seems, as in turbellarians, perfectly preadapted to a life as meiobenthos. This is the reason why the smallest, meiobenthic representatives (about 50 species) are essentially of the same organizational pattern as the huge ribbon worms (Fig. 34).

Many of the typical mesopsammic nemertines are relatively long (up to several centimetres!), but extremely flexible and thin. They are totally ciliated, have a unique ectodermal proboscis protrusible from a cavity lined by mesoderm, a "rhynchocoel". This apparatus is armed with viscid glands, and, in the order of Hoplonemertinea, with stylets used for catching prey. The proboscis works independently of the oral opening grasping prey. In some species it is also used as a locomotor organ. The head in some heteronemertines is flanked by two

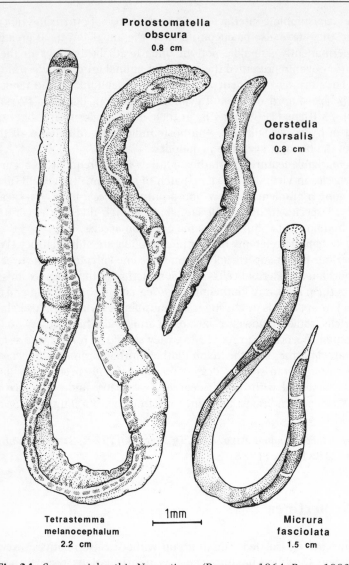

Protostomatella obscura 0.8 cm

Oerstedia dorsalis 0.8 cm

Tetrastemma melanocephalum 2.2 cm

1mm

Micrura fasciolata 1.5 cm

Fig. 34. Some meiobenthic Nemertinea. (Brunberg 1964; Riedl 1983)

deep lateral slits forming sensory organs. In many meiobenthic nemertines, the head has a pair of anterior statocysts associated with the brain. Nemerteans have four cerebral centres. The blood-vascular system is closed, numerous pseudosegmental protonephridia and gonads run in long rows along each side of the body. The complicated body musculature allows for an extreme flexibility and ever-changing body form and width: nemertines can contract their body to 1/12 of their length, can squeeze through tiny voids and even form papular swellings on their body surface.

The majority of the meiobenthic species lives interstitially in sand (30 species) and belongs to the class Enopla (i.e. "armed nemertines") of which the Hoplonemertinea are the main subgroup carrying stylets at the tip of their proboscis. Characteristic genera from sandy bottoms like *Arenonemertes* and *Ototyphlonemertes* demonstrate well the adaptations to their interstitial habitat: gliding ciliary movement, sticky areas through numerous epidermal glands, sensory cirri, paired statocysts, formation of a tail (in some species), turgescent protective tissues (*O. antipai*), suppression of the planktonic pilidium larva. There are reductive trends in the cerebral organs, in the intestinal diverticula, and the number of gonads and eggs.

Some nemertines, mostly belonging to the Anopla, lack ocelli, statocysts and any proboscis armature. Their mouth is ventral and they all are of meiobenthic size (the most common genus is *Cephalothrix* with a mouth far posterior to the cephalic tip; other genera are *Procephalothrix, Carinina*). The rather small species of the Anopla often live epibenthically on/in muddy bottoms.

While *Ototyphlonemertes* has a worldwide distribution, other genera like *Arenonemertes* are only known with a few species of restricted occurrence. Members of *Prostoma* and *Potamonemertes* can be considered stygobiotic freshwater forms.

Nemerteans are rather insensitive to factors like sand grain size, oxygen and even food supply. They are bolting predators, eat also carrion, but survive long phases of starvation while reducing their body size.

For investigation, their sluggish, extremely delicate body has to be handled with great care. However, the most gentle method, extraction by deterioration of the (oxic) environment when keeping a sediment sample for some time, is little effective in this group.

More detailed reading: interstitial species – GERNER (1969); taxonomy *Ototyphlonemertes* – KIRSTEUER (1977); key for interstitial species – NORENBURG (1988)

Group of Nemathelminthes (= Aschelminthes)

This combination of animal taxa is not unanimously considered a natural, monophyletic unit. The various groups, traditionally ranked as classes within the phylum Nemathelminthes, are therefore often also considered more or less independent phyla. They differ in characteristics mostly considered fundamental, e.g. the nature of their body cavity. In some nematodes, gastrotrichs and priapulids, the body cavity is completely lined by a mesodermal coelothel, i.e., a coelom is present in the cases studied, in other forms studied this mesothelium was absent, incomplete or reduced. Consequently, the body cavity was characterized as a "pseudocoelom", and Nemathelminthes are also termed in some text books "Pseudocoelomates". In many groups, the body cavity is more or less densely packed with the organs (numerous nematodes and gastrotrichs), in others it is spacious and filled only with body fluid and some mesenchymatous cells (kinorhynchs, rotifers). The free-living Nemathelminthes are non-segmented, bilateral animals, mostly with a mouth and an anus, and often with a complex cuticular lining covering the epidermis. Direct development after internal fertilization of the gonochoristic species is prevailing. There is a trend towards a defined and fixed cell number (eutelic development) and little regenerative potential. Asexual development is not known.

5.6 Nematoda

Nematodes usually dominate all meiofauna samples both in abundance and biomass, and represent the most frequent metazoans. In meiofauna samples, 90–95% of individuals and 50–90% of biomass are usually made up by this group (Fig. 35). Occurring in each substrate and sediment and in all climatic zones, they are of considerable ecological importance. Nevertheless, they are relatively little investigated which, in part, is referred to problems of identification.

All the free-living nematodes are of meiobenthic size. In their general body structure they are in many respects preadapted to live in sands and muds. "Contrary to popular opinion, marine nematodes do not 'all look the same' and it is time that this myth was finally put to rest" (PLATT and WARWICK 1980). Today, there are valuable illustrated keys at hand or to be published respectively

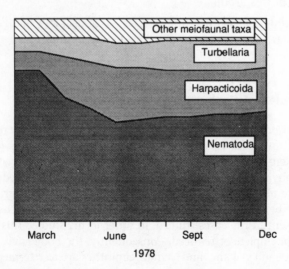

Fig. 35. Dominance of nematodes in a typical set of meiofauna samples. The main meiofauna groups from the shallow sublittoral of the Belgian North Sea coast are indicated in their relative abundance. (After HERMAN et al. 1985)

(PLATT and WARWICK 1983, 1988, in prep.). They demonstrate that it is not so much lack of valid and conspicuous diagnostic features which deters meio-benthologists from dealing with nematodes, but rather their exorbitant species richness and often minute size. About 4000 to 5000 species of free-living nematodes have been described so far, and it has been estimated that about 20,000 are as yet unknown. The marine biotopes around the British Isles alone harbour 41 nematode families with 450 species. Another estimate for the North Sea sediments is 800 species, a figure which even in this well-studied area will easily increase. In meiofauna samples, the species number of nematodes often exceeds that of all other groups by an order of magnitude.

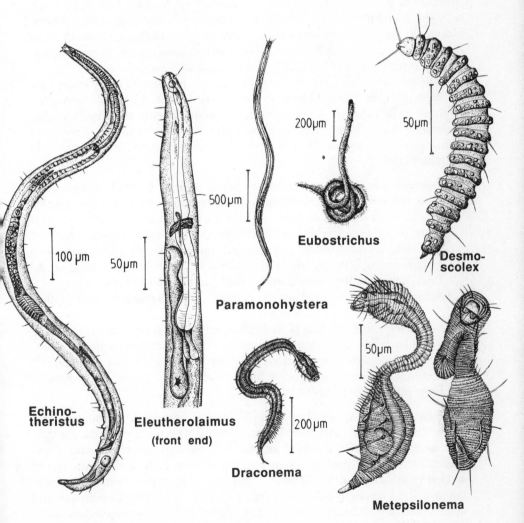

Fig. 36. Various Nematoda of different appearance. (Various authors)

Identification of nematodes is based mainly on characters visible without dissection in a good compound microscope (Nomarski contrast is very helpful). The general body form of nematodes is by no means invariable (Fig. 36) and often already helpful in restricting the variety of possible groups. Most of them are 0.5–3 mm long, the smallest being 0.2 mm, their length being 20–40 times their width.

External features relevant for taxonomical ordination are shape of the tail end, number and arrangement of sensory setae (particularly around the head), of caudal glands, and of gonads, position and shape of amphids (paired anterior chemical sense organs), epicuticular structures (e.g. annulation). The most important internal features are shape and cuticularization of the buccal cavity, structure of genital organs, and sperm morphology. In some taxa the general body shape and its external cuticula is of diagnostic value (e.g. some compress Chromadoridae, the Draconematidae and Epsilonematidae), while in most nematode groups the body is rather uniform with the anterior end more truncated and the posterior end slender, in males merging into an inflected tip.

The taxonomic scope of this book allows only mention of a few more comprehensive higher taxa and some extraordinary species. For details, perusal of the original literature is recommended. The systematics of the Nematoda has undergone much change and still is not generally recognized and stable. Most of the pertinent meiobenthic literature, however, is based on the system of LORENZEN (1981):

Adenophorea: setae and adhesive glands present, amphids conspicuous. Most meiobenthic nematodes belong to this subgroup. Preferably from marine and brackish water sites, but some 400 species of various families are also limnetic and even stygobiotic (Tobrilidae, Monhysteridae, Desmoscolecidae, Xyalidae: *Theristus*, many Dorylaimida). Important orders with numerous genera and species are the Chromadorida, Enoplida and Trefusiida. As Desmoscolecina the above family sometimes has also ordinal rank.

Secernentea: no setae and adhesive glands, amphids much reduced. Within the scope of this book, this taxon is of less relevance, since it contains the bulk of terrestrial, or parasitic and pathogenic forms. Among the meiobenthic forms there are those limnic groups that prefer a certain degree of pollution (Rhabditidae, Diplogasteridae). Only a few species of "halophilic" Secernentea occur in eutrophic biotopes of marine shores, albeit often in large numbers (e.g. *Rhabditis marina*).

Biological Aspect.   Nematodes move in a characteristic dorso-ventral wriggling due to their lack of circular musculature. As antagonists to the muscular contraction serve the rather stiff cuticle and the internal hydrostatic pressure. Some species can "jump" through rapid bending of their body with a subsequent sudden relaxation (e.g. in *Theristus*). Only the structurally aberrant Epsilonematidae and Draconematidae move by "looping", an alternative adhesion and detachment of their mouth and tail region (see Fig. 23). Desmoscolecidae move on their back by "contractive waves".

Although about 45% of all adenophoran nematodes are interstitial, there are few morphological features that occur regularly in mesopsammic species and can be considered typical for a mesopsammic life, although their function is not always clear (LORENZEN 1986): A strongly bent tail with adhesive glands (nematodes normally lie on their side when moving), additional adhesive organs (e.g. in Epsilonematidae), a flexible, tapering "neck" region, extremely long setae (*Thrichotheristus*), aberrant position of amphids (Epsilonematidae), flattened body (rare, e.g. *Neochromadora angelica*).

Life history studies on nematodes are badly lacking, the following observations refer to some Chromadorida and, considering their variance, it is questionable that they can be generalized. Adult females produce at least 50 eggs, after hatching the juveniles pass four moults before reaching maturity. All cuticular structures (also the armature of the buccal cavity) will be shed with each moult. Growth continues after the last moult. The average nematode has a fresh weight of 1 μg. Generation time is between 13 and 60 days, resulting normally in eight to ten annual generations (however, there are also species with just one reproductive period per year!). Each m^2 of sea floor is populated by about a million nematodes. The resulting high production is in contrast to a rather low biomass, the annual P/B-ratio is about 10 (see Chap. 9.2).

Ecological Aspects. It is not surprising that within the huge number of nematode species all trophic resources are utilized; generalists are represented as well as food specialists. WIESER (1953, 1959), in his basic papers on the ecology of nematodes from European and American littoral coasts, was the first to relate the multitude of trophic types to (a) the sediment structure and (b) the cuticular armatures of the nematode's buccal cavity (Fig. 37).

1. In mostly homogeneous muds and fine sand prevail non-selective deposit feeders with a well developed, but weakly cuticularized buccal cavity. Food particles, often larger bacteria, detritus, also diatom cells, are taken up with help of lips and anterior buccal cavity.
2. In more heterogeneous (fine) sandy substrates live selective or non-selective deposit feeders with a small or vestigial, not cuticularized buccal cavity. Their food particles (bacteria, small detritus) are soft and mostly obtained by suction of the muscular pharynx.
3. In sand with more microhabitats occur epigrowth feeders. They scrape off the (algal) surfaces of grains or pierce single cells. Hard cuticular ridges for scooping or pointed tips for piercing are well developed in the narrow buccal cavity. In coarser sandy sediments live mainly predators and omnivores with large and powerful pointed teeth and lancets as buccal armature; their buccal cavity is wide.
4. In the exposed phytal, algivores and predators/omnivores dominate. In more sheltered algal sites an increasing portion of epistrate feeders and selective deposit feeders scrape off and pick up particles (aufwuchs and detritus) from the thalli of the plants.

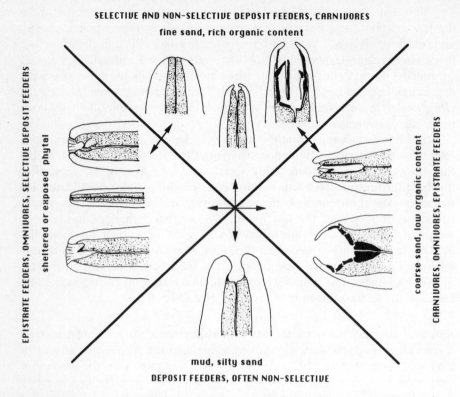

Fig. 37. Different eulittoral habitats and their nematode populations grouped as trophic guilds. These are indicated by schematic outlines of the buccal cavities of typical representatives. (Various authors)

WIESER's classification has been repeatedly found applicable also to studies from other areas (e.g. OTT 1972), but it has also been refined to meet the requirements of local nematode populations (JOINT et al. 1982; JENSEN 1983, 1987a). ROMEYN and BOWMAN (1983) pointed out that within one of the above ecological groups, e.g. epigrowth feeders, there may occur different buccal structures in different species depending how the food (diatoms) is actually being devoured. In a study on the nematodes of Australian mangrove muds, NICHOLAS et al. (1991) could parallel a pattern of different nutritive guilds with the shore profile. This grouping coincided well with the gradient of increasing exposure to the air: epistrate and diatom feeders prevailed at the low tide zone, the higher mangrove zone contained mostly selective microbial feeders while the sediments above high water neap were dominated by omnivores, predators and plant root feeders. This pattern is perhaps not generally valid; it contrasts to that described by ALONGI (1990a), who found an increase in omnivores and predators from the high intertidal mangroves to the sandflat.

By and large, there is substantial evidence that nematodes are mostly selective feeders with a restricted capacity of nutrient digestion. The high syntopic diversity of nematode species has been interpreted as a consequence of a much differentiated food partitioning. From the field distribution of nematodes in an American salt marsh, OTT (1972) separated four different associations related not only to physical parameters, particularly oxygen and hydrogen sulphide, but also to trophic aspects (based on analyses of the species' buccal armature).

While especially in tidal flats there are surprisingly many diatom eaters among nematodes, bacteria seem to play the outstanding role in the nutrition of the group. The worms accelerate bacterial digestion secreting bactericidal lysozymes which dissolve the thick cell walls. Decaying carrion or plant debris massively attracts nematodes, and even cultivation of bacteria has been suggested to occur (RIEMANN and SCHRAGE 1978) in the mucus threads excreted and subsequently ingested by the nematodes (gardening hypothesis). Nematodes regulate and stimulate bacterial growth. When (bacterivorous) nematodes were added to decomposing wrack, degradation times became significantly reduced (RIEPER-KIRCHNER 1989). Nematode mucus secretions and fine burrows modify the microtexture and, thus, the sediment climate (CULLEN 1973). *Ptycholaimellus* even constructs rather stable tubes of mucus and sediment particles with consequences for the aeration of the deeper layers, the depth of the oxic/anoxic interface, and also for the occurrence of other meiofauna (NEHRING et al. 1990).

Communities of nematodes and their diversity seem largely determined by sediment structure. It is probably the portion of the silt fraction which limits many stenotopic nematode species, while eurytopic species show a high affinity for silty or inhomogenous sediments. Since this cognition is rather generally applicable, it can be inferred that, in areas where this relation is not established, other factors like salinity gradients or pollution are modulating the distributional pattern (WIESER 1959; OTT 1972; VANREUSEL 1990, 1991). Sediment composition, in turn, often corresponds to food supply (organic content). The more micro-habitats there are, the richer the nematode communities. Fine sands in shallow sea bottoms with their varying and rich food supply harbour the highest species number; about 100 species per investigation area are not unusual. Extreme habitats with a more homogeneous sediment structure (and less food) are often populated by a rather monotonous community of characteristic specialists (VINCX et al. 1990). While the few typical mud inhabitants are mostly small and have short setae, sand specialists are usually elongated in shape, often with a well-sculptured cuticle, long setae and carnivorous (see above). Thus, in the interstitial of exposed sands typical representatives are Draconematidae, Epsilonematidae and Desmoscolecida (Fig. 36). Interestingly, epsilonematids have been found to occur frequently around deep-sea hydrothermal vents (DINET et al. 1988) and also in a remarkable diversity in the stem stratum of *Posidonia* stands. Other stenotopic psammobionts are *Daptonema gelana, Leptonemella aphanothecae* and *Theristus interstitialis* (VANREUSEL 1991). In contrast, the diversity of the nematode community on plant surfaces is reduced (NOVAK 1989).

Paralleling diversity, abundance also is highest in shallow flats. In the North Sea, up to $23 \times 10^6 \, m^{-2}$ have been found, decreasing towards the sublittoral to about 1 mio. m^{-2}, but with considerably lower values in greater depths. Deep-sea nematodes tend to be smaller in size than shallow-water species (see Chap. 8.3).

The intensive taxonomic radiation and ecological differentiation of nematodes already indicate a differentiated distribution pattern with many local and seasonal variations. Horizontally, from the shallow sublittoral towards the deeper bottoms of the continental shelf and the deep-sea, nematodes decrease only slightly in abundance. Also towards the eulittoral there is only a slight reduction in nematode dominance. The same is true for the nematode occurrence along the salinity profile in estuarine waters. Apparently, among the high number of species there are enough generalists plus specialists for each combination of factors to compensate for those nematodes that drop out.

Vertically, by far the most nematodes are restricted to the uppermost few centimeters of sediment. The vertical decrease is less expressed in well-oxygenated sands than in muds with their steep gradients in oxygen. Exposed beaches have been found to be populated by nematodes down to a depth of 1 m! Even in bottoms with a well-developed chemocline, specialized nematodes can regularly be found also in the anoxic depth. It was OTT (1972) who first underlined the occurrence of a distinct nematode community in these layers, specialized in many ways to live under sulphide-dominated conditions (see Chap. 8.5). Stilbonematinae live in symbiosis with "sulphur bacteria" around the oxic/sulphidic interface; but also the aposymbiotic *Siphonolaimus*, *Cyatholaimus*, *Terschellingia longicaudata*, *Monhystera disjuncta* and *Sabatieria pulchra* and some limnetic *Tobrilus* spp. have highest abundance around or underneath this chemocline. *Paramonohystera wieseri* from anoxic depth has been experimentally shown to survive extremes of temperature better under anoxia than in normoxic conditions and termed "obligate anaerobic" (WIESER et al. 1974). *Eudorylaimus andrassy* lives in oxygen-free zones of Lake Tiberias (Israel) for 8 months each year and can survive being placed in a sealed jar with completely anoxic sediment for 6 months (POR and MASRY 1968).

The vertical distribution varies, many nematodes migrate, depending on both the chemical stratification and diurnal and seasonal fluctuations. JENSEN (1984) suggested that in Finnish waters low temperatures might indirectly influence the vertical distribution of epigrowth feeding phytal nematodes: the animals change to the bottom-dwelling life when their habitat, the algal canopy, becomes destroyed by ice, and leave the sea bottom again when the vegetation is growing in spring. A substrate-mediated and food-influenced annual growth cycle was also found by NOVAK (1989) in the nematode fauna of a Mediterranean seagrass area reflecting the growth cycle of the plants. ALONGI (1990a) confirmed a food-mediated seasonal change in nematode dominance: deposit feeders and predators dominated during the cooler months in autumn and winter, while epigrowth feeders were most frequent in spring and summer.

Despite their considerable abundance, reports on an extended use of nematodes as a nutritional source are scarce. Only beside other items will crustaceans, small fish and carnivorous polychaetes also prey on nematodes. In tidal flats, recently metamorphosed *Carcinus* feed intensively and apparently selectively on the nematodes *Enoploides* and *Adoncholaimus* which, in turn, are attracted by dead animals. A good part of the nematode biomass does not seem to be transferred to higher trophic levels but is rather linked to the short trophic loops of destruents (see Chap. 9.3).

The high species number of nematodes even within small samples renders them good indicators for disturbance and pollution-induced changes. HEIP and DECRAEMER (1974) could relate a local decrease in diversity with the efflux of polluted river water. However, in nature influence of concomitant factors like changes in sediment structure, may interfere, and thus complicate the interpretation of diversity indices. A means of facilitating the assessment of community changes is avoidance of the tedious species identification in nematodes and introduction of summative parameters. BONGERS (1990) and BONGERS et al. (1991) contend that even on the family or genus level a "maturity index", relating the more r-selected "colonizers" to the more K-selected "persisters", is indicative for disturbance- or pollution-induced changes (see Chap. 9.1).

Far more simplified is the nematode-copepod index (RAFFAELI and MASON 1981) based on the observation that in general the more robust nature of nematodes compared to harpacticoids would lead to a superior persistence in gradients with increasing pollution. Also here, however, a refinement of identification and limitation was necessary to correct for changes caused by factors others than pollution. This, of course, limited the general range of applicability (see Chap. 8.6).

The fact that the freshwater nematode *Caenorhabditis elegans* (Rhabditidae) is so far the only metazoan in which the entire cell lineage has been completely determined and which is thoroughly known in its genome, renders this species after mass breeding a potentially powerful tool in aquatic toxicity testing, since it is particularly sensitive to many heavy metals (WILLIAMS and DUSENBERY 1990).

However, this does not invalidate the conclusive statement of HEIP et al. (1982b): "...free-living nematodes are ecologically the most important taxon in all marine sediments..."

More detailed reading: taxonomy and systematics – GERLACH and RIEMANN (1973/1974); HEIP et al. (1982); KEPPNER and TARJAN (1989); anatomy – CHITWOOD and CHITWOOD (1974); biology, life history – WHARTON (1986); HOPPER and MEYERS (1966); GERLACH and SCHRAGE (1971); TIETJEN and LEE (1977); HERMAN and VRANKEN (1988); ecology – WIESER (1959); ALONGI and TIETJEN (1980); PLATT and WARWICK (1980); HEIP et al. (1982, 1985a); JENSEN (1987a); OTT and NOVAK (1989); WARWICK (1989); books and reviews – FERRIS and FERRIS (1979); NICHOLAS (1984); HEIP et al. (1982b, 1985a); PLATONOVA and GAL'TSOVA (1985); WHARTON (1986)

5.7   Gastrotricha

Gastrotrichs are all of meiobenthic size; indeed, they are among the smallest metazoans (often only 0.1 mm). However, their characteristic gliding on their ciliated ventral side (derivation of scientific name!), their general body shape (head, thorax, tail), finally their armature of body spines and haptic tubes renders most of them microscopically well recognizable (Fig. 38). The haptic tubes contain rich adhesive glands which are enclosed by extensions of the body cuticle forming a stiff tubular sheath. With their steady gliding they differ from the more writhing gliding of turbellarians and gnathostomulids; another distinctive feature is their well-developed terminal mouth and anus.

There are almost 500 species described, classified in two orders with 13 families. From northern Europe, 180 species are presently known. Identification is mostly based on the shape of the caudal furca, the arrangement and shape of scales, spines and hairs on the cuticular surface, the position of haptic tubes and the structure of the radial pharynx musculature.

> *Macrodasyida*: discovered after the 1930s, this group contains about 180 more primitive, but apparently many radiating marine species with hermaphroditic sexual reproduction. The numerous haptic tubes extend in a symmetric arrangement anteriorly, laterally and caudally. *Macrodasys*, *Urodasys* (with a tail); *Turbanella*, *Cephalodasys* (Fig. 38).
>
> *Chaetonotida*: this mostly limnic group of approximately 350 species is more derived and has a less developed form variation; it contains only a few marine and brackish-water species. The cuticle is covered by conspicuous spiny or shindle-shaped sculptures, often with a circum-oral whirl of setae. There is only one pair of haptic glands on the furcate toes. Most species reproduce pathenogenetically. *Chaetonotus*, *Neodasys*, *Lepidochaetus*, *Xenotrichula* (a marine genus).

Biological and Ecological Aspects.   The cuticular hard structures as well as the occasional development of vacuolized epidermal cells are often interpreted as a protection against sediment agitation and pressure. With the very effective haptic tubes, glandular toes, and, in some species, also haptic "girdles", the animals can momentarily cling to sand grains. This reduces the effectivity of decantation considerably unless anaesthetization or freshwater shock is applied. Beside the typical continuous gliding, locomotion by crawling and alternating adhesion using the front end and the haptic toes (looping) occurs in gastrotrichs as well as sinuous swimming.

Within the mostly parthenogenetic Chaetonotida only in *Xenotrichula* and *Neodasys* have males been found that inject their sperm in a complicated way into the female partners (similar to Rotatoria, see Chap. 5.11). Typical for meiobenthic organisms, just a few large eggs are produced, from which juveniles hatch without passing through a larval stage. The life span of gastrotrichs is only a few weeks. They are microphagous, feeding on bacteria, protozoans etc.

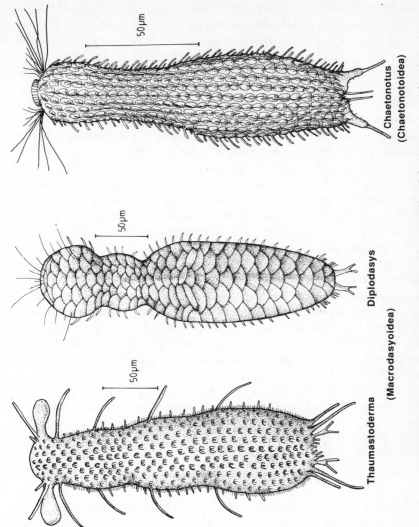

Fig. 38. Some representative Gastrotricha. (Combined from various authors)

Gastrotrichs are only occasionally common in meiofauna samples. Exposed beaches harbour just a few adapted species (e.g. *Xenotrichula*). Only in their preferred habitats, the fine lower tidal to subtidal coastal sands enriched with detritus, they can attain maximal densities of almost 7000 per 10 cm^2. On rare occasions, aggregations of more than 200 specimens per cm^3 have been found in a beach at the west coast of the USA (HUMMON 1972) and up to 400 specimens respectively in a tidal flat of the North Sea (POTEL and REISE 1987). In areas of coarse sublittoral sand, such as the entrance of the Skagerrak (North Sea), where meiofauna abundance is generally low, gastrotrichs can outnumber many other meiofauna groups (HUYS et al. 1992). It must be in these patches that gastrotrichs can ecologically compete with turbellarians and polychaetes (HERMAN, pers. comm.). In tropical beaches gastrotrichs were found to be particularly common after the decline of most meiofauna due to monsoonal rains (ALONGI 1990b). In contrast to abundance, diversity can be remarkably high, particularly in coarse, shelly sands. Up to 18 species have been described to occur syntopically in a few square centimeters.

Muds are seldomly populated, since most gastrotrichs prefer well-oxygenated sediments. BOADEN (1974) described some typical thiobiotic gastrotrich specialists, and the author found some 40 specimens in samples from North Sea methane seeps (unpubl. data). The brackish-water and freshwater gastrotrichs prefer zones of submerged or decaying vegetation, ephemeral pools and organic debris. Even within some congeneric species the variety of differentiated ecological demands results in a typical vertical and horizontal distribution pattern along a beach (RUPPERT 1977).

Gastrotrichs doubtless share numerous features with nematodes (e.g. triradiate pharynx), but the trend to reduce body structures, concomitant with diminution of size, aggravates conclusions on their phylogenetic position. Not even the nature of their body cavity (mesothelial coelom present?) is undisputed (TEUCHERT and LAPPE 1980).

More detailed reading: taxonomy – HUMMON (1971); structure, phylogeny – RIEGER (1976); TEUCHERT (1977); RUPPERT (1982); ecology – SCHMIDT and TEUCHERT (1969); physiology – HUMMON (1974); zoogeography – RUPPERT (1977); freshwater – KISIELEWSKI (1990); monographs – REMANE (1936a); D'HONDT (1971)

5.8 Kinorhyncha

The vermiform body covered with cuticular rings led DUJARDIN (1851), in the first description of a kinorhynch, to classify it intermediate "between worms and crustaceans". The striking "in-and-out movement" of the eversible anterior end of their body is so characteristic that later the scientific name Kinorhyncha has been derived from it ("animals with motile proboscis"). Today, their nemathelminth character seems established and the 150 purely marine, meiobenthic species are grouped into two orders distinguished by differences in the number of plates in the "neck" segment (the second of 13 segments) of cuticular rings, sometimes referred to as "zonites" (Fig. 39). The head can be withdrawn in the trunk and the plates of the neck then serve as a closing apparatus. The zonites bear spines in different arrangement and shape, an important taxonomic feature.

Cyclorhagida, mainly represented by the genus *Echinoderes* with 60 species – 2nd zonite covered with 14–16 plates or "scalids", round to oval in cross-section. *Semnoderes* is another better known genus from more sandy sediments.

Homalorhagida, mainly the genus *Pycnophyes* with about 35 species – 2nd zonite with only 6–8 scalids, triangular in cross-section. *Kinorhynchus* is a well-known genus typical for very fine sediments.

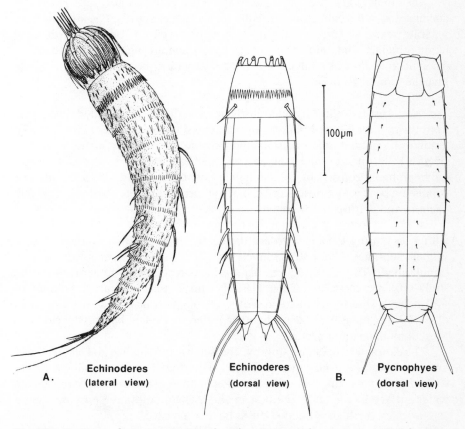

Fig. 39A, B. Some characteristic Kinorhyncha. **A** Natural appearance. **B** Schematic graph showing structure of body plates. (HIGGINS 1981, 1986)

Kinorhynchs are not often encountered in meiofauna samples, although their unique movements makes them immediately noticeable. First, they are not frequent and are very tiny (180 to 1000 µm), second, they occur mostly in fine sandy to muddy sediments which usually attract less attention in meiofauna studies, and third, they are not to be obtained quantitatively by the usual extraction methods since their body, covered by a strongly water-repellent cuticle, tends to adhere to the water surface once in contact with air.

The flexibility of the stiff body is possible by the articulation of the cuticle in 13 annular zonites which additionally attain a certain elasticity in lateral and medioventral furrows subdividing them in separate plates. Rhythmical inversion of the first zonite into the neck and trunk region creates the typical kinorhynch movement by which the scalids serve as anchoring devices so that the body can be slowly dragged forward. When retracted, the anterior scalids can completely cover the front end, when fully extended, the nine long oral styles of the 1st zonite point forward. They surround a small oral cone which often remains retracted in specimens from live samples.

The subdivision of the body, externally evident in the zonites, is also internally documented in the epidermis, musculature and nerve system. The animals have separate sexes, spermatophores have been observed on the females of some homalorhagids. The hatching juveniles have spine arrangement and structure different from that in adults. All cuticular structures become shed during the juvenile moulting phases.

Ecological Aspects.   Kinorhynchs mostly occur in fine sandy to muddy sediments from the eulittoral (e.g. *Echinoderes coulli*) down to the deep sea. They are also found in the phytal, and occasionally in coarse clean sand (*Cateria*, with very slender body, in intertidal high-energy beaches). The food of shallow water forms probably consists mainly of diatoms, which correlates with the marked population peak in summer recorded in some species. In deeper bottoms, probably bacteria and detritus are ingested.

Mostly, kinorhynchs are found in the range of only 15 specimens per $10 \, cm^2$, with a decreasing tendency toward the depth. In oil-polluted shores in Prince William Sound, Alaska, 130 specimens per $10 \, cm^2$ have been recorded (SHIRLEY, unpublished). That these figures require further perusal is underlined by the study of PFANNKUCHE and THIEL (1987), who found in Antarctic bottoms in 400 m depth kinorhynchs to represent 5–6% of all meiofauna with abundances of > 250 specimens per $10 \, cm^2$. In other Antarctic studies, kinorhynchs ranked third of all meiofauna groups.

The easiest way to obtain kinorhynchs is to use the "bubble and blot method" for muds, because even when preserved, kinorhynchs tend to adhere to the surface film; for coarse sand, decantation with subsequent inspection of the water surface in the jar is the best method. For quantitative purposes, more sophisticated methods (HIGGINS 1988) have to be used.

More detailed reading: anatomy – KRISTENSEN and HIGGINS (1991); mono-graphs – ZELINKA (1928); REMANE (1936a)

5.9 Priapulida

Just half of the known priapulid species (altogether 16) are of meiobenthic size; however, these are rather heterogeneous in appearance and of wide distribution. Certainly best known is *Tubiluchus corallicola*, which is relatively frequent in sublittoral coralline sands in the Caribbean. Other meiobenthic forms are the interstitial *Meiopriapulus* fijiensis, found in the eulittoral of Pacific islands and *Maccabaeus* (= *Chaetostephanus*), a tubicolous form from muds in the Mediterranean. As temporary meiobenthos, the larvae of the common macrobenthic *Priapulus caudatus* can be encountered in fine sediments.

The body of priapulids is covered by a chitinous cuticle, often bearing tegumental spines, setae and papillae. In its anterior part the body is structured as an eversible and retractable proboscis, the "introvert", which is studded by various diagnostically relevant teeth and scales, the "scalids" (Fig. 40). A post-anal extension of the body can be developed as a long tail (*Tubiluchus*) serving as an anchoring organ. Beside the well-developed dermal-muscular layer, two strong longitudinal muscle strands traverse the body inserting in the proboscis to serve as retractors. It has been argued whether generally in priapulids the body cavity is coelomic and, thus, has a mesodermal lining. In *Meiopriapulus* this is certainly not the case. In *Tubiluchus* spp., which are sexually dimorph, the males have a stronger ventral setation. Except for *Meiopriapulus*, which may have a direct development (Higgins and Storch 1991), priapulids have charac-

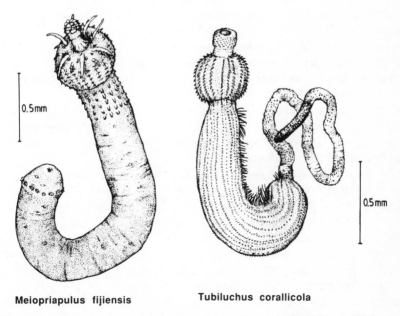

0.5mm

0.5mm

Meiopriapulus fijiensis   **Tubiluchus corallicola**

Fig. 40. Characteristic meiobenthic Priapulida. (Combined from various authors)

teristic larvae differing from the adults in having a fortified cuticle, the "lorica" (Latin: "coat, case") consisting of solid plates (dorsal, ventral, lateral) equipped with scalids (Fig. 40). The larvae of *Tubiluchus* spp. lack the tail characteristic of the adults. The chitinous cuticle of priapulids is periodically moulted.

The meiobenthic priapulids probably feed on bacterial films and other small organisms which they scrape off or sieve out of the sediment with their relatively wide, scoop- or comb-shaped anterior scalids.

Priapulids are a very old group, known from fossils since the Cambrian. Their relation to kinorhynchs and especially to the Loricifera (see below) suggests their relationship to the nemathelminths.

> *More detailed reading*: *Maccabaeus* – POR and BROMLEY (1974); *Tubiluchus corallicola* – KIRSTEUER 1976); *Meiopriapulus fijiensis* – MORSE 1981); ultra-structure – HIGGINS and STORCH (1989); monograph – VAN DER LAND (1970)

5.10   Loricifera

This new nemathelminth phylum was first described by KRISTENSEN (1983) on the basis of *Nanaloricus mysticus* from the coast off Roscoff (France). In the meantime, these bizarre, minute animals have increased in species number up to about 50 (KRISTENSEN 1991a), grouped in several families, and at present > 60 species still await formal description (KRISTENSEN 1991b). While the first specimen described lived in shell gravel in only 25 m water depth, members of the group have also been found in deep-sea muds. Reasons for the late discovery of this group might be their minute size (250–300 µm), their similarity (when fixed) with contracted rotifers and their rare occurrence, which, in turn, might be a methodological problem, since their viscid surface badly adheres to sand grains. Loriciferans were first noted in 1974 by HIGGINS and 1975 by KRISTENSEN, and probably also by other meiobenthologists. However, it needed fresh material to recognize their unique structures separating them from all other nemathelminths (Fig. 41). Today, sufficient specimens have been found to suggest a worldwide distribution of the taxon.

The dorsoventrally flattened body of the Nanaloricidae is covered by a solid armour, the lorica, divided by longitudinal furrows in six plates. These are studded by about 230 spiny scalids, often of bizarre shape, arranged in nine transverse rows. The head region has a non-eversible, but telescoping mouth cone which can be retracted (e.g. when fixed) along a flexible neck into the trunk, which then recalls certain rotifers. The anterior proboscis contains an internal stylet apparatus with a complicated triradiate muscular pharyngeal bulb. The Pliciloricidae (*Pliciloricus*) have a round body covered with a set of plates similar to the priapulid larva. The separate sexes of Loricifera differ in the structure of the scalids (Fig. 41). The complicated larva has numerous long

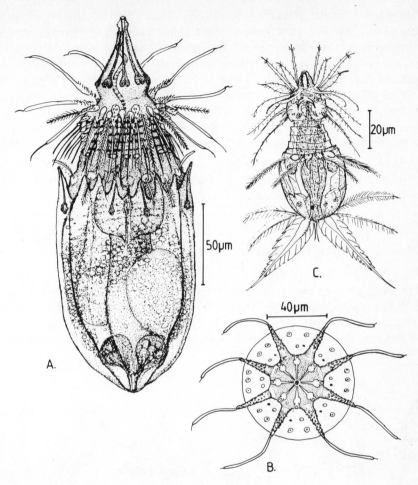

Fig. 41A–C. The first described representative of Loricifera, *Nanaloricus mysticus.*
A Female, ventral view. **B** Female, front end with mouth opening. **C** Larva with flosculi.
(KRISTENSEN 1984)

spines around its head end, and, at least in *Nanaloricus*, at its hind end a pair
of toes which serve as adhesive organs. Three leaf-shaped, locomotory flosculi
may be present, probably propelling and pushing the animal forward through
the sand. The larva passes several moults shedding the lorica. Some deep-sea
species of *Rugiloricus* develop progenetic larvae. Beside normal males and females
with internal fertilization, parthenogenesis also occurs in *Rugiloricus* (KRISTENSEN
1991b).

The systematic position of Loricifera within the various groups of Nemathelminthes is not yet established. Despite their minute size, loriciferans have a
considerable number of small cells (> 10,000) which is in contrast to nematodes

and rotifers. Singular similarities with some rotifers (lorica) and kinorhynchs (scalids) are believed convergences or old nemathelminth symplesiomorphies. The resemblance in some features to tardigrades (stylet apparatus) may indicate convergences, but basal ties between arthropods and aschelminths are being discussed (KRISTENSEN 1991b). There is a clear structural relationship to the larva of priapulids, so that a neotenic origin of loriciferans from priapulids has been suggested, but for more clarity we need more details from both new loriciferans and the priapulid larvae. KRISTENSEN (pers. comm.) contends that the loriciferans are morphologically the most complicated meiobenthic animals.

More detailed reading: first description of taxon – KRISTENSEN (1983); taxonomy, anatomy – KRISTENSEN (1991a); biology and phylogeny – KRISTENSEN (1991b)

5.11   Rotifera, Rotatoria

With about 1000 species, rotifers are a dominant group in the freshwater meiobenthos, but in marine samples the taxon is less frequent and little documented. The structural diversity of rotifers is remarkable, facilitating identification in live samples, at least to the generic level. With maximally 3 mm, the bulk of

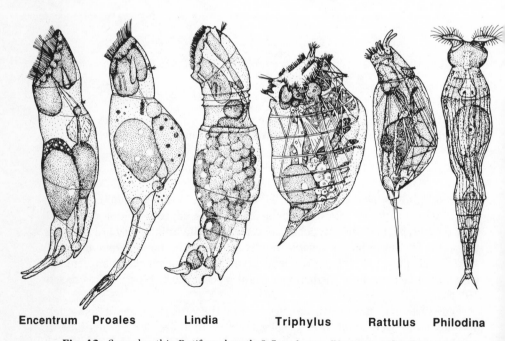

Encentrum   Proales   Lindia   Triphylus   Rattulus   Philodina

Fig. 42. Some benthic Rotifera, length 0.5 to 1 mm. (Various authors)

the group is of meiobenthic size, the smallest rotifers (about 50 μm) easily reach protozoan size range and represent some of the smallest metazoans.
There are two important benthic orders (Fig. 42):

Order Bdelloidea – *Rotaria, Mniobia, Philodina*. They live mostly in phytal habitats, e.g. between mosses; reproduction is parthenogenetical, there are no males.
Order Ploima – *Trichocerca, Lepadella, Lecane, Proales, Notomma, Notholca, Ploima*; common marine species are members of the genera *Encentrum, Trichocerca, Proales*.

Most meiobenthic forms belong to the Ploima. They live predominantly in the well-oxygenated and water-saturated sandy shores. The males are short-lived dwarf males without intestine. Reproduction is by heterogonous alternation of parthenogenetic summer generations and one bisexual winter generation. Sperm transfer is by injection into the females, brooding of the few eggs is frequent.

The wheel organ is the characteristic structure from which the Latin group name is derived. Originally a ciliated plate for creeping and a simple mode of food uptake from the substratum, it has modified and complicated its function by the formation of separate ciliary whirls (resembling rotating wheels) which produce strong incurrent water eddies for food ingestion.

The body is covered by a rather thick syncytial epidermis which can locally be reinforced to form plates (e.g. in *Notholca*). The head with the wheel organ is retractable into the trunk so that, in fixed samples, this diagnostically important organ is often not visible. Internally, the pharynx is equipped with a very complex grasping and chewing apparatus, the mastax, supported by its own muscle strands. Its structures are an important basis for rotifer identification.' The body often tapers into a flexible and eversible foot which mostly ends in a pair of toes. Carrying adhesive glands, the tail with its toes serves as anchoring organ, but in some species it participates also in a looping locomotion where body contractions and extensions alternate while the wheel organ or the foot respectively are fixed.

Typical interstitial forms have been termed by Wiszniewski (1934) "psammobiotic", species less strictly bound to the mesopsammon were named "psammophilous". Stygobiotic rotifers live preferably in the groundwater, e.g. *Encentrum subterraneum* (*Wierzejskiella subterranea*), *Paradicranophorus* (see Chap. 8.2.2). The psammobiotic rotifers have typical morphological adaptations to an interstitial life: flattened body, enlarged adhesive glands, long toes, no pigment and eyes (e.g. many Dicranophoridae like *Encentrum*).

Rotifers are strictly eutelic; with about 1000 cells, their cell number is definite and remains constant. Since mitoses terminate already after the first 5 h of life, rotifers, in contrast to other nemathelminths with a thick lorica, do not shed their cuticle during growth. Convergent to tardigrades and some nematodes, many rotifer species can develop cryptobiotic stages. The diploid winter eggs of the Ploima represent resting stages for overwintering. Mostly described from planktonic species, the regular change in form and structure of the body, known as cyclomorphosis, occurs also in some meiobenthic forms (e.g. *Mniobia*). The adaptive significance of this interesting phenomenon, described for meiobenthos also from a tardigrade (see Chap. 5.15), is not understood.

Ecological Aspects.   Resting stages together with parthenogenesis contribute to the extreme capacity of rotifers to quickly colonize and populate new and often ephemeral habitats. They are also the basis of drastic population fluctuations typical for rotifers. This renders quantitative data such as the average population density of rotifers (about 100 specimens per 10 cm^3) sediment problematical. In optimal zones, a sample contained up to 11,500 individuals per 10 cm^3 sediment (PENNAK 1940).

These figures underline that in limnic meiofauna samples rotifers often exceed nematodes in abundance. Nevertheless, few ecological details are known about them. Their preferred habitat are phytal environments, but also the shores of lakes, where sand and some detritus accumulates, are populated by rich and often typically adapted rotifers. Along a shore profile of a lake, the permanently water-saturated sandy surface sediments in shallow reaches near the water line ("hydropsammon" according to WISZNIEWSKI 1934) would contain most rotifers, higher up in the moist, aerated shore ("hygropsammon") they would occur in lower numbers and in deeper sediment layers. Data on freshwater rotifers from deep water bottoms are scarce.

While brackish-water biotopes are well populated by meiobenthic rotifers, a marked reduction is noticeable in fully marine habitats. TURNER (1990), without giving quantitative figures, described from the marine psammon of Florida 11 species of *Lecane*, *Trichocerca*, *Lindia* and *Colurella* as psammobiotic forms. From European beaches, REMANE (1949) and TZSCHASCHEL (1979, 1980) found in sandy North Sea shores predominantly species of *Proales*, *Lepadella*, *Colurella* and *Encentrum*, often in densities between 30 and 60 specimens per 100 cm^3. Mostly preferring the upper subsurface horizons in the vicinity of the highwater mark, marine rotifers do not colonize the lower reaches of tidal flats where oxygen supply may limit their distribution. However, in well-oxygenated sediments (coarse sand, shell-hash) rotifers have been regularly found also sublittorally down to 300 m water depth (HIGGINS, pers. comm.). In winter, the populations from shallow reaches seem to perform temperature-dependent vertical migrations in somewhat deeper layers. In the void system of Arctiç sea ice close to its lower surface, rotifers seem to be fairly frequent and accounted in samples for > 20% of the overall meiofauna (GRADINGER et al. 1993).

Both structurally and biologically, rotifers are a much derived group. Their nemathelminth relationship is often concealed by highly apomorphic features. The body cavity is not lined by a mesothelium. The phylogenetic origin of the complicated mastax is probably a typical triradiate pharynx typically known from nematodes. In psammobiotic forms the depressed body shape, well-developed adhesive glands in the foot, and very flexible toes may be considered adaptations to the sandy habitat.

More detailed reading: ecology – REMANE (1949); TZSCHASCHEL (1980); PENNAK (1951) (freshwater), WISZNIEWSKI (1934) (freshwater); taxonomy – TZSCHASCHEL (1979); monograph, mainly taxonomy – VOIGT and KOSTE (1978)

5.12 Sipuncula

So far only two species of this isolated phylum of unsegmented marine coelomates can be considered as permanent meiofauna. The body of Sipuncula is divided in a posterior trunk into which a narrower anterior introvert can be retracted by ventral muscles. At the anterior end of the body, the terminal mouth opening is surrounded by a tentacular crown of varying structure. In close vicinity to the anterior tip of the body is the anal opening. Thus, what seems to be the posterior body end is morphologically a median "sac". Internally, the intestine is coiled with an escending and ascending portion.

While macrobenthic Sipuncula live mostly in crevices and holes of porous rocks, in animal tubes and shells, the tiny and almost transparent meiobenthic species have been found in shelly coarse sand in the Atlantic sublittoral off Florida (*Phascolion* sp.) and in the brackish interstitial of intertidal sands of some Caribbean islands (*Aspidosiphon exiguus*). These species are about 4 mm in length, but their flexible body can squeeze through narrow voids in the sand. As a characteristic feature, *Aspidosiphon* has two "shields", hardened areas of the skin.

More detailed reading: EDMONDS (1982); GIBBS (1985)

5.13 Mollusca

After metamorphosis most molluscs pass through a first benthic phase during which they belong to the "temporary meiofauna" and can become ecologically important (ELMGREN 1978), but later they grow to macrobenthic size.

However, there are a number of about 100 mollusc species, belonging mainly to the Opisthobranchia, which remain minute and permanently of meiobenthic size. These are mostly animals of worldwide occurrence, living preferably in the interstices of well-oxygenated medium to coarse sand (Fig. 43).

Due to their remarkable deviation from the typical mollusc body plan, a mixture of "idioadaptations" and regressive features, it is sometimes not easy to recognize the small and vermiform bodies as molluscs, especially when fixed. They may be confused with turbellarians (*Rhodope*) and nemerteans (*Philinoglossa*, *Platyhedyle*). While most of them retain a (simple) radula as the main synapomorphic structure of molluscs, they have reduced their gills, the ctenidia, and in many cases their shell, suppressed the planktonic veliger larva, simplified the gonads and digestive gland, and are whitish-opaque without pigments. On the other hand, most meiobenthic gastropods developed rhinophors and tentacles

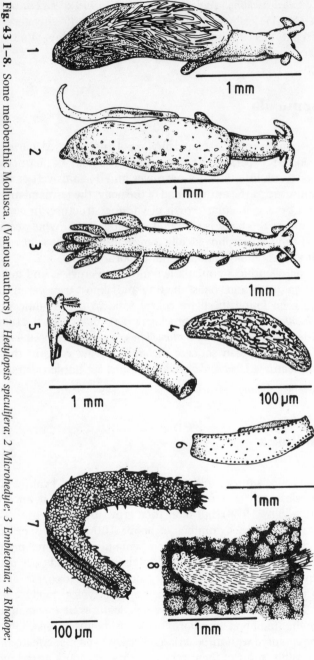

Fig. 43 1–8. Some meiobenthic Mollusca. (Various authors) 1 *Hedylopsis spiculifera*; 2 *Microhedyle*; 3 *Embletonia*; 4 *Rhodope*; 5 *Caecum*; 6 *Philinoglossa*; 7 *Meioherpia*; 8 *Prochaetoderma*

in a specific arrangement at their anterior end, some have subepidermal spicules to stabilize their body (*Hedylopsis, Rhodope*, see Fig. 43). Most interstitial gastropods have an enhanced number of adhesive glands in their epidermis. The aplacophoran molluscs are externally characterized by their coat of cuticular spines and needles to which are attributed not only a protectice and reinforcing function but also a locomotory relevance.

The small number of meiobenthic genera allows for an introduction to some of the more common or characteristic forms:

5.13.1 Gastropoda

Opisthobranchia
Acochlidioidea: no shell and ctenidia, two pairs of anterior tentacles
Hedylopsis: six species, with long and broad posterior tentacles, with spicules in their dermis (e.g. *H. spiculifera*), hermaphroditic. Members of the genus occur only sublittorally. *Strubellia*, related to *Hedylopsis*.
Microhedyle, Unela: with short, slender posterior tentacles (rhinophores), with spermatophores and integumental fertilization (see Fig. 43). They live eu- and sublittorally in sand
Cephalaspidea: some species with internal rudimental shell and ciliated creeping sole (Philinidae), and a small visceral hump, The ctenidia are sometimes reduced (Philinoglossidae)
Philine, Philinoglossa, Pluscula: the vermiform body is square in cross-section; Nudibranchia: with clusters of dorsal processes (cerata), mostly containing cleptocnides (functional cnidocysts adopted from their cnidarian prey)
Aeolidiacea: seven meiobenthic species, some of them, as in their macrobenthic relatives, with cleptocnides in processes of the midgut gland (preying on interstitial hydroids). Pseudovermidae: several species of *Pseudovermis*, head end swollen, set off from body, devoid of appendages; continuously probing the sand when creeping. Embletoniidae: several species of tiny *Embletonia*, similar to the macrobenthic relatives
Rhodopidae: this group has a doubtful position (between opisthobranchs and pulmonates), hermaphrodites without a shell. *Rhodope* has numerous verrucose spicules (Fig. 43), similar to a turbellarian since a foot is not delimited and redula is reduced. Discovered in 1847 as the first interstitial mollusc.

Prosobranchia
Omalogyra: with "normal", typically coiled, operculated shell, about 1 mm in height. *Caecum* spp: with characteristic curved tubular shell, but juvenile conch coiled, adults about 1.2 mm long; common in calcareous sediments of warm water regions, may be confused with juvenile tusk shells (Scaphopoda)
Meiobenthic gastropods are characteristic for water-saturated intertidal and subtidal sands with a well-developed interstitial system of high permeability

and a rich oxygen supply. Their typical adaptations prevent them from being displaced or damaged by sand agitation. Only a few species are found in sands with somewhat higher organic content or in polluted areas. The shape of the grain particles seems to play a major role for their occurrence, since densely packed sediments are only little inhabited, while shell hash with its loose structure of often flat fragments is preferred.

5.13.2   Aplacophora (Fig. 43)

These peculiar marine molluscs have a vermiform, bilateral-symmetrical body. The subgroup Neomeniomorpha (= Solenogastres s.str.), is characterized by a longitudinal pedal groove (raphe) with mucus glands and by many iridescent calcareous spines and plates. These can be hollow (*Biserramenia*) or solid (*Lepidomenia, Meiomenia, Meioherpia*), their shape and arrangement are taxonomically relevant features as are the position and form of the radula teeth. They lack tentacles and eyes, and exibit many reduction phenomena.

In contrast, the Chaetodermomorpha (= Caudofoveata) have a frontal shield, a set of long circum-cloacal spines and a pair of posterior ctenidia; they lack a pedal groove. While most Aplacophora are beyond meiobenthic size, about 20 species are only 1–2 mm long and are considered meiobenthic. The Neomeniomorpha among them are anatomically well adapted to live mostly in the interstitial of coarse sand and shell hash in shallow sublittoral depths. They occur also on colonies of hydrozoans, where they feed on the polyps. The deep-water forms (below 100 m depth down to the deep sea) burrow slowly in muddy bottoms. The meiobenthic Chaetodermomorpha (Prochaetodermatidae, genera *Prochaetoderma, Chevroderma*) have a rasping radula and live on foraminiferans and detritus. *Lepidomenia hystrix*: described in the last century by MARION and KOVALEVSKI from the Mediterranean; *Meiomenia, Meioherpia, Psammomenia*: found by SWEDMARK in detritusrich coarse sand ("Amphioxus sand") off Roscoff, reported also from the North Sea.

> *More detailed reading*: taxonomy and biology of meiobenthic gastropods – SWEDMARK (1968); POIZAT (1985); ARNAUD et al. (1986); Aplacophora – SALVINI-PLAWEN (1985)

5.14   Annelida

5.14.1   Polychaeta

In each meiofauna sample, polychaetes belong to the most striking and beautiful animals due to their multitude of structures, sizes and movements. Although there are only about 200 to 250 polychaete species of meiobenthic size, belonging to approximately 25 families, their abundance is fairly high, usually ranking fourth in meiofauna samples. Many of the meiobenthic forms have been classified earlier as a separate group, the "archiannelids", a term coined by HATSCHEK. This group was characterized by rather aberrant features which were held to be archaic. The structure often regarded the most valid to unify the "Archiannelida", the ventral pharyngeal bulb, is probably a convergent adaptation to the uptake of microphagous food: they evert this bulb by complicated muscular contractions for pipetting bacterial and diatom aufwuchs from sand grains. It has been shown, however, that the various structures forming the ventral pharyngeal bulb are not homologous, and thus defy a synapomorphy for all the archiannelids (PURSCHKE 1988; but see RIEGER and RIEGER 1975).

Today, it is well established that there is not a single synapomorphic character unifying the "Archiannelida", a fact which abolishes this taxon as artificial. Archiannelids are rather a convergent assembly of about 60–100 species of meiobenthic, aberrant polychaetes belonging to approximately 12 families with numerous reductive, neotenic or highly modified features (see below). It is, however, not without reservations to group all of them into polychaetes, since polychaetes as a taxon are problematical, not to be defined by clear synapomorphies. There certainly exist so many aberrant forms that a classification as "Annelida incertae sedis" (e.g., Parergodrilida, Aeolosomatida, *Lobatocerebrum*, see Chap. 5.14.3) would be more correct than a forced lumping into the traditional annelid subgroups polychaetes and oligochaetes. Some of these strange annelids play a central role in phylogenetic discussions on annelids.

Among polychaetes many characteristic convergent adaptations typical for meiobenthic and particularly interstitial life are beautifully realized (Fig. 44):

– mostly very small (mature *Nerillidium* only 250 µm long!), with only a few segments,
– parapodia often reduced and not protruding (*Protodrilus*), sometimes even setae reduced (*Polygordius*),
– often ciliated, with ciliary rings or a ciliated "creeping sole" (with gliding locomotion; e.g. *Dinophilus, Trilobodrilus, Ophryotrocha*),
– no circular musculature, no peristaltic movements.

Some of the features mentioned above clearly relate to the progenetic nature of many "archiannelids", a contention supported by other structures: epithelial nerve system, simplification of the dermal ultrastructure, no vascular system nor coelomic cavities. Other morphological features seem secondary adaptations to the void system of sediments:

– threadlike form, then with many segments (*Polygordius*, up to 10 cm long, with up to 185 segments!),
– flattened body with a ventral "creeping sole" (*Protodrilus*),
– numerous adhesive glands on parapodia and caudal appendages (*Hesionides*),
– eyes and pigments reduced,
– no planktonic trochophora larva,

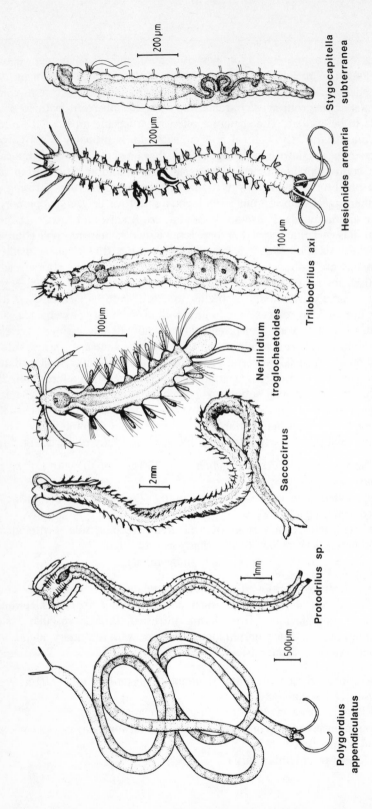

Fig. 44. Some typical meiobenthic Polychaeta. (Various authors)

200 μm

Stygocapitella subterranea

200 μm

Hesionides arenaria

100 μm

Trilobodrilus axi

100μm

Nerillidium troglochaetoides

Saccocirrus

2mm

Protodrilus sp.

1mm

Polygordius appendiculatus

500μm

– complicated genital organs, often hermaphroditic (*Ophryotrocha*), with copulatory structures (*Hesionides, Microphthalmus, Questa*).

Many meiobenthic polychaetes are extremely euryoecious and well adapted to
live under the fluctuating conditions of eulittoral beaches. Here, *Protodrilus* and
Hesionides can become locally abundant. Their abundance and hardy nature
made meiobenthic polychaetes to favorite animals for experimental work with
meiofauna: *Protodrilus* was shown to "recognize" its natural sand due to preference for its bacterial colonization (BOADEN 1962; GRAY 1966). *Ophryotrocha*
and *Dinophilus* have long served as convenient culture animals for biossays and
studies on speciation and sex determination (ÅKESSON 1975; PFANNENSTIEL
1981). Respiration experiments have been performed with *Protodrilus* and
Ctenodrilus (BOADEN 1989a). The latter is being studied at present for ontogenetic
determination processes of general relevance on the level of gene sequences and
expression (DICK and BUSS 1993).

Despite the lack of planktonic larvae, the distribution of many genera and
species of meiobenthic polychaetes is fairly wide. Identical genera, sometimes
even species, can be found amphi-atlantically and occasionally they have a
cosmopolitan occurrence. Considering the often highly apomorphic features in
almost all organs, it seems doubtful that this uniformity in distribution can be
interpreted as a slow radiation of a very old animal group (see Chap. 6.2).

Many meiobenthic polychaetes occur in brackish-water biotopes from where
some entered the continental groundwater system (*Meganerilla, Thalassochaetus*),
others became troglobitic (*Troglochaetus*). *Parergodrilus* even lives in moist soil
of forests far away from the sea.

Despite their often euryoecious nature, the interstitial polychaetes populating
a beach have a very distinct distribution pattern (Fig. 45) with clear preferences
for certain ecological gradients.

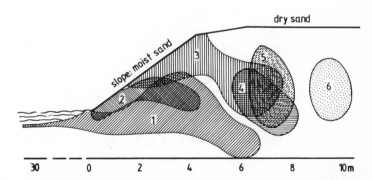

Fig. 45. A differentiated distributional pattern of some interstitial polychaetes along a
beach profile. *1 Hesionides gohari; 2 Diurodrilus* sp.; *3 Hesionides arenaria; 4 Diurodrilus
benazzi; 5 Protodriloides* sp.; *6 Stygocapitella subterranea.* (WESTHEIDE 1972)

The wealth of structural details useful as diagnostic features renders it possible to identify meiobenthic polychaetes in most cases to the generic level; often even the species is recognizable without further dissection (WESTHEIDE 1990).

Below, some of the more frequent and interesting genera are introduced and commented, their classification into families follows WESTHEIDE (1988, 1990).

Polygordiidae: *Polygordius* (with 15 species) occurs worldwide in sublittoral sands. Together with *Nerilla* (see below) the genus was one of the first "archiannelids", discovered in 1848 by O. SCHMIDT in sublittoral sand near the island of Helgoland. The long, thread-like body with its smooth surface devoid of any appendages or setae makes the genus easily recognizable. The numerous (up to 185) segments are not visible externally. The two prostomial tentacles are short and stiff; the pygidium with adhesive glands is set off.

Protodrilidae: two genera, among which *Protodrilus* with 30 species. They are distributed worldwide (NORDHEIM 1989), occurring mainly in the sublittoral, but are also found in sandy tidal flats. The slender body consists of many discernible segments without parapodia, but with the setae visible. In contrast to *Polygordius*, the two head appendages are flexible and longer. After extended and rough extraction procedures they tend to break off. The pygidium has two to three lobes which are of diagnostic relevance. The animals are hermaphrodites, some species have larval dwarf males which attach to the partner. They produce spermatophores and fertilization of the eggs is internal. The larva is known to metamorphose only in its domestic sand probably due to a genuine bacterial composition (GRAY 1966). The two species of *Protodriloides*, characterized by numerous refringent epidermal glands, lay their eggs in cocoons; fertilization is external, possibly within the cocoons. *Parenterodrilus* is a gutless protodrilid from coral sand in the Pacific Ocean (JOUIN 1992).

Saccocirridae: *Saccocirrus* with 17 species, is distributed worldwide in eulittoral and sublittoral coarse sand. They are thin and very active worms with short parapodia and setae, one pair of eyes and two sticky pygidial appendages. The long and flexible tentacles are used as tactile probes. In some species, the males use eversible papillae in their genital segments as copulatory organs; the females have corresponding spermathecae. In other species of *Saccocirrus*, spermatophores are attached to the female partner. *Saccocirrus* is probably predaceous.

Nerillidae: most of the > 30 species (classified in 12 genera of which *Nerilla* and *Nerillidium* are best known) are small polychaetes (< 1 mm long) with only seven to nine segments. They are ventrally ciliated, have characteristic, slightly lobate anterio-lateral palps, and relatively long parapodia and setae. The pygidium has anal cirri; all appendages are very flexible. Development is direct, eggs are sometimes brooded in an epidermal "hood". Ecologically,

Nerillidae are very diverse, some (*Mesonerilla*) are limnic or troglobitic (*Troglochaetus*). Often Nerillidae are brought in experimental seawater systems. Some species have a worldwide distribution. Although many species possess jaws, they are microphagous bacteria and diatom feeders.

Dorvilleidae: *Ophryotrocha, Parapodrilus, Apodotrocha*. Previously often considered as belonging to separate (archiannelid) families, WESTHEIDE (1984) clarified their relationship to dorvilleid polychaetes. The approximately 70 meiobenthic species, gliding on a ventral ciliated "creeping sole", are mostly progenetic with only a few segments of larval character. As with nerillids, also dorvilleid species (particularly *Ophryotrocha*) often are of unclear origin and distribution, suddenly "occurring" in seawater aquaria. Parthenogenesis is considered possible in this group. *Ophryotrocha* species are protandric hermaphrodites protecting their eggs in a cocoon. They have conspicuous jaws. *Parapodrilus* is only 0.6 mm long when adult.

Dinophilidae: the mostly clumsy body has ciliary rings but lacks tentacles and appendages. *Dinophilus*, discovered in 1948, with about ten, sometimes pigmented species. After a phenotypical sex determination, the dwarf males inject their sperm into the females. This copulation is performed directly after hatching from the eggs, sometimes even still in the egg-cocoon. *Trilobodrilus*, is one of the frequently occurring meiobenthic polychaetes in North Sea tidal sands.

Hesionidae: *Hesionides* and *Microphthalmus* are common meiobenthic genera within this polychaete family. Both have long but flexible parapodia and setae. *Hesionides* is moving over the sand grains resembling of chilopod centipedes with the parapodia orientated ventrally. The genera and species differ in the arrangement of prostomial tentacles and parastomial cirri and the formation of anal lobes equipped with adhesive glands. They are hermaphroditic animals with complicated copulatory organs, spermatophores and encapsulation of eggs in cocoons.

Syllidae: this large family of Polychaeta contains numerous small, meiobenthic forms belonging to genera such as *Sphaerosyllis, Streptosyllis, Petitia, Exogone, Brania*. Most of them have a conspicuous barrel-shaped proventricular musculature. The parapodial tentacles often have a characteristic articulated structure. Some species have eyes. Representative of meiobenthic syllids occur both in sand and mud, but most often between coral rubble and encrusting epiphytic algae.

Pisionidae: the mostly small, psammobiotic species (best-known genus *Pisione*) occur worldwide, also in the continental groundwater. They have many adaptations to the interstitial system: adhesive glands, complicated genital organs and reproductive patterns with numerous copulatory appendages (males with penes, females with spermathecae, see Fig. 46).

Psammodrilidae: an isolated group of aberrant, neotenic polychaetes (*Psammodrilus, Psammodriloides*) with complete body ciliation, no appendages, but with three pairs of long cirri. They live interstitially in a transparent housing glued to sand grains.

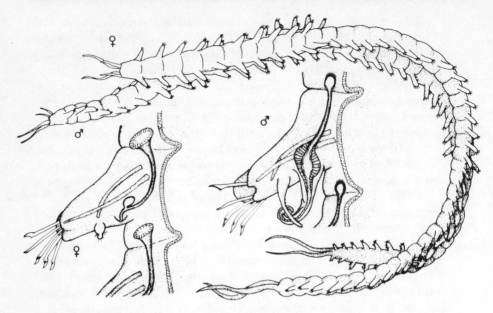

Fig. 46. Copulatory structures in the meiobenthic polychaete *Pisione remota*. (Ax 1969)

Parergodrilidae: an isolated group of aberrant, neotenic? polychaetes. *Stygocapitella subterranea* without head appendages and parapodia, but with characteristic setation (long hair setae in segment V) and a conspicuous S-shaped intestinal curvature in the posterior body (see Fig. 44). Originally described and for a long time considered as an enchytraeid oligochaete, this species occurs regularly in the supralittoral, rather dry zone of beach sand. Its slow movements and the features given above make it easily recognizable. *Parergodrilus* is a rare genus occurring in soil of Middle European beech forests.

More detailed reading: taxonomy and reproductive biology, phylogeny – WESTHEIDE (1984, 1985, 1986, 1987a); monograph on archiannelids – REMANE (1932)

5.14.2 Oligochaeta

This group, usually considered as primary freshwater inhabitants, contains about 600 meiobenthic species, of which about 300 have intensively radiated in marine habitats, some of them even occurring in the deep sea. Most of the numerous interstitial species belong to the Tubificidae, followed by the Enchytraeidae. Distinction of the families is largely by position of genital segments in combination with features of setation (Fig. 47). Genera and species are mostly

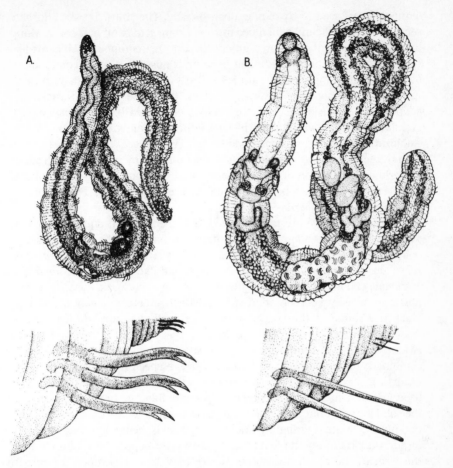

Fig. 47A, B. Some typical meiobenthic Oligochaeta and their setae (*below*) **A** Tubificidae (*Aktedrilus monospermathecus*). **B** Enchytraeidae (*Enchytraeus* sp.)

defined by setation and details of the genital organs. Potamodrilida and Aeolosomatida do not have genuine oligochaete features; they are covered here as "Annelida incertae sedis".

If sufficiently small, the body organization of oligochaetes does not require spectacular idioadaptations for a meiobenthic life: oligochaetes never have any head tentacles or protruding parapodial appendages, their setation is retractable or flexible. All of them are hermaphrodites with complicated genital structures and copulatory modes. Reduction phenomena are frequent: planktonic larvae are not developed, they mostly lack eyes and a marked coloration. In contrast to polychaetes, they have only fairly simple, never "composite" setae (Fig. 47).

Tubificidae: marine tubificids in numerous genera and species belong to the meiobenthos occurring mostly in sandy sediments, while the frequent limnetic forms are mostly of macrobenthic size. The characteristic bifurcate setae (see Fig. 47) can be modified into pectinate setae and hair setae. Many have salient "penial setae" in segment XI, some are equipped with penis-like structures for sperm transfer. Identification is possible using a combination of metric, setal and genital features (mostly mature specimens required). Among the numerous sand-living species, there are many characteristic adaptations to the interstitial: thread-like, elongated (*Olavius longissimus*), also flattened body, with tail (*O. planus*), adhesive glands, turgescent cells for stabilization (*Aktedrilus monospermathecus*). Two genera (*Inanidrilus, Olavius*) with numerous species are adapted to a thiobiotic life by symbiosis with sulfur bacteria and reduction of gut and nephridia (see Chap. 8.5). Other frequent marine genera of meiobenthic tubificids are *Heterodrilus, Limnodriloides, Thalassiodrilus, Phallodrilus* spp.

Enchytraeidae: there are an estimated 200 species from aquatic biotopes, some of them with adaptations to interstitial habitats: the usually simple-pointed setae occasionally become more or less reduced (*Achaeta, Marionina subterranea, M. achaeta*), and the body is thread-like elongated (*Grania* sp.). *Marionina* can attach to sand grains by secretion of coelomic fluid. Identification is possible by a combination of general metric, setal and genital features. Since the shape of the spermatheca is of particular importance, species determination is possible only in mature specimens. In *Enchytraeus* and *Henlea*, terrestrial, limnic and marine species from the upper shore are combined. In *Lumbricillus* and *Marionina* the species are mostly marine, occurring predominantly in the upper eulittoral. Species of *Grania* have been found in sublittoral and even deep-sea sediments. In this latter genus, the solid longitudinal musculature reduces peristaltic movements and leads to a more nematoid wriggling. Most enchytraeids, being extremely tolerant to fluctuations of abiotic factors (except oxygen supply), are abundant in the wrack zone of the sea shore, where they have an important role in degrading the debris washed ashore (GIERE 1975). Most members of the family occur in terrestrial biotopes.

Naididae: setae of naidids are conspicuously bifurcated with a nodulus. Long hair setae occur frequently in dorsal bundles. Since genital organs are rarely developed, identification has to be done by form, arrangement and metric relations of setal size. The predominantly asexual reproduction by budding, fission or fragmentation leads to sudden outbursts of naidid populations, often in correspondence with blooms of their favourite food, benthic diatom algae. *Chaetogaster* is predominantly predaceous. Mostly as epibenthos in freshwater (*Nais, Dero*), a few species entered the brackish tidal flats (e.g. *Nais elinguis, Paranais, litoralis, Amphichaeta sannio*).

Lumbriculidae: with a body length of < 10 mm, the subterranean freshwater genera *Dorydrilus* and *Trichodrilus* contain a few truly meiobenthic, albeit rare species belonging to the continental groundwater fauna, the stygon (see Chap. 8.2.2).

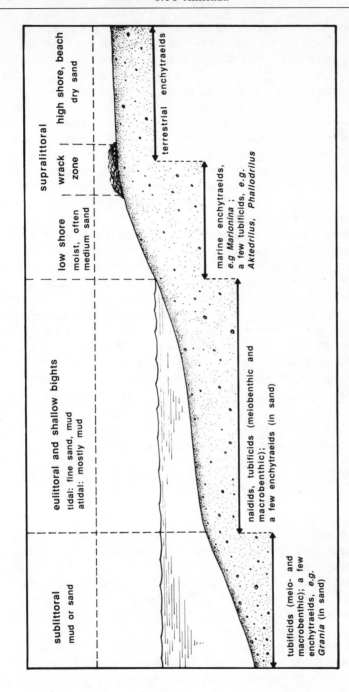

Fig. 48. The general distribution pattern of marine meiobenthic Oligochaeta. (After GIERE and PFANNKUCHE 1982)

Biological and Ecological Aspects. Most naidids have an annual life span, while the other families do not reproduce before their second year. Reproduction in the limnetic and boreal eulittoral forms is mostly seasonally restricted (often in summer/autumn), while the numerous marine species from subtropical to tropical habitats apparently reproduce continuously. In tropical beaches, oligochaetes were found to recover particularly rapidly after the massive decline of meiofauna through monsoonal rains (ALONGI 1990b). Despite the existence of some deep-water forms, marine oligochaetes are typical eulittoral and sublittoral meiofauna. In limnetic habitats, naidids will be encountered as the predominant meiobenthic oligochaetes, while sandy sea shores in their upper (less water-saturated) reaches mostly harbour enchytraeids. In deeper horizons or further out in tidal flats, particularly in fine sediment, tubificids take over. They represent the typical interstitial marine oligochaete (Fig. 48).

Freshwater oligochaetes usually are an important member of the benthos in all limnetic habitats (see Chap. 8.2.3), their juveniles belonging to the temporary meiofauna. In the continental groundwater and hyporheic system (see Chap. 8.2.2), about 60 species of small stygobiotic species have been recorded belonging mainly to lumbriculids, haplotaxids and rhyacodrilids. Vertical distribution is mostly limited by oxygen supply to the uppermost layers, except from the gutless thiobiotic tubificids which accumulate in slightly sulphidic sediments around the oxic/sulphidic interface (GIERE 1975; GIERE et al. 1991). Their abundance makes oligochaetes locally a significant food source for other meiobenthos (e.g. some halacarids, turbellarians) and for small fish.

More detailed reading: taxonomy, identification – BRINKHURST (1982a); ERSÉUS (1980, 1984a, 1990b); phylogeny – BRINKHURST (1982b, 1984, 1991, 1992) ERSÉUS (1984b, 1987, 1990a); review on ecology of marine Oligochaeta – GIERE and PFANNKUCHE (1982)

5.14.3 Annelida "incertae sedis"

Progenetic trends may have played a major role in the development of these rare, but phylogenetically possibly relevant "microannelids". Potamodrilida and Aeolosomatida (Fig. 49) do not match the usual definitions of the above larger groups and are also grouped as "Aphanoneura" (BRINKHURST 1982b).

Their systematic links remain unclear, particularly since miniaturization (body length only 200–1000 µm) and progenesis have led to various reductive phenomena. The few segments are not always clearly delimited, the nerve cord is epithelial, a ventral ciliation (particularly of the prostomium) is responsible for the gliding movements, the setation consists of hair setae only, etc.

Potamodrilida live stygobiotically in the gravel and sand of river banks. *Potamodrilus* fluviatilis, a hermaphroditic species which merely reproduces sexually, has seven segments only and a typical "archiannelid-like" protrusible ventral pharyngeal bulb (see Fig. 49). Aeolosomatida consist of about 25 species

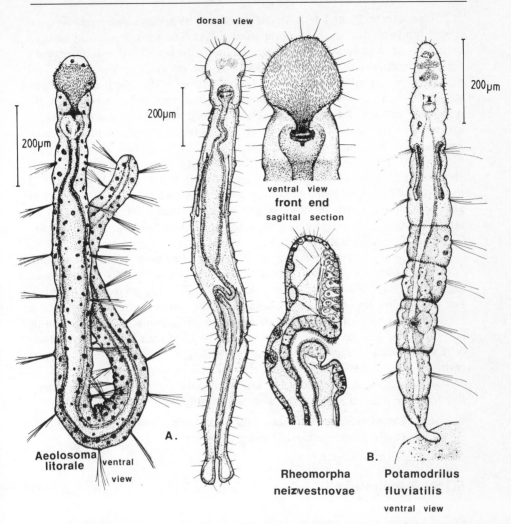

dorsal view

200µm

200µm

200µm

ventral view
front end
sagittal section

200µm

A.

Aeolosoma litorale ventral view

B.

Rheomorpha neizvestnovae

Potamodrilus fluviatilis
ventral view

Fig. 49A, B. Representatives of the annelid groups Aeolosomatida (**A**) and Potamodrilida (**B**). (Combined from various authors)

characterized by numerous epithelial "oil glands" of different colour. Aeolosomatids occur preferably in the phytal and epibenthic detritus layer of more or less stagnant, oxygen-rich freshwater habitats (*Aolosoma hemprichi*). *A. litorale* is common in brackish coasts; *A. maritimum* was described from the interstitial of a Tunesian beach. *Rheomorpha neizvestnovae* is a rare species without setae from the hyporheos of boreal rivers and lakes; Fig. 49). Aeolosomatids normally have an asexual reproduction by paratomy leading to a sudden mass multiplication and chains of two to eight zoids. In the few mature specimens the spermatheca is formed by just one cell, a clitellum is not developed.

Lobatocerebrida: in 1980, Rieger described the genus *Lobatocerebrum* from coarse sands off the North Carolina coast (USA). The vermiform, fully ciliated animal lacks any setation and segmentation. Its body cavity is filled with mesenchymal cells, a mesodermal lining is not discernible. The mouth and pharynx are ventrally located, the genital system is hermaphroditic. Thus, superficially, it can be characterized as turbellariomorph. However, a well-developed hindgut and anus, remnants of a blood vessel system and a strange inversion of the muscular layers contradict any body plan of turbellarians. Additionally, many autapomorphic histological details do not allow an alignment into the Plathelminthes. They rather suggest putting the peculiar Lobatocerebrida with their three species as "secondary acoelomates" (Rieger 1991) to the annelid stock in a wide definitory sense of "Annelida". Westheide (1988) groups the family in the polychaeta. Rieger argues that Lobatocerebrida might attain a high relevance for the evolutionary understanding of acoelomate flatworms from coelomate ancestors.

Jennaria pulchra: this peculiar slender worm has been found in fine sand of the highwater line at the US-Atlantic coast (Rieger 1991). Its ciliated, non-segmented body of only about 2 mm length is divided by grooves in a rostral and caudal part set off from the trunk. The body musculature consists almost exclusively of unicellular, obliquely striated longitudinal muscle cells. The mouth is ventral, the anus terminal. The body cavity is filled with large enchymatous cells and contains some protonephridia. The genital system is possibly gonochoristic. Like *Lobatocerebrum*, also this turbellariomorph worm should be considered a "secondary acoelomate" which originated as an isolated evolutionary line from the annelid stock or its close relationship.

More detailed reading: Aeolosomatida and Potamodrilida – Bunke (1967); phylogenetic considerations – Rieger (1991)

5.15 Tardigrada

This taxon is a characteristic member of the meiobenthos, present both in marine and freshwater habitats, in sand and mud, in the supralittoral and the deep see. The high group diversity (most genera with a few species only) point to a marked radiation. Today, there are about 680 species known; however, especially from marine habitats, more are being continuously described. Description of the first interstitial tardigrade, *Batillipes mirus*, was in 1909. The minute size of all tardigrades or "water bears" (50–1500 µm) often requires a double check at high magnification when sorting a sample.

The five-segmented body of tardigrades is of sometimes bizarre appearance (Fig. 50). Its head segment has numerous sense organs on cirri whose arrangement is of diagnostic relevance (Heterotardigrada). The following three thoracal segments and the caudal segment have flexible legs without arthropod joints, ending mostly in claws, sometimes fixed on long toes (subgroup Arthrotardigrada;

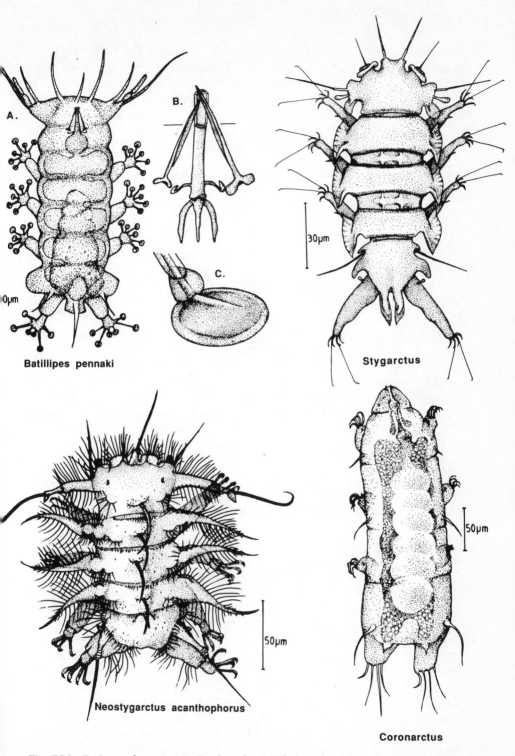

A.

B.

C.

0μm

Batillipes pennaki

30μm

Stygarctus

Neostygarctus acanthophorus

50μm

50μm

Coronarctus

Fig. 50A–C. Some characteristic Tardigrada. **A** Whole animal from dorsal. **B** Piercing stylet apparatus. **C** Viscid digital end plate. (Various authors)

Fig. 50). The distal part of the extremities is retractable like a telescope and eversible by turgor pressure exerted by the mixocoelom.

The pharynx is equipped with a complicated, diagnostically important stylet apparatus for piercing mainly plant cells. Its specific muscles act as a suction pump, recalling many namatodes. As in nematodes, circular musculature and ciliated epithelia are absent. The minute size allows reduction of a respiratory, circulatory and excretory (nephridial) system.

> *Heterotardigrada*: this more primitive group today consist of more than 250 species (which all have head cirri). In most heterotardigrade species the body is well segmented, often carrying plates, Most live in marine sediments and in algae.

Arthrotardigrada: legs have toes from which claws arise. *Stygarctus, Halechiniscus, Batillipes*, are common species from marine eulittoral sands; altogether 35 genera with 110 species.

Echiniscoidea: the claws insert directly on the legs, there are no toes. With altogether 14 genera and 190 species, most echiniscoid tardigrades are terrestrial forms; however, *Echiniscus* is a typical marine genus; the rather vermiform *Carphania* lives in hyporheic freshwater. Echiniscoid tardigrades survive short periods of desiccation without problems.

> *Eutardigrada*: this more advanced group consists of 34 genera with 380 mostly limnetic and terrestrial species. They never have head cirri and lack a distinct segmentation. Their mouth is surrounded by a whorl of lamellae. *Macrobiotus, Hypsibius, Milnesium* are characteristic representatives. Many occur in mosses and lichens, but also in mesopsammal habitats. Some have secondarily re-adapted to marine biotopes (e.g. Halobiotus)

Biological and Ecological Aspects. The aquatic heterotardigrades survive adverse periods through cyst formation, while the eutardigrades have developed a remarkable anhydrobiosis (cryptobiosis) during which they dehydrate their water content down to 98% and survive most inhibitory life conditions. During this phase, stabilization of membrane structures is possible through synthesis of glycerol and trehalose metabolized from lipids and glycogen which replace the integral water of lipid bilayers. Occasionally, cyclomorphotic changes (i.e. regularly occurring alterations of certain structures within a population) in the stylet apparatus have been found in *Halobiotus* (KRISTENSEN 1982).

Little is known about the developmental biology of tardigrades. They usually have separate sexes and are sometimes sexually dimorphous. Males have a shorter maturation period than females. Reproduction is through copulation; eggs are often unusually shaped and ornated; the cleavage pattern is irregular. Growth of the juveniles, which have less legs and claws than the adults, is through molts. All hard structures (claws, stylet apparatus) consist of the chitinous material covering the whole body as a cuticle. Thus, they have to be molted during growth. Mitosis is limited to a few tissues and possible only during the juvenile phase. In the adult body, which consists of relatively few cells, the cell number remains constant (eutelic) in most organs. *Batillipes pennaki*, one of the most frequent interstitial tardigrades around the Mediterranean, has a bivoltine developmental cycle with (in Italy) maxima in spring and autumn related probably to favourable food conditions.

Tardigrades feed on the contents of bacteria and algal cells (both macroalgae and diatoms) which they pierce with their stylet apparatus. Only a few species are carnivorous or detritivorous. Many aquatic species can be found in mosses, among algae, in seagrass meadows, even in floating *Sargassum*. These phytal forms use their legs and claws for clinging to the fine thalli.

Sandy biotopes represent the other major habitat for (marine) tardigrades. Here, they crawl characteristically and slowly on the sand grains (group name derived from Latin: "slow walkers"). *Batillipes* with its viscid digital plates instead of claws is fairly active, quickly "running" up and down the grains. *Halechiniscus* with its clumsy, barrel-shaped body pushes its way through the finer sand. In the mud-inhabiting *Coronarctus*, found in greater depth and even in the deep-sea, the body is more worm-like and the legs are very short. The semi- or epibenthic forms attain the most bizarre body shape: *Neostygarctus, Zioella.*

The overall abundance of tardigrades, even in favourable sites, is never very high. Densities of > 500 ind. per 100 cm^3 of sand must be considered extremely unusual. Patches up to 3500 *B. pennaki* per 100 cm^3 (= 32,500 under 100 cm^3) have been counted in the low water line of a Portuguese beach (THIERMANN, unpubl.). On the other hand, their number is often underestimated due to inadequate sorting procedures and methodological shortcomings. A good method for obtaining (interstitial) tardigrades fairly quantitatively is the "freshwater shock" decantation method which causes them to release their immediate grip on sand grains (see Chap. 3.2.1).

While in coarse sand most species live interstitially, in fine sand with detritus the tardigrade community tends to live epibenthically and becomes less diverse. Apparently, sediment porosity and structure are relevant distributional factors. In sediments with a good oxygen supply, vertical occurrence of tardigrades down to 150 cm and deeper is not unusual, but the food source in these depths remains unknown.

A typical distributional profile along the sea bottom (Fig. 51) will have the most eurytopic forms, common even in various types of sediment, in the eulittoral (*Stygarctus, Batillipes*). Genera occurring in the sublittoral can clearly be differentiated into mud- and sand dwellers and the deep-sea ooze harbours rather specific tardigrade species of very limited distribution.

However, many genera, although of ubiquitous occurrence, seem to have a well-developed potential of differentiating the ecological conditions in a beach, resulting in a remarkably heterogeneous colonization with distinct population centres (Fig. 52).

Despite the limited active distribution potential of tardigrades, many tardigrade genera and species are widely distributed, some of them are cosmopolitans occurring in all climates. Their extreme survival capacity in resting stages may have contributed to this ubiquity. On the other hand, depressed forms with flat,

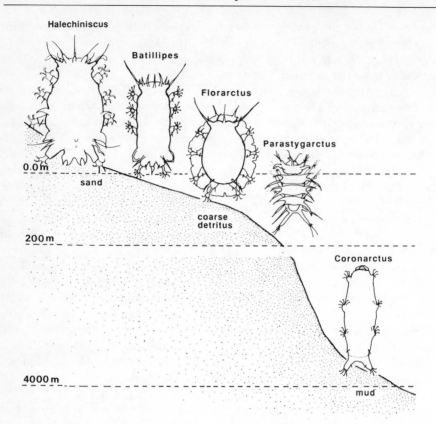

Fig. 51. A horizontal distribution pattern of tardigrades along the sea floor. (After GRIMALDI DE ZIO et al. 1984)

extending body plates have been considered particularly well adapted for a passive transport clinging to sand grains (GRIMALDI DE ZIO et al. 1983).

Tardigrades with their segmented body and chitinous cuticula, their meta-meric nerve system and characteristic legs are phylogenetically linked to the "annelid-arthropod line of development." Parallels to Nemathelminthes, such as the structure of the pharyngeal pump, eutelic fixation of cell number, lack of ciliated epithelia and circular musculature, are interpreted today as convergences. The coastal zone of the sea is often considered the original habitat of primitive tardigrades from which they entered both the freshwater and terrestrial biotopes as well as the deeper marine bottoms.

More detailed reading: marine species, taxonomy and faunistics – SCHULZ (1963), KRISTENSEN and HIGGINS (1984); phylogeny – GRIMALDI DE ZIO et al. (1987); biology, ecology – DE ZIO and GRIMALDI (1966); GRIMALDI DE ZIO and D'ADDABBO GALLO (1975); GRIMALDI DE ZIO et al. (1983); RENAUD-

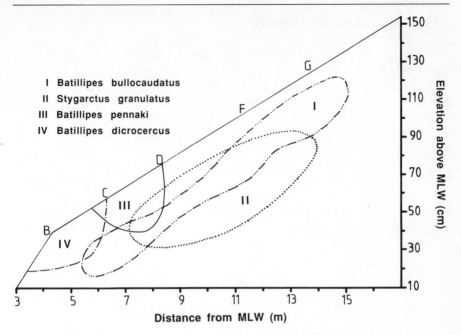

I Batillipes bullocaudatus
II Stygarctus granulatus
III Batillipes pennaki
IV Batillipes dicrocercus

Distance from MLW (m)

Elevation above MLW (cm)

Fig. 52. Occurrence of various tardigrades in a beach profile. (After POLLOCK 1970)

MORNANT (1982); freshwater species – IHAROS (1975); SCHUSTER et al. (1980); cryptobiosis – WRIGHT et al. (1992)

5.16 Crustacea

Next to nematodes, crustaceans are the meiobenthic group dominating in abundance and species richness in most meiobenthic studies. There are, however, many taxonomic lines in crustaceans with groups often consisting of a few isolated species only and living in refuge zones (e.g. groundwater, caves). These rarer forms are also considered here because of their high phylogenetical and zoogeographic significance. Progenesis is of particular importance in the development of meiobenthic crustacean groups. They seem "prone" to this evolutionary abbreviation leading to taxa such as ostracodes and cladocerans which are generally of small size and rapid generation turnover. In these groups, as well as in the species-rich copepod suborder Harpacticoida, the majority is of meiobenthic size. But there are other, normally macrobenthic crustacean groups, e.g. isopods and amphipods, in which by structural deviation and miniaturization only some few aberrant taxa are adapted to the requirements of meiobenthic life.

5.16.1   Cephalocarida (Anostraca)

This meiobenthic group, discovered by SANDERS, consists of three, perhaps four, genera only with a few species. *Hutchinsoniella* (Fig. 53) was found in the sublittoral down to 300 m depth off Long Island (east coast of America), *Lightiella* in soft bottom off the American Pacific shore, other Pacific areas, and off Venezuela (HIGGINS, pers. comm.). *Sandersiella* occurs in mud bottom off Japan.

The small animals (about 3 mm body length) have a slender body with numerous, homonomous segments (nine thoracal and ten abdominal metamers). They lack a carapace and a caudal furca. Unlike other anostracans, Cephalocarida have articulate phyllopodia and a maxilla morphologically identical with the thoracal legs. The group is found in the superficial detrital layer of the sea bottom, flocculent enough to filtrate their food from it when moving (the row of legs form suction chambers).

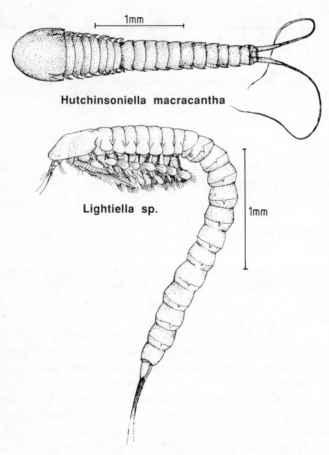

Hutchinsoniella macracantha

Lightiella sp.

Fig. 53. Some anostracan crustaceans: Cephalocarida. (Combined from various authors)

Little is known about this rare group. The disjunct geographical distribution, the primitive structure of the maxilla and, most important, the Devonian fossil *Lepidocaris* which is very similar to recent Cephalocarida characterize this order as "living fossils".

▓ *More detailed reading*: SANDERS (1959)

5.16.2 Cladocera (Branchiopoda)

There are about 270 species of typical meiobenthic cladocerans or water fleas living in all kinds of freshwater from puddles to lakes, most of them belonging to the Chydoridae (Fig. 54). As phyton they climb and swim among the macrophytes of shores (*Eurycercus, Chydorus*), as epibenthos they dig through the surface of muddy bottoms (*Ilyocryptus, Macrothrix, Pleuroxus*), some few species represent even typical subterranean stygobiota (*Alona* spp.).

Although some adult water fleas exceed meiobenthic size, the bulk of the benthic forms do not surpass 1 mm in length or represent small pre-reproductive instars.

Diagnostic features in cladocerans are the general shape and sculpture of the large carapace (the "valves") which completely encloses the unsegmented and short thorax, certain pores on the headshield which often has the shape of a helmet, and the setation and spines of the terminal claws which are mostly bent ventrally.

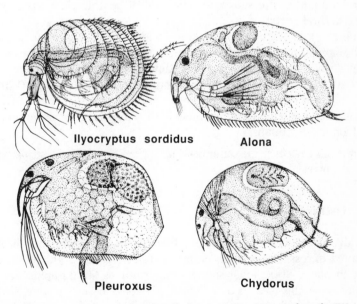

Ilyocryptus sordidus   **Alona**

Pleuroxus   **Chydorus**

Fig. 54. Some representative meiobenthic Cladocera; 0.5 to 1 mm in length. (WESENBERG-LUND 1939)

All cladocerans moult five times before reaching maturity thereby shedding their large carapace. The population consists almost exclusively of females which predominantly reproduce parthenogenetically by diploid eggs. These are brooded in a chamber underneath their carapace. Each egg deposition is preceded by a moult; development of the juveniles hatching from these "summer eggs" is very rapid. With deterioration of life conditions, e.g. sinking temperatures, males are produced and, after normal meiotic division, haploid eggs that need fertilization by the males. The resulting thick-shelled resting eggs are stored in a chamber of the carapace, the ephippium. Here they can survive extremes of temperature and salinity even for years. During this resting phase of embryonic development, the eggs can also dry out completely. Moreover, the ephippium is an excellent raft for distribution drifting in the water, adhering to plants and bird legs. Depending on the physiographic situation, the production of fertilized eggs occurs during winter only. However, in species typical for ephemeral pools which repeatedly dry up and fill again after new rain falls, several sexual cycles occur independently of the season. This heterogonous generation cycle, resembling that of rotifers, ensures on the one hand genetic mixis and survival (of eggs) during adverse life conditions, on the other hand, through its parthenogenetic phase, it renders a huge potential to rapidly populate new areas and isolated ponds.

The meiobenthic cladocerans dig, punt, rake and climb with their large and muscular locomotory antennae and the following thoracic appendages rather than using them as swimming and filtering legs. The number of filter chambers is correspondingly reduced to just one or two. Also the terminal claws can be used for scraping off the substrate. The food of benthic cladocerans is small algal and detrital particles.

The highest diversity of meiobenthic cladocerans occurs in the phytal zone where the biotopical complexity is rich. Due to their parthenogenetic development, they can develop dense populations in almost all types of freshwater (except streams). Abundances of 1 million per m^2 substrate are not unusual (WHITESIDE et al. 1978). Considering their high consumption of bacteria, algae and detritus, cladocerans represent a relevant member of the limnetic meiobenthos important also as food for macrofauna.

More detailed reading: taxonomy – FREY (1987); ecology – GOULDEN (1971); WHITESIDE et al. (1978)

5.16.3   Ostracoda

The bivalved carapace characterizing this crustacean group has been well documented in fossils since the Cambrian. Today, about 5000 recent species contrast to about 8000 fossil species. Important as stratigraphic indicators, ostracodes were classified in a geological grouping based on hard structures only. This developed separately from the zoological taxonomy. It has been attempted to overcome the inherent confusing consequences by inclusion of both

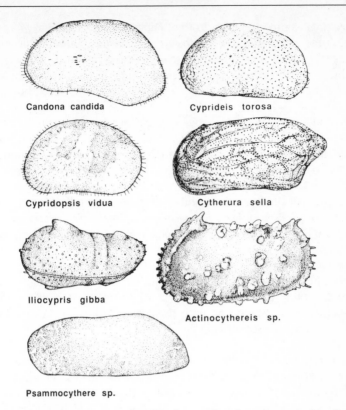

Candona candida

Cyprideis torosa

Cypridopsis vidua

Cytherura sella

Iliocypris gibba

Actinocythereis sp.

Psammocythere sp.

Fig. 55. Some representative meiobenthic Ostracoda; about 1 mm in length. (Various authors)

hard structures and soft-body features into taxonomical considerations (HARTMANN 1963).

Ostracodes were first described in 1722 as "bivalved insects." Their short body, consisting of only a few segments, is completely enclosed into the two valves of a carapace joined by a complicated hinge that is of taxonomical relevance (Fig. 55). Other diagnostic features for species identification are the shape and sculpture of the shells, structure of their closing muscles, and position of spines on the short legs.

The problems of basing the taxonomy merely on shell structures becomes obvious considering the fact that almost identical valve types have been developed convergently in several independent lines, that shell structures can be subject to ontogenetic changes and sex dimorphism. Moreover, shells can also very in their ornamentation (*Leptocythere*; HARTMANN and KÜHL 1978) and their pattern of lobes depends on environmental conditions (e.g. salinity and Ca-content of the ambient water in *Cyprideis torosa*). In many interstitial forms (e.g. Xestoleberidae) the 2nd antennae are equipped with a complex of spinneret glands releasing adhesive fibres from openings in long setae. The myodocopid ostracodes have compound eyes, in the Podocopa only nauplius eyes are developed. The huge spermatozoa are transferred by large copulatory organs. Some ostracodes (*Cyprideis*) reproduce parthenogenetically. The nauplius is already covered by a two-shelled carapace; after five to eight moults the adult stage is reached; the first naupliar stage is sometimes brooded between the shells.

Order Podocopa: the ventral edge of the shells forms a straight to concave line. The shells of these marine (Cytheracea) and freshwater (Cypridacea) forms often have marked sculptures (Fig. 55), but no rostral incisure. The mostly small species of meiobenthic size often show reduction phenomena like the lack of the abdominal furca, heart, and compound eyes. *Cyprideis*, *Paradoxostoma* (interstitial; phytal), Bairdiidae (phytal), *Microcythere* (only 0.2 mm long), *Leptocythere*, *Limnocythere*, *Cythere*, *Xestoleberis* (both interstitial and phytal species, also in brackish and freshwater water), *Cytherissa* (limnetic).

Order Myodocopa: the ventral edge of the shell is convex, there is often (but not in Cladocopida) a rostral incisure in the anterior edge of the shell. Except from the suborder Cladocopida, species of Myodocopa mostly exceed meiobenthic size. Cladocopa with their small, roundish body swim freely in the marine interstitial; *Polycope*, *Polycopsis*.

Biological and Ecological Aspects.	Ostracodes are microphagous feeders on bacteria, detritus and sometimes carrion; some of them are phytophagous, piercing plant cells with their modified mouth parts (*Paradoxostoma* spp.) It has been observed in some species that the detrital material fixed by secreted fibres is ingested.

Only occasionally is there a clear adaptive correlation between structural or biological features of ostracodes and their habitat. There are 60–70 species only which are structurally adapted for a life in the interstitial of sands. Their small body (only 0.2×0.3 mm) has a smooth carapace of elongated, almost cigar-like or conical shape (Fig. 55) which seems advantageous for moving their way between sand grains. The shells of interstitial species may be laterally compressed (*Paradoxostoma*) or ventrally flattened (Xestoleberidae). While usually crawling on the grain surface or swimming in the interstitial system, they can immediatly clutch the grains by their well-developed spinneret complex on setae of the antenna. Their eyes are often reduced (HARTMANN 1973).

Most ostracodes, however, live in fine or shelly sand, burrowing and pushing through the sediment with strong legs often armed with claw-like setae. Many Bairdiidae with their specialized mouth parts and hairy carapace live in the phytal zone. Ostracodes prefer the surface sediment layers of shallow bottoms where they can attain abundances well beyond 200 specimens per $10 \, \text{cm}^2$ (when sorting, it is a difficulty to count the fresh individuals only and to omit empty shells). In the beaches of the Galápagos Islands, ostracodes were found (WESTHEIDE 1991) to rank second after the nematodes and sometimes even to dominate (2850 specimens per $100 \, \text{cm}^{-3}$). However, usually, their share is much lower. While ostracodes, like harpacticoids, are generally restricted to the well oxidized sediment layers, *Cyprideis torosa* survives mild sulphidic environmental conditions. The lacustrine *Cytherissa lacustris* is one of the best-known ostracodes and a preferred object of studies on the limnetic meiobenthos in general. Its ecological demands and life cycle are studied in detail in an interdisciplinary "*Cytherissa* project" (DANIELOPOL 1990a). Ostracodes populate the lower tidal flats as well as sublittoral coralline sands, but they also occur in the deep sea (DINET et al. 1988) and have even been found in sediment around hydrothermal vents (FRICKE et al. 1989). Their distribution is both marine and limnetic, some even live in moist mosses and litter. Phylogenetically, ostracodes are considered to have originated very early by progenesis from unknown ancestors.

More detailed reading: taxonomy – ELOFSON (1941); HARTMANN (1963); monograph – HARTMANN (1966–1989)

5.16.4 Mystacocarida

This entirely interstitial, marine group belonging to the Super-Order Maxillopoda consists of some 12 species comprising only two genera (*Derocheilocaris, Ctenocheilocaris*). Although locally occurring in considerable numbers in well-accessible beaches, the group is scientifically young, originally found at the New England coast of America (*D. typicus*, PENNAK and ZINN), and at the French Mediterranean beaches north of Banyuls-sur-Mer (*D. remanei*, DELAMARE DEBOUTTEVILLE and CHAPPUIS).

Despite the close structural relations which bind the animals to the interstitial system, and the lack of any distributional stages, *D. typicus* is found conspecifically on both sides of the Atlantic. Meanwhile, other species have been discovered along the African Atlantic coast. Recent findings of *D. remanei* from Portuguese beaches (unpubl.) do not fit the original description and underline the need for a taxonomic revision of the group. *Ctenocheilocaris* has been found both in Atlantic and Pacific beaches of South America. An undescribed species has been found at the Australian coast (HUYS, pers. comm.). Hence, the early amphiatlantic distribution pattern can not longer be maintained.

All mystacocarids are small (max. 0.5 mm) crustaceans with 11 free body segments which render the vermiform body high flexibility (Fig. 56). Equipped with long antennae, a primitively branched mandible and conspicuous claw-like caudal rami, a certain convergence to some interstitial harpacticoids is only very superficial.

In a detailed study, LOMBARDI and RUPPERT (1982) revealed the high degree of specialization in the swift movements (fifty times their body length = 2.5 cm min^{-1}) of these interstitial animals (Fig. 56). The long exopodites of the antenna and mandible are held upwards, supporting and pushing the animal dorsally, their endopodite counterparts have the same function ventrally. All other

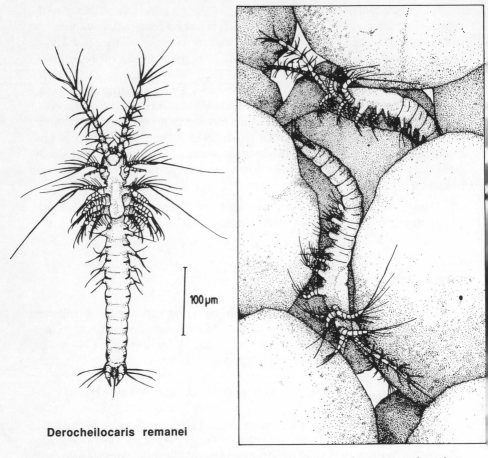

100 µm

Derocheilocaris remanei

Fig. 56. The mystacocarid *Derocheilocaris*. *Left* body structure; *right* two animals in their natural environment. (After LOMBARDI and RUPPERT 1982)

appendages are non-locomotory; the antennules are merely tactile; the caudal claws serve perhaps in copulation.

Thus, locomotion is fully dependent on an interstitial void system of a given width; in other surroundings the animals move helplessly. Their occurrence is local and patchy, but in favourable sites they have been found in frequencies which ranked them third in overall abundance of the meiofauna sample (70 ind. 100 cm^{-3}, THIERMAN, unpubl.). They apparently prefer water-unsaturated median sand above the (high) water line poor in detritus, where they are most common in the permanently moist subsurface layers above the groundwater horizon. Only rarely have they been found sublittorally. In accordance with this occurrence high up in the shore is their euryhaline and eurythermal nature (down to 10‰ S; 7–25 °C; JANSSON 1966b; KRAUS and FOUND 1975).

Reduction of vascular and respiratory organs, of egg number and of eyes can be interpreted as consequences of the minute body size. Secondary adaptations to life in the interstitial system seem to be (a) specialized locomotory organs, (b) development of antennules as long tactile organs, and (c) abbreviation of ontogenesis omitting the naupliar stage. Conservative features reflecting a long and isolated phylogenesis are the biramous mandibles with maintenance of their original function in locomotion, and primitive traits in the nervous and cerebral system. These features justify the separation of Mystacocarida as a distinct, albeit small crustacean order, perhaps of progenetic origin.

▒ *More detailed reading*: bibliography – ZINN et al. (1982)

5.16.5 Copepoda: Harpacticoida

Within the approximately 17 species-rich, relevant families of harpacticoids (about half the total family number of this taxon), there are about 4000 to 4500 meiobenthic species described, of which live 20% in freshwater biotopes (MARCOTTE 1983; WELLS 1988). In the North Sea, an area usually considered well investigated, a recent survey yielded 278 species whereof 121 were new to science (HUYS et al. 1992). The term interstitial fauna was coined for the rich harpacticoid populations in a sample from a British sandy beach (NICHOLLS 1935).

The slender, usually rather linear body of harpacticoids ranges in length from 0.2–2.5 mm, the thorax is not well set off from the abdomen. Harpacticoid copepods are distinguished from calanoid copepods by their short antennules, and the position of the articulation between metasome and urosome: in calanoids it lies between the last thoracic and the first abdominal segment, in harpacticoids, however, the last thoracic segment is included in the urosome (Fig. 57).

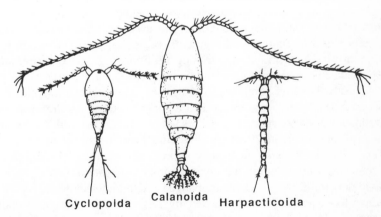

Cyclopoida   Calanoida   Harpacticoida

Fig. 57. The general body structure of the three suborders of Copepoda (SIEWING 1985)

Identification of harpacticoid species mostly depends on examination of individual appendages, which for the beginner may be a difficult task due to their minute size. The flattened pereiopod 5 is of particular taxonomic significance. Compared to other meiofaunal groups, in harpacticoids there are carefully prepared systematic keys which provide a very good basis for taxonomic work (e.g. LANG 1948; WELLS 1976). The body size and shape varies, often in accordance with the preferred biotope (Fig. 58). A description of the spherical naupliar stages (Fig. 58) from several species is given by DAHMS (1990).

General features, evident even to the non-specialist, and valid for entire families, hardly exist in harpacticoids. Hence, the following short characterization of some more common harpacticoid families serves for a rough orientation only.

Tisbidae: body shape cyclopoid (see below), more or less depress, about 120 mostly benthic and phytal species, occurring worldwide from littoral to deep-sea bottoms; *Tisbe, Scutellidium.*

Ectinosomatidae: body spindle-shaped or vermiform, no marked demarcation between thorax and abdomen; some 200 benthic and phytal species occurring worldwide from littoral to deep-sea bottoms; *Ectinosoma, Arenosetella.*

Diosaccidae: body elongate, somewhat tapering towards the end, cephalo-thorax extending into a marked rostrum, females with two egg sacs; largest family of harpacticoids with about 350 mostly benthic, but also phytal species occurring worldwide from littoral to deep-sea bottoms; *Amphiascus, Stenhelia.*

Cletodidae: body dorsoventrally depress, with a rostrum, segments well separated by deep incisions; the roughly 200 benthic species prefer muddy sediments rich in detritus, occurring worldwide from littoral to deep-sea bottoms; some species also in freshwater; *Cletodes, Metahuntemannia, Enhydrosoma.*

Paramesochridae: body elongate-cylindrical to vermiform, the small species are well adapted to the marine interstitial; about 90 species, by far the most dominant harpacticoids in sandy shores, *Paramesochra, Scottopsyllus.*

Cylindropsyllidae and Leptastacidae: mainly slender, interstitial forms of major importance in sandy habitats. *Leptastacus* excretes mucopolysaccharides in caudal glands of its terminal segments, the mucus strands becoming subsequently ingested (mucus trap feeding, see Chap. 2.2.2.1); *Cylindropsyllus.*

Ameiridae: body shape similar to Paramesochridae and Diosaccidae, but in contrast to the latter family with one egg sac only; mostly interstitial species; about 230 species, some living in the phytal; *Nitocra, Ameira, Ameiropsis.*

Peltidiidae: body short and broad, dorso-ventrally depress, with large epimeral (lateral) plates, about 70 species live mostly among algae; *Alteutha.*

Tegastidae: body short, laterally compress and of unusual shape for har-pacticoids; cephalothorax has conspicuous ventro-lateral flanks, very small abdomen, some 50 species occur on algae and muddy sediments; *Tegastes.*

Ancorabolidae: body characterized by large dorsal spines and epimeral "wings" (chitinous extensions), live in the mud of deep-sea and polar bottoms, some 30 species; *Laophontodes, Ancorabolus, Echinopsyllus.*

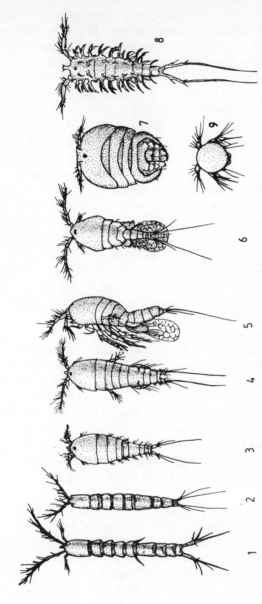

Fig. 58 1–9. Some Copepoda Harpacticoida with different body shape adapted to various biotopes; between 0.2 and 2 mm in length. (Various authors) **1** *Leptastacus*. **2** *Arenosetella*. **3** *Tachidius*. **4** *Parathalestris*. **5** *Harpacticoides*. **6** *Tisbe furcata*. **7** *Porcellidium*. **8** *Ancorabolus mirabilis*. **9** A typical nauplius larva of harpacticoids

Biological and Ecological Aspects.   Many mesopsammic species have a thin, almost vermiform body with minute legs, and short, non-protruding appendages and setae, their uniform segments give them a high flexibility. They seem to swim rapidly through the interstices of the sand, but their movement is, in fact, a very quick writhing of the whole body, not just the legs (*Arenosetella, Stenocaris, Parastenocaris*). Species living in the phytal have a stout, sometimes depressed body with richly setose, often sturdy legs well adapted to clinging to plants, but also to swimming (e.g. *Thalestris, Porcellidium*, Fig. 85). This habitat correlation is less obvious in the harpacticoid species that live in areas of fine sands or muds. Their body is often tear-drop-shaped, almost cylopoid and of fairly large size. Their stout legs help to dig in the mud, they prefer the surficial sediment and live mostly epibenthically (*Cletodes, Tachidius, Nitocra, Microarthridion*). Some epibenthic deep-sea forms (e.g. Ancorabolidae) have developed bizarre dorsal spines to anchor mud balls stabilized by mucus as camouflage (Fig. 58). Some species of *Stenhelia* and *Pseudostenhelia* are tube builders living in blind-ending tubes constructed of sediment (CHANDLER and FLEEGER 1984).

Previously, harpacticoids have been considered mainly "detritus feeders". Recent studies, however, revealed selective grazing on single food particles like diatom cells, bacteria, protozoans, which the animals strip with their mouth parts from detritus, algae and sand grains (MARCOTTE 1983, 1984; BOUGUENEC and GIANI 1989). In diatom-feeding species a close distributional correlation has been found between the microphytobenthos and harpacticoid occurrence. Exudates of bacteria and plants are also included in the preferred trophic spectrum of these animals (DECHO and FLEEGER 1988). However, in some species the food demands seem fairly flexible, so that they can be easily cultivated for experimental work (e.g. *Tisbe* spp., *Nitocra* sp.). A differentiated feeding style might explain the distributional pattern of many harpacticoid species.

Temperature and food supply are the prime determinants for the development of harpacticoid populations. Hatching and growth of the six naupliar and five copepodite stages in the field is mostly linked to increasing annual temperatures (NODOT 1978). In winter, migration into deeper sediment layers has been observed, provided there is enough oxygen available. On the other hand, populations of *Tisbe furcata* have been reported to migrate during the winter months into the sea-ice layers where they feed on the rich supply of algae (GRAINGER 1991). *Drescheriella glacialis* from Antarctic sea-ice seems to compensate for its extreme ambient temperatures by having a life cycle similar to r-strategists of more temperate habitats (BERGMANS et al. 1991). Most harpacticoids are sensitive to reduced oxygen supply, which restricts their occurrence to the upper sediment layers and favours epibenthic life. One marine and many limnetic species can develop cysts for survival in adverse conditions. From other species an environmentally induced retardation of naupliar development or a diapause phase during the copodite stage has been reported. Species found in crevices of Antarctic ice contained resting stages in their life cycle (DAHMS et al. 1990).

Nutritional selectivity and a marked adaptation of species to physico-chemical conditions of the sediment (see above) favour distinct horizontal

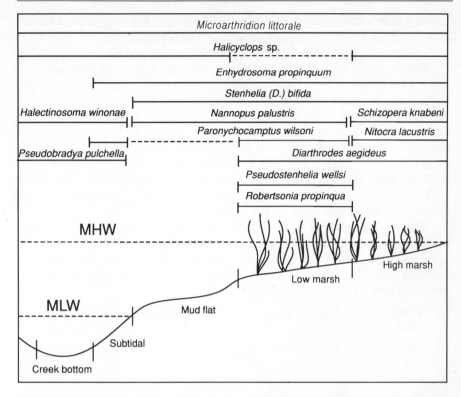

Fig. 59. The horizontal distribution pattern of harpacticoids along the shore of a salt marsh in the southeastern United States. (After COULL et al. 1979)

distribution patterns (Fig. 59). Disturbance by (tidal) currents, agitation of superficial bottom layers during storms and through burrowing activity of macrobenthos have been shown to considerably affect harpacticoids. Passive suspension (BELL and SHERMAN 1980; PALMER 1988) combined with active emergence (ARMONIÉS 1989a) are regular phenomena and lead to rapid dispersal and recolonization of disturbed habitats. The fact that mainly diatom-eating species have been recorded to perform excursions in the water column, may imply an active adaptation to a diet of planktonic diatoms (see Chap. 7.1.2).

Despite this periodical "hyperbenthic" life style, a fairly consistent small-scale distribution pattern has been observed. This suggests that beyond a random distribution some biological and physical factors are responsible for an active and selective reentry into the sediment, leading to aggregations and patches in harpacticoids. In an extensive survey of the North Sea meiofauna (HUYS et al. 1992), five well-defined groups could be discerned confined mainly by sediment structure and the impact of pollution.

In meiobenthos samples, harpacticoids after nematodes rank usually second in overall abundance. However, it occurs that they even represent the dominant taxon (e.g. in some areas of the southern North Sea). The highest harpacticoid abundance is encountered in shallow flats with fine sand and some mud enriched with detritus (from several hundred to > 1000 specimens per $10\,cm^2$). While the species composition can be rather monotonous in the eulittoral (about 20 species per sampling area), similar bottom types in the deeper sublittoral tend to have an increased diversity (60–70 spp.), yet with decreasing abundance. The Southern Bight of the North Sea harbours about 140 species. In deep-sea sediments there often live only one to ten specimens per $10\,cm^2$, but a high diversity is maintained. The frequent epibenthic occurrence of harpacticoids makes them a preferred prey for many small demersal fishes, carnivorous crustaceans (e.g. *Crangon*), and polychaetes. In sandy tidal flats, particularly where small annelids and other prey objects are rare, harpacticoids play a decisive nutritional role for fish (GEE 1987; SERVICE et al. 1992).

The great diversity of copepods with the numerous parasitic families is far from being grouped into a natural system. According to NOODT (1971), harpacticoids of a more cyclopoid type with a roundish and stout body shape (e.g. *Tachidius*) represent the more primitive type and the vermiform sand dwellers are considered secondarily derived.

> *More detailed reading*: systematic monographs and identification keys – LANG (1948); WELLS (1976); ecology – NOODT (1971); GEE (1989); ecological review – HICKS and COULL (1983)

5.16.6   Copepoda: Cyclopoida

This copepod subgroup is widely held to be restricted to a planktonic life in freshwater. This is misleading, since the Halicyclopinae and most Cyclopininae (altogether about 30 genera) live in and on littoral bottoms of the sea. They are quite common in the southern North Sea (HUYS et al. 1992). Many of the smaller forms represent typical interstitial or epibenthic meiobenthos (*Cyclopina, Metacyclopina*). They are structurally well adapted to the mesopsammal with their vermiform body (e.g. *Psammocyclopina*) and reduced egg number and show some convergence to the interstitial harpacticoid family Paramesochridae. The genus *Halicyclops* can be abundant in brackish-water sediments.

Most freshwater cyclopoids live epibenthically among macrophytes, with all transitions towards an endobenthic life. About 25% of all limnetic cyclopoids (i.e. about 160 species) can be considered endobenthic, subterranean or even troglobitic species. Beside the reduction in body size, many of these species have also reduced their egg number (e.g. the famous *Eucyclops teras*, the first cave-dwelling cyclopoid discovered). The most specialized species have even lost the typical egg sacs carrying their few eggs on long filaments (*Speocyclops, Graeteriella*). Some genera are restricted to the phreatic sediments of river beds and shores (*Haplocyclops, Speocyclops racovitzai*). In contrast to harpacticoid copepods, most Cyclopoida are predaceous carnivors.

> *More detailed reading*: see citations in the text of this section

5.16.7   Malacostraca

More commonly known by their macrobenthic representatives (Decapoda and also Peracarida), there are also meiobenthic forms in almost all orders of malaco-stracan crustaceans. In some they represent isolated, specialized miniatures of a speciose group of larger-sized animals (e.g. Isopoda, Amphipoda). In others, they represent the sole survivors of rare relict groups often living in refuge biotopes (see Chap. 8.2). Members of the first group have mostly typical convergent features secondarily adapting them to a meiobenthic, and particularly to an interstitial live. The latter forms, however, represent a composite of specialized-derived and archaic-primitive features. Combined with their often enigmatic zoogeography, this makes the few existing species a particularly rewarding study object for the phylogenetically interested zoologist.

The malacostracan crustaceans mentioned here, often exceed in size the operational limits which separate meio- from macrobenthos; but they live in habitats typical for meiobenthic and interstitial fauna, their ecology is typical meiobenthic in many respects, and they have been subject to adaptations held characteristic for meiobenthos. Therefore, they are here considered as an "ecological meiobenthos" which justifies their inclusion in this book. Their zoological, geographical and phylogenetical relevance (particularly in the fresh-water realm) would make this book incomplete had they been omitted.

5.16.7.1   *Syncarida*

This superorder contains two orders, the Bathynellacea and the Anaspidacea. Today almost restricted to freshwater habitats, their fossilized ancestors are known from marine strata of the Carboniferous. It is believed that the immigration route was via the coastal, brackish groundwater. While the Anaspidacea are still very little known, the Bathynellacea are better investigated and some details beyond the few data on morphology and zoogeography are described. Syncarid morphology (Fig. 60) contains both primitive features and derived characters adapting them convergently to the freshwater interstitial: metamerization is homonomous; the carapace, normally typical of each malacostracan, is reduced as well as the vascular system. The originally present peduncle of the eyes is often abandoned, resulting in sessile eyes; sometimes the eyes become altogether rudimentary.

Today, almost all syncarid crustaceans are found in the old supercontinent of Gondwana, corresponding to the recent southern hemisphere. Here, they are apparently still radiating. With further studies, the number of species described is expected to increase rapidly. The few species found in other areas seem to have an almost ubiquitous distribution.

– Bathynellacea

They consist of two families with about 135 species, some of which represent with 0.5 mm body size the smallest malacostracans (Fig. 60). Eyes and statocysts are reduced. The uropodes are styliform and sometimes also

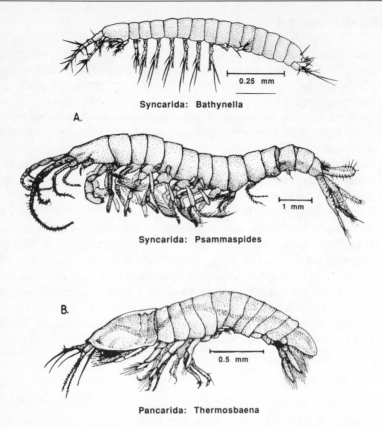

Fig. 60A, B. Representatives of the Syncarida (**A**) and Thermosbaenacea (Pancarida) (**B**). (**A** SCHMINKE 1986; **B** DELAMARE DEBOUTTEVILLE 1960)

termed a furca. Bathynellaceans occur worldwide in the groundwater system, in wells and in river sands, where they seem to feed on bacteria and fungi colonizing the sand grains and detritus particles.

Bathynella: 1–2 mm long, circum-mundane in groundwater.

Hexabathynella: cosmopolitan, mostly in rivers.

SCHMINKE (1981) could show that the group has clear homologies to the zoea/protozoea stages of the Eucarida (Penaeida), and thus suggested that it has developed by progenesis from common ancestors of these primitive malacostracans. Like their penaeid relatives, bathynellids were originally of larger size (there still exists one species of 50 mm length in Tasmania) and probably had free larval stages. It is thought that miniaturization and progenetic development enabled them to enter the mesopsammal of river mouths and from there the groundwater system of the (southern) continents, thus, reducing their competition against more "modern" crustaceans.

– Anaspidacea

Psammaspides, *Stygocarella*: stygobiotic in New Zealand and Australia respectively.

Micraspides (0.8 mm) and *Koonunga* (some species up to 8 mm) from crayfish burrows and moist mosses in Australia, caudal furca absent.

Stygocaris Noodt, 1.5 mm long, in the mesopsammal of South America and Tasmania, caudal furca only rudimentary retained.

New reports of stygocarid crustaceans from Australia and New Zealand indicate that this group, originally established as a separate syncarid order, are to be assigned now as a family "Stygocaridae" within the Anaspidacea.

More detailed reading: taxonomy, zoogeography – NOODT (1965); PENNAK and WARD (1985); ontogeny, phylogeny – SCHMINKE (1981)

5.16.7.2 *Thermosbaenacea, Pancarida*

These rare, meiobenthic malacostracans (largest species 4 mm, Fig. 60) are known from 16 species within six genera and three families (FOSSHAGEN, pers. comm.). They are related to the Peracarida (see Chap. 5.16.7.3).

Monodella: about ten species with an "amph-Northatlantic" distribution from the Caribbean to Italy and the near East. The vermiform species are very euryoecious, occuring in sands and muds of both marine and freshwater biotopes.

Halosbaena (1.8 mm): the only marine form of "amphi-Southatlantic" distribution, mostly in coral rubble.

Thermosbaena mirabilis, a monotypic species found in the sediment of thermal springs (42 °C) in Tunisia; with five pairs of pereiopods only.

Instead of typical peracararid oostegites, Pancarida have developed a dorsal brood pouch or "marsupium" underneath the posterior part of the wide carapace which covers the dorsum up to the fourth free thoracic segment; the animals are blind.

Apparently, the distribution centre of Pancarida today are refuge biotopes such as marine caves, sometimes also the groundwater system, in the Caribbean. Probably several independent lines have emigrated from the Oligocene shore lines of the Tethys Sea via brackish groundwater and colonized the subterranean freshwaters. Ontogenetically, parallel to the Bathynellacea (see before) they have homologies to the mysis-stage of penaeid shrimps (Decapoda) which indicates the progenetic origin of Pancarida.

More detailed reading: bibliography, biogeography – STOCK (1976); monograph – MONOD (1940)

5.16.7.3 *Peracarida*

In several phylogenetic lines peracarid crustaceans have independently developed meiobenthic forms, either the whole taxon (Mictacea), or a good part of it (Tanaidacea), or just some specialized forms (like in Isopoda and Amphipoda).

– Mictacea

First discovered in 1985, there are now two monotypic genera. *Mictocaris halope* Bowman and Iliffe (1985), 3–3.5 mm long, in marine caves of Bermuda (Fig. 61); and *Hirsutia bathyalis* Sanders et al. (1985), from the deep-sea benthos.

The homonomously segmented crustaceans have, like all true peracarids, ventral oostegites for brooding the eggs. Their general body organization is a combination of features not fitting into any existing peracarid order, but they are distinguished by one unique character, their eyestalks, which lack functioning visual elements. The isolated occurrence of Mictacea in disjunct biotopes points to a long, independent development. *Mictocaris* has been observed to live epibenthically, swimming with the exopodites of its pereiopods.

– Spelaeogriphacea

An isolated monotypic taxon; *Spelaeogriphus lepidops*, 6–8 mm long, blind, from a cave near Cape Town, South Africa, where it lives in pools and in a stream (Fig. 61). This troglobitic group is phylogenetically close to Tanaidacea (see below). It is distinguished by the well-developed exopodite on all but the last thoracopods. Females have a ventral marsupium.

– Tanaidacea

A good part of this speciose and widespread marine group (for figures see general zoology textbooks) contains numerous small species which have a typical

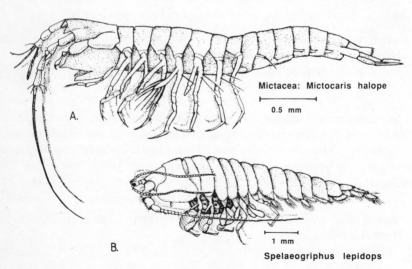

Mictacea: Mictocaris halope

0.5 mm

1 mm

Spelaeogriphus lepidops

Fig. 61A,B. Primitive meiobenthic Peracarida. **A** Mictacea. **B** Spelaeogriphacea (A Bowman and Iliffe 1985; B Botosaneanu 1986a)

meiobenthic size of a few mm length. Distributed worldwide, they prefer muddy sediments and the phytal, only a few live in sand. In deep-sea bottoms they occur frequently and are here among the most abundant crustaceans.

Tanaissus lilljeborgi, 2.5 mm, *Heterotanais oerstedti,* 2 mm, occur in muds of the North Sea; *Anarthrura simplex,* 1.5 mm, Atlantic; *Psammokalliapseudes,* from the mesopsammal off Brazil.

The carapace of tanaidaceans is so much reduced that it does allow for a high flexibility of the body, a trend enhanced by the numerous free and rather homonomous segments. This is important for their digging of U-shaped tubes in muddy sediments. Especially in sandy bottoms and among algal mats, they stabilize their tunnels by spinning silk from glands at the tips of pereiopods 1–3. The exopodite of all thoracopods is absent. The 2nd thoracopod has a large distal chela. In some tanaidaceans, sex determination is phenotypic and complicated. Here heteromorphic genders occur. In some families, different types of males and a protogynous hermaphroditism are described (a rare exception in malacostracans!). Most representatives are detritivores and scavengers, feeding also on diatoms, however, some predaceous species grasp nematodes and harpacticoids with their chelate legs. In shallow waters abundance of tanaidaceans can attain 100 to 1000 individuals per 100 cm², but the distribution is extremely patchy. In many areas, tanaidaceans are common food for polychaetes, malacostracans and small fish. Hence, they represent an important member of the marine benthic food chain.

– Isopoda

The large and diverse order of isopods has developed numerous meiobenthic forms through different adaptational lines (Fig. 62). Structurally divergent, they belong to various suborders, super-families and families which have adapted to an epibenthic, interstitial-mesopsammic or endobenthic life. Especially the groundwater species are often only 1 mm long (which is an unusually small size for the complex body organization of Malacostraca); but many species are 2–3 mm long, and yet well adapted to live in the void system, particularly of river gravels (see Chap. 8.2).

Most meiobenthic isopods belong to the superfamily Janiroidea, the genus *Microcharon* alone contains 55 species. Microcerberoidea represent about 50–60 spp. of *Microcerberus,* all are meiobenthic in size. Within the Cirolanidae (Flabellifera) there are about 45 stygobiotic freshwater species.

Most meiobenthic isopods from sand or gravel have modified the typical dorso-ventral depression of the body into an almost round shape. They have reduced tagmatization becoming fairly vermiform with rather uniform metamers. The resulting slender body is highly flexible. Reduction of eyes, pigments and long appendages occurs frequently in the numerous troglobitic and interstitial species, which all have a strong thigmotactic behaviour. In *Microcerberus,* having reduced oostegites, eggs are deposited directly into the voids. Another

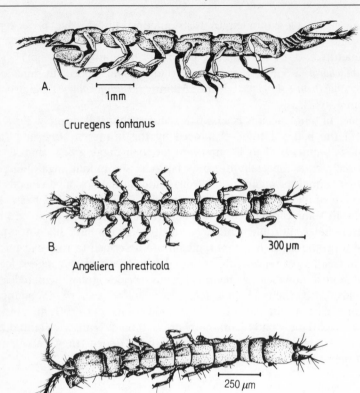

A. |——| 1mm

Cruregens fontanus

B. |——| 300 μm

Angeliera phreaticola

C.   Microcerberus sp. |——| 250 μm

Fig. 62A–C. Characteristic representatives of meiobenthic Isopoda. (Various authors)

developmental line is evident in the epibenthic forms. With their flattened body (large epimeres) and long legs they can easily move over soft surfaces.

Microparasellidae: the five genera with numerous species are all only a few mm long. Their general distribution pattern indicates the region of the former Tethys Sea. *Microparasellus* lives in the shallow marine interstitial as well as in freshwaters of European karstic regions; it is not present in America. *Angeliera phreaticola* occurs in the brackish coastal groundwater of Mediterranean beaches, but also in surrounding freshwaters; it was also found off Madagascar. *Microcharon* is mostly known from the groundwater system, but some archaic forms are also marine with an apparently worldwide distribution: they have been found off Roscoff (coast of Brittany), along the Mediterranean coasts, and at the Galápagos-Islands. *Microjaera* lives sublittorally off Banyuls-sur-Mer (Mediterranean) and off Roscoff. Other taxa: *Protocharon, Paracharon*.

Microcerberoidea: the genus *Microcerberus*, with often just 1-mm-long species, occurs worldwide in the continental groundwater system, but also in caves and aquifers of karstic bottoms. It contains also interstitial marine and coastal-brackish species. This subterranean taxon is probably related to the epigean group Stenasellidae (Asellota).

Phreaticoidea are restricted to subterranean freshwaters of the southern hemisphere, their body is strongly vermiform and attenuated. *Neophreaticus*, from New Zealand

Anthuridea: the centres of this group seem to lie in the Caribbean and the Indo-Pacfiic; it is lacking around the Mediterranean. The species are mostly marine or live in freshwater in the vicinity of the oceanic coast lines. With their slender body they dig in the sediments, occur often in tubes of polychaetes and in the culms of sea grasses, where they live as predators. The stygobiotic, often blind groundwater forms are sometimes only 1 mm long. Most-known genus: *Cyathura*

Flabellifera, Cirolanidae: these are mostly stygobiotic species of circum-Mediterranean and circum-Caribbean distribution. With their flattened body and reduced size (often only 2–3 mm long), they are typical for karstic habitats, where they live mostly as detritivores or scavengers. *Cirolana*, *Speocirolana*, *Arubolana*

Oniscoidea: *Nannoniscus* is a small (1–2 mm long), flattened deep-sea isopod living in the surface layer of the sea bottom.

The striking phenomenon that most of the circum-mediterranean findings of meiobenthic isopods occur both in marine and limnetic (groundwater) habitats, led to the discussion whether they represent relict forms of the former Thethys Sea which principally kept their old distribution, but, after regression of the Thethys, slowly adapted in several independent lines to the brackish and finally to freshwater conditions. The alternative would be that the marine forms migrated independently and actively into the groundwater via the brackish mesopsammon of many beaches (see Chap. 6).

More detailed reading: Faunistic and zoogeographic aspects – DELAMARE DEBOUTTEVILLE (1960); BOTOSANEANU (1986a)

– Amphipoda

In accordance to similar trends in Isopoda (see before), within many amphipod orders dwarf forms developed independently. Even the morphological adaptations are fairly convergent to isopods, sometimes modifying the latero-compressed body shape of Amphipoda into a roundish, slender body (Fig. 63). Some of the morphological species acquired an extreme euryhalinity enabling them to live in marine as well as limnic habitats (different haline capacities in divergent physiological species or just various populations?; compare KINNE 1964).

The data on the occurrence of meiobenthic Amphipoda do not yet allow drawing a realistic picture of their geographical distribution. There seems

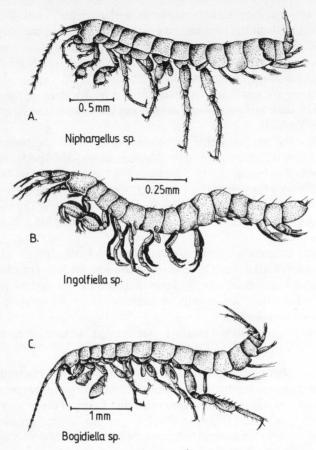

A.

0.5 mm

Niphargellus sp.

B.

0.25mm

Ingolfiella sp.

C.

1mm

Bogidiella sp.

Fig. 63A–C. Characteristic representatives of meiobenthic Amphipoda. (Various authors)

to be a trend that most marine meiobenthic amphipods live in the southern hemisphere. Many freshwater forms are found in identical genera or even species in the old and new world, suggesting that these groups are very old, originating on Gondwana or at least on the Laurasian archaic continent before its breakup, which formed the Atlantic Ocean. Meiofaunal amphipods live both in the sea and in freshwater, either interstitially in sands (rarely in mud), or, particularly often, in rubble from encrusting algae and corals. Many occur also in the culms and holdfasts of sea grass and algae and belong to the phyton. In freshwater, a large portion of small amphipods lives worldwide in subterranean biotopes like groundwater, springs, caves and river beds.

Although the size of small Amphipoda often somewhat exceeds the traditional limits which separate meio- from macrobenthos, they often beautifully exemplify the typical adaptations for meiobenthic, in particular interstitial, animals (Fig. 63): minute size, vermiform body with reduced epimeres, only one to two eggs, reduced eyes, no pigmentation, high flexibility of the homonomous body. Although ecological studies on meiobenthic amphipods (as well as isopods) are scarce, it can be inferred that most of them feed on small detritus particles on and in the sediment.

Gammaridea:

Niphargidae: mostly in subterranean waters of the holarctis
Niphargus and related spp.: within this large species group, widely known from European groundwaters, most species have kept the typical amphipod morphology. But there are some forms of small size (only 2–3 mm) living in the interstices of sand, which have a typically modified body organization: the slender, vermiform type with short legs (e.g. *Niphargellus*), or the stout, short type which can roll up its body. Numerous species are adapted to brackish water conditions.
Microniphargus, 2 mm, from wells in western parts of Germany and in Belgium
Psammoniphargus, from the groundwater of Yugoslavia and Brazil
Crangonyctidae: prevalently of nearctic distribution, these amphipods are mostly beyond meiobenthic size, however some typical stygobionts have an attenuated body of only 1–2 mm length. All species are cold-stenothermal, photonegative and mostly lack eyes and pigments. They occur in springs and caves in karstic areas of the southern and central United States, but also in Eurasia. The stygobiotic forms are clearly derived from epigean ancestros.
Crangonyx spp., mainly from the United States.
Synurella spp., hypogean as well as epigean forms, mostly from eastern Europe, smallest species 1.5 mm long; in springs, subterranean waters, in the mud of creeks *Stygobromus* spp, mostly subterranean from the United States.
Bogidiellidae: *Bogidiella* spp. These typical subterranean animals of often only 2.5 mm length are found in caves, springs, river beds and groundwater, but occasionally also in the marine interstitial (*B. chappuisi*), mainly in Europe and South America.

Ingolfiellidea:

Ingolfiella: about 25 species of this strongly modified group of amphipods are described. Many are only 1–3 mm long and have a vermiform body which lacks oostegites in the female gender. They are found in refuge biotopes like caves, continental and coastal groundwater, but also in deep-sea bottoms. Their vermiform body demonstrates best the typical adaptation to the *lebensform* of the mesopsammon.

Within many other amphipod families, both from marine and freshwater habitats, there exist species of meiobenthic size, e.g. *Uncinotarsus pellucidus* (Aoridae) which is 1.5 mm long. Found in the shallow sublittoral off Roscoff, it is a typical interstitial form with the characteristic adaptations to the mesopsammal. *Salentinella* (1.6 mm) is a characteristic element of the continental groundwater fauna.

> *More detailed reading*: faunistic and zoogeographic aspects – DELAMARE DEBOUTTEVILLE (1960); BOTOSANEANU (1986a); monographs – BARNARD (1969) (marine species); BARNARD and BARNARD (1983) (freshwater species)

5.17 Acari

The small body size of mites enabled them to enter the realm of meiobenthal in several independent subgroups (Fig. 64). The most successful invasor in marine habitats is the suborder Halacaroidea with the family Halacaridae, which even sent some 50 species back into freshwater biotopes (Limnohalacaridae). Another small family of the Halacaroidea, the Hydrovolziidae, occurs in brooks and subterranean waters. The other suborders of mites live mainly in the bottom substrates of lakes and rivers, most of them belong to the Hydracarida and Gamasida. Some other mite groups can be encountered with some representatives in marine sands: the Oribatei, with their uniform brownish colour and heavily sclerotized dorsal shield represent a suborder that normally lives in terrestrial soils. The famous vermiform trombidiform mites Nematalycidae have developed remarkable convergences to adapt to the interstitial environment (COINEAU et al. 1978), *Nematalycus nematoides* has been found in a beach near Algier (Fig. 64). Within the Rhodacaridae (Gamasida) there are two genera with species regularly found in marine intertidal sand (e.g. *Rhodacarellus*). The slender body in some of these interstitial species seems well adapted to move in the interstices of the sediment. In marine sediments of warm water areas, Pontarachnidae, related to freshwater mites, are characteristic representatives.

5.17.1 Halacaroidea: Halacaridae

800 known species, most of meiobenthic size, are inhabitants of marine bottoms. The family Halacaridae alone harbours > 200 species. About 50 additional halacarid species are specialized inhabitants of subterranean groundwater. Halacaroid mites are easily recognized by the division of their body in the "gnathosoma" carrying the chelicerae and pedipalps and the "idiosoma" with the two first pairs of legs directed anteriorly and two pairs of posterior legs directed backwards. The distance between these groups of laterally attached legs can become rather wide, and the body attain an elongate shape, adapting the species to a life in the interstices of sand. The chitinous body cuticle can develop a pattern of sclerotized plates ornated with numerous setae. Position of setae (which also arise from the legs) and shape of the plates are reliable diagnostic features, as is the articulate structure of the legs. Interstitial

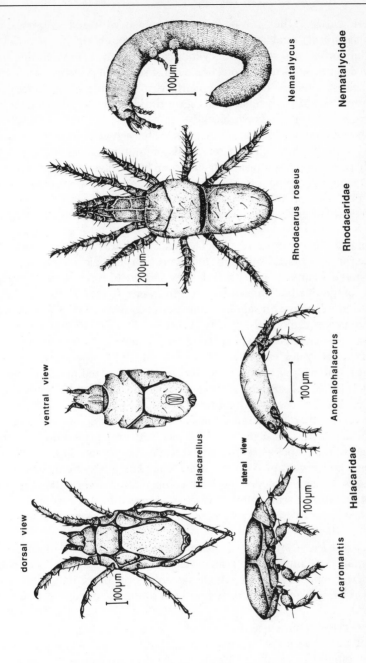

Fig. 64. Some typical meiobenthic representatives of the Acari, belonging to three different families. (Various authors)

forms are either rather soft-bodied and slender with reduced body plates (*Anomalohalacarus*) enabling them to squeeze through the narrow voids; or interstitial forms are stout and cylindrical, strongly armoured by heavily sclerotized plates and legs as protection against sediment pressure (*Acaromantis*). In unfixed specimens, the gut content, visible through the more or less transparent body of many halacaroids, gives the idiosoma a characteristic colour (dark green, reddish, brownish black).

Despite the relatively clearcut diagnostic characters, halacaroids have been little investigated, although they represent common members of the meiofauna regularly occurring in samples from almost all biotopes, whether littoral or deep sea (see below). The genera *Halacarellus* and *Copidognathus* alone comprise about 24% of all halacaroid species. Other genera which are frequently encountered in boreal shores (tidal flats) are *Lohmannella, Acarochelopodia, Actacarus, Rhombognathus*.

Biological and Ecological Aspects.   Halacaroids are hardy creatures, able to live in a wide range of biotopes without too many morphological variations of their general body organization. Although they proved to have a clear preference reaction for gradients of moisture, pH values etc. (BARTSCH 1974), they can withstand in full activity a salinity range from freshwater to 30‰ S. A premise for this amazing ecological capacity, however, is good oxygen supply, since most mites are sensitive to hypoxic conditions and do not occur in hypoxic and sulphidic muds. Mites can also survive extremes of temperature, desiccation and (hyper-)salinity in a resting stage during which they reduce their respiration to negligible amounts. After fixation of meiofauna samples halacarids tend to move their legs for a wearily long time!

Like many other chelicerates, halacarids have piercing and succing mouth parts and an extra-oral digestion. Most of them are carnivorous, feeding on crustaceans and oligochaetes, the phytal forms (see below) on the soft parts of hydrozoan and bryozoan colonies. *Rhombognathus* is phytophagous, piercing diatom cells. In some biotopes, a certain competition with oribatid mites can be expected. In turn, marine mites have been observed to serve as prey for some small fish, also for hydrozoan polyps, but altogether there is not much predation pressure on them.

Halacarid females take up a spermatophore deposited by the males. The tiny animals have only one to a few eggs and a relatively long developmental time. There is one larval stage which has only three pairs of legs, followed by one to two nymph stages occurring mostly in summer. Average life span is 5–9 months.

Mites crawl characteristically slowly and somewhat awkwardly. The abundant phytal forms climb with particularly strong clinging appendages, while the often very small (200 µm) and slender mesopsammic forms have a rather flexible, concave body shape that better attaches to the sand grains (e.g. *Anomalohalacarus*; Fig. 64). Those species living in exposed and agitated coarse sand protect their roundish body with armoured, solid plates, with their legs tightly pressed to the body in depressions of the cuticle.

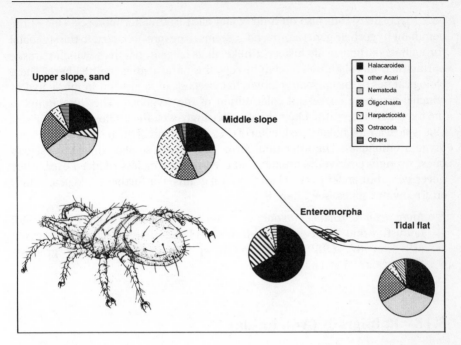

Fig. 65. Portion of halacarids in the meiofaunal spectrum along a beach profile. (After BARTSCH 1982)

Halacaroida, especially the rhombognathids, will be most frequently encountered in the phytal of boreal regions. In algal mats and *Enteromorpha* canopies they can make up 90% of all meiofauna, in the mesopsammal of medium sand still 15%, while in fine sand and muds with their limited supply of oxygen their portion is reduced to about 5% (Fig. 65). Here, they never enter the deeper horizons and tend to live epibenthically. In the epigrowth of larger animals (crustaceans, gastropods), mites are regular phoretic guests.

The set of preference reactions (see above) allows one to allocate certain species of marine mites in a well-defined distributional pattern to the various regions of the shore line. However, their hardy nature has not resulted in the development of many endemic species, e.g. the halacaroid fauna of the Baltic Sea and the North Sea is more or less identical and the identity of amphi-North Atlantic species still remains 45% (BARTSCH 1974).

5.17.2 Freshwater Mites: "Hydrachnellae", Stygothrombiidae and Others

Mites from many orders and families (altogether 5000 spp.), combined in the artificial group "Hydrachnellae", live among the phytal, in the sand of river beds (hyporheos), but also on mud of stagnant freshwater biotopes. One of the dominant hyporheic genera, *Atractides*, seems to be (pre-)adapted to the stygobial through its particular life history. Unlike all its epigean relatives, which parasite during their larval phase on adult insects, this parasitism is omitted here. These biological ties of the epigean relatives to insects seem to prevent in many hydrachnellid groups a successful colonization of the stygobial. The inhabitants of the hyporheic interstitial often show surprisingly well the analogous specialisations typical for the habitat: reduction of eyes, elongation of the body (Wandesiidae, Stygothrombiidae). The interstitial forms are clearly smaller than the epigean ones, strongly positive thigmotactic and with more or less reduced eyes. They never swim but prefer to crawl on the sand grains. For further ecological details on freshwater mites, see Chap. 8.2.

> *More detailed reading*: taxonomy – Viets (1927); Bartsch (1972, 1979); faunistics and zoogeography – Bartsch (1982, 1989); ecology – Bartsch (1974, 1982, 1989); Pugh and King (1985a, b); freshwater groups – Schwoerbel (1961b, 1967)

5.18 Palpigradi (Arachnida)

It is with hesitation that representatives of this most primitive and rare group of Arachnida are included here. While most Palpigradi live in moist terrestrial soil and caves, some few species of three genera have been found in the eulittoral of tropical beaches and shallow coral sand. Through their small size (less than 2 mm) and slender shape, the long flagellum and the fairly thin and flexible appendages, they seem to match the requirements of an interstitial life. The most interesting features which give convincing arguments for an ordination into truly marine meiobenthos are of ecological nature: specimens of *Leptokoenenia scurra*, the best-studied species, normally crawl on and clutch the sand grains. When extracted, they successfully escape back into seawater. While all terrestrial palpigrades have a hydrophobic cuticle which would trap them at the water surface, these marine forms can easily pass the surface film. The occurrence of marine palpigrades evoked new considerations on the marine origin of Arachnida and on the colonization of terrestrial habitats from the sea shore.

> *More detailed reading*: Condé (1965); Monniot (1966)

5.19 Pycnogonida

Among the meiobenthos extracted from sublittoral marine sand samples, there are occasionally also some minute representatives of the usually macrobenthic Pycnogonida or Pantopoda. They belong to about ten species of *Anoplodactylus*,

Nymphonella and *Rhynchothorax*. Except for their small size (the smallest about 1 mm in body length) and, in some species, reduction of ocular tubercles and eyes, there are no characters which differ essentially from the general body organization of this strange animal group, which usually is associated with the chelicerates.

▩ *More detailed reading:* CHILD (1988)

5.20   Terrigenous Arthropoda (Thalassobionts)

There is a heterogeneous assembly of normally terrestrial arthropods of meiobenthic size, which so regularly occur in marine biotopes that they should shortly be mentioned here. These animals are more characterized by biological properties than by indigenous morphological adaptations (perhaps their normally well-developed clinging legs can be mentioned in this context). The strong affinity of these arthropods to the marine realm seems to be based on an ecological niche realized in marine, (occasionally) moist sand and in aufwuchs of supra- and eulittoral hard bottoms.

The typical and highly specialized forms, although mostly wide-spread, have been little investigated. Most of them belong to chelicerate groups, followed by Tracheata:

1. Mites: Hydracarida, Gamasida (*Hydrogamasus*), Oribatei, Uropodidae
2. Pseudoscorpiones
3. Aranea
4. Chilopoda (centipedes)
5. Insects: Collembola (Anurida, Archisotoma), Diptera (larvae of Ephydridae), Coleoptera (Staphylinidae)

Contrasting to the normal meiobenthos, the problem of thalassobiotic terrigenous arthropods is not lack of moisture in their environment, but inundation by the sea that might last too long. On the other hand, adapted to live in genuinely marine habitats like regularly flooded cliffs and islets, their resistance to survive flooding by the sea is highly developed, sometimes up to several months. During this time they remain, of course, air-breathing, adopting mostly the principle of "physical lungs" (plastron respiration). Moreover, many of the thalassobiotic arthropods have a resting stage with only minimal oxygen requirements. Osmoregulation remains effective through their coxal glands. These adaptations are so specific that survival after inundation is only possible in seawater, not in freshwater. Indicating their terrestrial descent, nutrition of this ecologically defined mixture of animals is not of marine origin. Terrestrial detritus, fungi, lichens and carrion is washed ashore even in the most isolated islands, apparently in sufficient amounts to sustain a small outpost of strange terrestrial life in the marine biome with highly interesting physiological and ecological adaptations.

▩ *More detailed reading:* SCHUSTER (1962, 1965, 1979)

5.21 Tentaculata

5.21.1 Brachiopoda

Among this group of macrobenthic, sessile animals there is one species of meiobenthic size (only 1 mm in diameter) described from the coasts of the Channel and Great Britain. Known to live in sublittoral shell-gravel sediment attached to shell and tube fragments, *Gwynia capsula* Jeffreys was described already in 1859, but was only much later recognized as an aberrant brachiopod of meiobenthic size (SWEDMARK 1967; Fig. 66). *Gwynia* lives together with psammobiotic foraminiferans. Its simple lophophore and articulation of the shells may indicate a progenetic nature of the animal, the few eggs are brooded for a long period, the larva remains benthic.

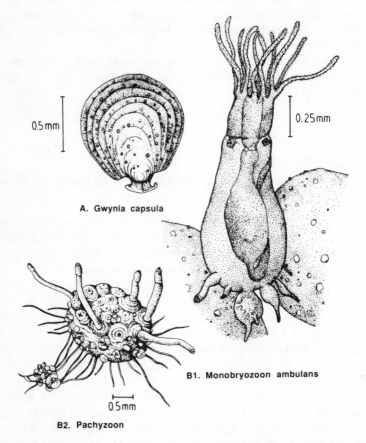

0.5 mm

0.25 mm

A. Gwynia capsula

B1. Monobryozoon ambulans

0.5 mm

B2. Pachyzoon

Fig. 66A, B. Meiobenthic Tentaculata. **A** Brachiopoda. **B** Bryozoa. **B1** Gymnolaemata. **B2** Lunulitiformes. (**A** SWEDMARK 1967; **B1** REMANE 1936b; **B2** COOK 1988)

5.21.2 Bryozoa

This group of usually sessile and colonial tentaculates of microscopic size contains some meiobenthic, hapto-sessile species belonging to different families and living in sublittoral sediments. The more conspicuous *Monobryozoon* and its relatives and the recently described lunulitiform bryozoans will be mentioned here.

The zoologically famous *Monobryozoon ambulans*, found by REMANE (1936b) in sublittoral coarse sand ("Amphioxus-sand") off the island of Helgoland (German Bight, North Sea) was the first bryozoan described to live as a non-sessile, solitary organism (2 mm length), anchored by viscid stolons (rhizoids) in the sand (Fig. 66). Although a rare animal, it has been recorded today also from other European coasts. *M. bulbosum* and *M. sandersi* are congeners from the American east coast. Another, more slender form was originally described as a new European *Monobryozoon* species (FRANZÉN 1960), but is now classified as *Nolella limicola*. It is found in relatively high densities (240 ind./sample) in deep-water muddy sediments of some West Scandinavian fjords (BERGE et al. 1985). Belonging to the Gymnolaemata, the above bryozoans have a non-calcified, but still fairly compact cystid which is separated from the soft-bodied retractable polypid by a deep annular furrow. With its semi-transparent yellowish colour, the animal is well hidden among the sand grains. The mouth opening ("orifice") is surrounded by a circle of ciliated tentacles. The numerous characteristic stolons are hollow ambulatory processes well supplied with musculature for (slow) contractions pulling the animal through the sand. Terminally they are equipped with adhesive glands and sensory hairs. Stout, non-contractile buds, structurally different from the rhizoids, attach directly to the cystid. This is in contrast to *Nolella*, where buds arise from slender processes of the cystid, probably representing "kenozoids', i.e. thread-like, modified bryozoan individuals. Consequently, *Nolella limicola* must be considered a small mobile colony and not a solitary bryozoan like *Monobryozoon*.

Lunulitiform bryozoans: today, the numerous small lunulitiform colonies (COOK 1963, 1966, 1988) of variable, but mostly conical to discoid shape are grouped in various genera belonging to several families of vagile bryozoans (Fig. 66). Together with other soft-bottom fauna they occur circum-mundane in little exposed deep-water sand and mud (Atlantic Ocean, Red Sea, Indian Ocean). With a size of 1.5 to 9 mm, they are found, at least during their initial stages, when sampling meiobenthos.

Long and flexible setate vibracularia help to transport food, remove foreign particles and serve as the locomotory organs of the colony. Although food can be taken up in each position, the colonies, when turned upside down, follow a distinct vertical polarity with the vibracularia pulling them back in an upright position. If buried too deep, they can also be brought back by the vibracularia in more superficial layers.

As usual, reproduction in these bryozoans is both asexual (by marginal buds) and sexual. In the conical species the apex is formed by a foreign particle, e.g. a sand grain to which the first zooid attaches, subsequently forming a new colony by budding some zooids. If broken away, parts of the colony will be regenerated.

Despite their wide distribution, the lunulitiform bryozoans have not been discovered until recently. This is partly because of their restriction to deeper bottoms, but, like in many benthic foraminiferans, they have certainly also been overlooked due to their variable and often irregular shape.

▨   *More detailed reading*: see citations in the text of Sections 5.21.1 and 5.21.2

5.22   Kamptozoa, Entoprocta

Among this isolated group of sessile animals which usually form colonies of individuals connected by stolons, there are some strange solitary representatives of meiobenthic size. *Loxosoma isolata*, described by SALVINI-PLAWEN (1968) from the northern Adriatic Sea, is less than 1 mm in size (Fig. 67) and has only 16 ciliated tentacles. At the basis of its stalk there is a glandular disc for temporary fixation on shell fragments in coarse, sublittoral sands.

Other, yet undescribed meiobenthic species of kamptozoans have been found off North Carolina, and Florida (USA), and Roscoff (Britanny), and additional forms will probably be discovered if unfixed samples of coarse shell hash are scrutinized for them. Asexual reproduction by budding has been observed.

▨   *More detailed reading*: see citation in the text of this section

5.23   Echinodermata

Directly after metamorphosis and settlement, most echinoderms pass a temporary meiobenthic phase. Judging from their abundance, particularly the numerous species of juvenile Ophiuroida can become an ecologically important fauna element in the crevices of coarse sand and shelly sediments. However, almost nothing is known of their ecological role during this stage of their life. This holds true also for the other echinoderms during their meiobenthic life span. Only within the holothurians have permanently meiobenthic species with characteristic adaptations to a mesopsammic life evolved.

5.23.1   Holothuroidea

Within this group of usually large animals, in several apodous families some species have developed which remain as adults only one to a few mm long (Fig. 67). Most of them belong to the family Synaptidae (*Labidoplax*, *Leptosynapta*,

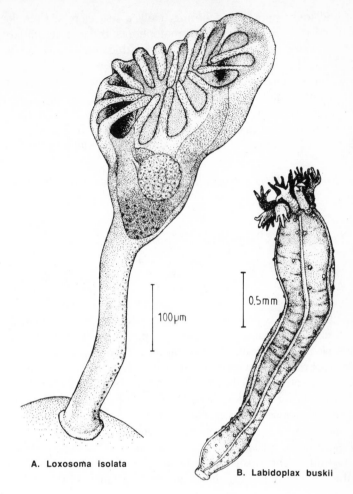

A. Loxosoma isolata

B. Labidoplax buskii

Fig. 67A, B. Interstitial representatives of Entoprocta or Kamptozoa (**A**) and Holothuroidea (Echinodermata) (**B**). (**A** SALVINI-PLAWEN 1968; **B** SWEDMARK 1971)

Rhabdomolgus?, Psammothuria), but also some small Chiridotidae (*Chiridota; Trochodota*) and Myriotrochidae can be grouped as meiobenthos.

Apart from their size, also other characters adapt these tiny sea cucumbers to meiobenthic life: the often transparent body is rather tough and in some species seems mechanically well protected by calcareous platelets against abrasion. Other species like *Rhabdomolgus* lack these skeletal ossicles completely. Conspicuous statocysts occur regularly. The mouth tentacles are strongly adhesive and their contractions pull the body slowly forward through the sediment. These large and sticky branched podial appendages are also used for food uptake (sediment particles) and for temporary anchoring. Only a few eggs are developed and brooded after fertilization through spermatophores; the Pentactula-larva is not planktonic as would be typical for the macrobenthic forms. *Leptosynapta minuta* is a hermaphroditic species.

Among the few meiobenthic species described so far, the best known are the synaptids living in shallow sublittoral coarse sand; *Rhabdomolgus ruber* occurs also in more silty bottoms, the Myriotrochidae have been found in the deep-sea. The regular occurrence of meiobenthic holothurians along the European, American and Indian coasts probably indicates a distribution wider than documented today.

▨   *More detailed reading*: SWEDMARK (1971); SALVINI-PLAWEN (1972)

5.24   Tunicata (Chordata)

5.24.1   Ascidiacea

Among the usually sessile and macrobenthic ascidians there are some 20 species of 3–4 mm size which can be found among meiofauna from sand samples and have the typical adaptations of mesopsammic animals. These small tunicates originated apparently convergently from different orders and families, often through progenesis (Fig. 68). Described by WEINSTEIN (1961) from "Amphioxus-sand" off Banyuls-sur-Mer (Southern France), *Psammostyela delamarei* was the first tunicate of meiobenthic size. However, its first encounter may date back to 1924, although then it was not recognized as an ascidian. Other genera: *Heterostigma, Psammascidia, Polycarpa, Dextrogaster, Heterostyela, Molgula* sp.

Contrasting to the sessile, macrobenthic ascidians, the body wall of the meiobenthic species is mostly glassy and lacks pigments. The sticky tunic is usually covered with sand grains or foraminiferan shells. Often the meiobenthic ascidians have the openings for ingestion and egestion currents at opposite ends. Corresponding to their minute size, the characteristic branchial filter system is often simplified. The planktonic ascidian tadpole larva is often suppressed. These features indicate that meiobenthic ascidians have originated by neotenic processes. Typical adaptations to a hemi-sessile life are the numerous, mostly ventral rhizoids anchoring the body temporarily in the soft substrate. The animals move by slow contractions of the rather tough muscular body wall.

In contrast to the other species representing solitary individuals, *Arenadiplosoma migrans* MENKER and AX (1970) is a colonial aggregate of only 1–3 mm length found in the North Sea. It moves by contractions of its rhizoid-like "tunic vessels", which have a viscid terminal end.

Meiobenthic ascidians tend to avoid sediments in which their delicate filtering apparatus might become clogged and their supply with oxygen-rich water limited. Thus, they are found sublittorally in the well-exposed surficial

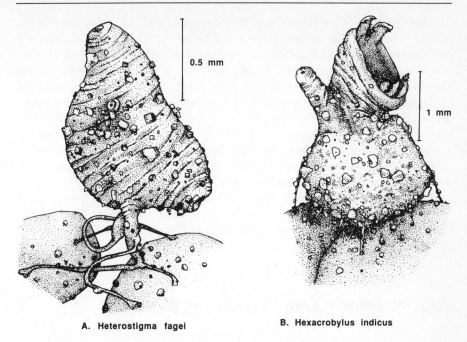

A. **Heterostigma fagei**

B. **Hexacrobylus indicus**

Fig. 68A, B. Meiobenthic Tunicata. A Ascidiacea. (MONNIOT and MONNIOT 1988). B Sorberacea. (MONNIOT and MONNIOT 1984)

layers of coarse, silicate sands ("Amphioxus sand"), but seldomly in muds and fine calcareous sediments. However, recently, meiobenthic ascidians have been encountered in fine deep-sea sediments, either with ventral anchoring rhizoids or with "setate" rhizoids covering the tunic. They are oviparous and their simplified internal organs indicate progenesis. So far, details on their ecology are lacking.

Distributional Records. Meiobenthic ascidians are mostly found in temperate and boreal latitudes, in deep bottoms of the Atlantic Ocean from northern to southern latitudes and in the Mediterranean Sea. A few littoral species are reported worldwide.

5.24.2 Sorberacea

This recently established fourth class of tunicates (MONNIOT et al. 1975) with unique morphological and ecological characters contains about 20 not permanently sessile, solitary species, many of which measure only 1–5 mm, and, thus, have been found in meiofauna samples (Fig. 68); *Hexacrobylus*; *Hexabranchus*; *Sorbera, Hexadactylus.* Encased in a tunic, the body is densely covered by adhering sediment particles. It is anchored in the sediment by rhizoids. Diverging from all other tunicates, the oesophagus is not filtering. Instead, these tunicates are carnivorous, they have a huge mouth opening surrounded by six prehensile lobes or "fingers", well suited to grasp prey and fold over the oral cavity. These muscular organs are well innervated, their function is efficiently coordinated by the well-developed dorsal neural system. All animals have one large excretory organ (kidney), they are hermaphroditic and oviparous. Their ontogeny is as yet unknown, vegetative regeneration and colonial forms have not been observed. The intestine of these predatory tunicates has been found filled with nematodes, harpacticoids, acarids and polychaetes, but also ophiurids and gastropods. Although predominantly found worldwide in deep-sea samples, more careful examination revealed Sorberacea also from the continental shelf. In fact, it now seems that the first sorberacean tunicate was found in only 92 m depth off the English coast, but was described as an ascidian (BOURNE 1903).

▓   *More detailed reading*: MONNIOT (1965); MONNIOT and MONNIOT (1990)

5.25   Meiofauna Taxa – Concluding Remarks

The examples of meiobenthic Sorberacea and lunulitiform Bryozoa suggest that the deep-sea is the region where further discoveries of new and exciting meiobenthos can be expected. However, the recent findings of Mictacea, Cephalocarida, and, of course, Loricifera, underline that even shallow bottoms harbour spectacular new animal groups, many of which belong to the meiobenthos. With the closer examination of calcareous sands from tropical regions, with the better accessibility of polar sediments and thorough investigation of caves, discovery of a rich number of new animal forms can be expected, many of which will be of meiobenthic size. It does not need much prophecy to state that the classification of higher animal taxa is by no means concluded and will much be influenced by research on meiofaunal groups.

CHAPTER 6

Phylogenetic Aspects in Meiobenthology

Discussions on evolutionary links between animal groups often center around the meiobenthos, particularly from exotic and undisturbed habitats like the deep-sea, caves, and the groundwater system. Phylogenetic considerations mostly refer to morphological features of the small-sized animals, often in combination with an unusual, disjunct zoogeographic pattern. This chapter will deal with (1) structural and (2) distributional aspects of meiofauna studies relevant for evolutionary aspects.

6.1 Structural Considerations

The relatively stable physiographic nature of the prevailing habitat for meiobenthos, the subsurface layers, favours long-lasting selective trends. This often results in an "orthogenetical" evolution: the structural "answers" of the meiobenthic inhabitants are seemingly directed towards an adaptive end point. This becomes particularly evident in the numerous cases of regressive evolution within the stygofauna (Chap. 8.2).

On the other hand, the uniformity of the physiographic milieu also favours convergent evolution resulting in analogy, (1) since numerous morphological and physiological features are subject to an identical selective pressure, and (2) since the minute body size of meiofauna organisms may already limit the options for a diverging functional design. (However, considering the elaborate structural complexity of Loricifera, this argument is questionable.) A similar adaptational advantage, effective during long periods of time, often results in analogous structures within most diverse and unrelated meiobenthic groups (e.g. adhesive organs, caudal appendages, ciliation; see Chap. 4.1). It is only at a lower taxonomic level that these synapomorphies, particularly if identical in details of structure and hierarchial arrangement, may indeed indicate a natural relationship (Ax 1963).

As the basic prerequisite for every phylogenetic argument, the nature of each structure considered has to be carefully examined, whether based on true homology or on a deceptive convergence, whether plesiomorphic and archaic or apomorphic (secondarily derived) and simplified. The interstitial fauna offers good examples for the complex structural mosaic of features differing in their evolutionary significance: in cases where the whole taxon is restricted to the interstitial of sediments and has entirely evolved in this ecological refuge ("plesiotope"), we frequently find a combination of archaic and highly specialized and derived features: the fairly homonomous body segmentation and the biramous mandible of Mystacocarida are plesiomorphic features, while the furcal claw is clearly derived. Similar examples can be given from Gnathostomulida, Gastrotricha, Loricifera, Bathynellacea etc.

The situation is different and more straightforward in those cases where a few specialized members are reduced in size and live interstitially while their "normal" relatives are of macrofaunal size and belong to large and diverse animal groups. Combined with their minute size and often also reduced vagility, some trends have evolved which results in a simplified, "primitive" body organization. These represent cases of an apomorphic and phylogenetically rather recent development: *Halammohydra* among cnidarians, *Psammostyela*, *Heterostigma* among ascidians, *Monobryozoon* among bryozoans, *Labidoplax*, *Leptosynapta* among holothurians, *Dinophilus* among polychaetes, *Microhedyle* among molluscs.

The situation is more confusing in animals such as nematodes, harpacticoids, and turbellarians. They belong to ubiquitous taxa whose members are mostly of meiobenthic size. They possibly have some other features, e.g. a slender body, preadapting them to an endobenthic life. Here, subgroups of lower taxonomic rank living in an identical biotope, e.g. the interstitial, may be characterized by special evolutionary trends that result in a characteristic appearance or behaviour (e.g. Epsilonematidae in nematodes, Otoplanidae in turbellarians). These similarities may express a natural relationship based on common ancestors.

Examining natural relationships in meiobenthic animals, ultrastructural investigations have often been helpful. In order to ascertain homology, REMANE (1952b) summarized general guidelines in his "criteria of homology". These have been specified for ultrastructural features by RIEGER and TYLER (1979) and RUPPERT (1982) working on meiobenthic animals. Below, some examples of (ultra-)structural analyses on meiobenthos are given contributing to a better assessment of natural relationships between phyla, but also within subgroups of higher taxa, e.g. the fairly well-studied polychaetes.

On the basis of number and arrangement of cilia, RIEGER (1976) concluded that the cells of the more primitive phyla (at least of their larvae) such as Placozoa, Cnidaria, Gnathostomulida, and Gastrotricha (partim), are "monociliated", i.e. flagellated, while the Spiralia with their typically multiciliated arrangement were considered more derived. On the basis of the cuticular ultrastructure, RIEGER and RIEGER (1976) further derived the annelids from plathelminths. RUPPERT (1982) considered a myoepithelial pharynx to be a symplesiomorphic character important enough to interrelate gastrotrichs, nematodes, but also bryozoans and tardigrades. On the basis of the ultrastructure of adhesive glands, phylogenetic considerations have been made on those groups conventionally regarded as acoelomatic, such as plathelminths and nemertines, and on their links to typical coelomates (TYLER 1977; RIEGER and TYLER 1979; RIEGER 1985).

Turning now to annelids, the ultrastructural details of the nephridia and the pharyngeal bulb have revealed that these structures, traditionally used as criteria of natural relationships, do not represent a valid basis for a polychaete classification. According to BARTOLOMAEUS and Ax (1992), metanephridia arose several times independently from protonephridia which, in turn, are homologous structures connecting various primitive animal groups. Similarly, the analogous nature of the pharyngeal bulb emerging from ultrastructural studies, proved that archiannelids represent an artificial group of miniaturized and often neotenic worms belonging to various polychaete families (WESTHEIDE 1985, 1987a; PURSCHKE 1988). Already in 1974, FAUCHALD postulated that neither the archiannelids, nor the typical errant polychaetes are close to the hypothetical primitive annelid. Instead, he envisaged as the probable annelid ancestor a worm of meiobenthic size with simple setae and head appendages.

For considerations on phylogenetic relationships within meiobenthic groups, the role of larval structures is very important. Persistent retention of larval structures, i.e. progenesis, is a characteristic phenomenon of probably considerable relevance for the phylogeny of meiobenthic taxa. According to WESTHEIDE (1987a), loss of planktonic stages and adaptation to a permanent settlement in the voids of the sediment have led to the evolution of an interstitial meiobenthos (Fig. 69).

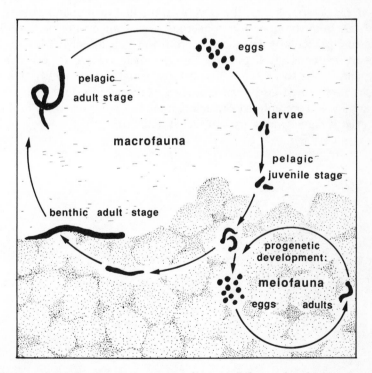

Fig. 69. Hypothetical progenetic origin of interstitial meiobenthos evolving from macrobenthos with a meroplanktonic ontogenetic phase. (After WESTHEIDE 1987a)

Larval ultrastructure documented the relations of some "archiannelids" to various polychaete families. The study of nauplius larvae clarified the position of syncarid crustaceans as primitive Malacostraca. An intensive future study of larval structures may also allow for conclusions on the connections between Loricifera and other nemathelminth groups, as for the assumed relationships of aberrant "secondary acoelomates" to annelids (RIEGER 1991).

In a strictly cladistic analysis and on the basis of numerous morphological characters, EHLERS (1985) refuted the classical plathelminth grouping and showed that "Turbellaria" are an artificial taxon of several convergent lines which should better be combined under the term "free-living plathelminths."

These examples underline that studies of structural details render a fruitful field for phylogenetic conclusions on meiobenthic groups. Naturally, they must remain hypothetical since they are all based on the study of recent animals. Positive evidence can only be given if connections between present-day animal groups are based on meiobenthic fossils. However, most findings of meiofauna fossils are rare and fortuitous due the small size and delicate consistency of the remains. While hard-shelled meiobenthos like ostracodes, despite their minute length, are fairly well documented to reach back into early Cambrian times and are even of economic relevance as stratigraphic indicators in oil drilling, information on other fossil meiobenthos is available only from scanty records: the earliest copepod, *Lepidocaris*, is from the Devonian, the first nematode has been found in horizons from the upper Carboniferous. An exception are the meiobenthic arthropods, particularly crustaceans, from the rich Swedish "Orsten fauna". which are extremely well preserved despite their old age (Upper Cambrian) (MÜLLER and WALOSSEK 1985, 1991; WALOSSEK and MÜLLER 1990). While some of these phosphatized fossils can doubtless be related to extant higher groups, there are some representatives which cannot be clearly allocated to a known taxon even if higher ranks are considered (Fig. 70).

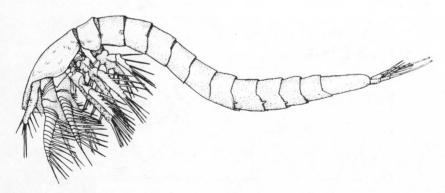

Fig. 70. A Cambrian meiobenthic crustacean from the Orsten formation: *Skara anulata* (total length 1.2 mm). (MÜLLER and WALOSSEK 1985)

It is postulated that all phyla existing today were already present at the Precambrian/Cambrian boundary. The microfossils from the Burgess and similar shales give some hope of unraveling the numerous unknown and important interrelationships between higher animal taxa whose members were apparently often of meiobenthic size. It has been hypothesized (JENKINS 1991; RUNNEGAR 1991) that even in the early Ediacaran, a primitive, probably small and simply structured benthos existed under hypoxic, perhaps anoxic conditions (see Chap. 8.5).

The "thiobios hypothesis" (BOADEN and PLATT 1971; BOADEN 1975, 1977; for details see Chap. 8.5) contends that the archaic animals belonged to a "thiozoic" meiofauna without a planktonic phase. It is reasoned that some of the more primitive meiobenthic taxa existing today can be derived directly from these thiozoic roots. Still today they live in anoxic and sulphidic habitats corresponding to life conditions in the primaeval (benthic) biotopes. However, the thiobios hypothesis has been seriously doubted: today, it is acknowledged that many of the recent thiobiotic animals certainly represent highly derived specialists. The frequent occurrence of progenetic trends in meiobenthos would speak rather for a primitive benthic fauna possessing a planktonic phase which increased their distributional potential. Secondarily then, this phase became abbreviated and eventually omitted (WESTHEIDE 1987a; HADZI 1956; see Chap. 4.1.4; Fig. 69).

More detailed reading: Ax (1963); RIEGER (1976); RIEGER and TYLER (1979), CONWAY MORRIS et al. (1985)

6.2   Distributional Implications

Whether of primary origin or secondarily acquired, restriction to a sedimentary life with limited vagility put meiobenthos in the focus of evolutionary considerations which base on horizontal distribution patterns in a zoogeographical scale. In order to assess the relevance of phylogenetic arguments using meiofauna, one has to consider the scenario which probably led to colonization of subsurface habitats.

It is widely accepted that the marine littoral and particularly the epibenthal represent one of the earliest and preferred habitats for animal life due to its rich food supply. On the other hand, the astatic, often rigid conditions typical for this zone forced the fauna to continuously adapt and respond to the biotopical stress with evolutionary versatility. Those species not able to cope with the demands of this competitive and radiative centre became displaced. Many forms of less radiative and adaptive capacity are thought to have evaded vertically away from the surface into the endobenthal and, if small enough, into the interstitial habitat with its lower number of competitors and more balanced physiography. In this refuge areas of enhanced stability, many meiobenthic forms could maintain their original features and ecology. In subsurface horizons, the protective habitat character not only enabled survival of "old" forms, but allowed for reduced reproductive rates, limited distributional means via propagative stages, and highly specialized adaptations: it favoured development of K-strategists. These features are particularly evident in the freshwater stygofauna.

REMANE (1952a) termed the littoral mesopsammal a zone of radiation into adjacent habitats and DELAMARE DEBOUTTEVILLE (1960) repeatedly underlined the brackish coastal groundwater to be a predominant route of immigration into the continental groundwater system. The characteristic ecological stability of groundwater habitats and their typical isolation often favoured a fauna with relictary characters or long-lasting evolutionary lines, frequently of regressive nature. This coastal-brackish pathway of colonization would explain why many of the stygobiotic amphipods (*Bogidiella*; *Ingolfiella*) and isopods (*Microcerberus*, *Microcharon*) have marine rather then freshwater relatives. This relationship is particularly striking in the freshwater polychaetes *Troglochaetus* and *Hesionides riegerorum* (WESTHEIDE 1979), the nematode genus *Desmoscolex*, some macrodasyoid gastrotrichs (KISIELEWSKI 1987) and some turbellarians (see Chap. 8.2.2 for a detailed account).

Another pathway of refuge, apparently taken by many meiobenthic groups, led away from the shallow seashore into the deeper and/or more protected regions of the sea. A stable and conservative environment is typical for marine and anchihaline caves and for the deep-sea. Both are often considered to represent refuge biotopes (ILIFFE et al. 1984). The interrelations between troglobitic and abyssal fauna have been explained by the continuous pathway connecting deep bottoms via the sheltered interstitial and crevicular system of marine caves (WILKENS et al. 1986; ILIFFE 1990), from where immigration routes led to the freshwater stygobios (WARD and PALMER 1992). Although this paradigm is still much disputed (STOCK 1986; DANIELOPOL 1980), it can be generalized that it is the refuge character of many meiobenthic habitats which renders the study of meiofauna distribution a valuable approach for drawing zoogeographical conclusions. In small soft bottom coves bordered by rocky areas, in coastal lagoons and ponds, geographic isolation may lead to accidental disparity of the founder populations and to the development of aberrant meiofaunal assemblages with the enhanced probability of an unusual speciation due to genetic drift.

Turning now to biological features: what are the prerequisites that make meiofauna good zoogeographical indicators for possible evolutionary connections? The dominant formative condition in meiofauna is miniaturization. It is mainly attained by reduction of cell number, not so much of cell size. Rotifers consist of only about 1000 cells. The nematode *Caenorhabditis elegans* is so small that every cell lineage is thoroughly known. Yet the sometimes only 50-µm-long loriciferan species (perhaps the smallest known metazoans) reportedly consist of as many as 10,000 and more cells (KRISTENSEN 1991a).

Miniaturization of body size, in turn, favours geographical isolation because of a reduced potential of dispersal and speciation. These characteristics are often linked to a set of features frequently occurring in meiobenthos of extreme habitats:

– low food consumption,
– simplification of organs (saving material),
– conservation of energy by developing highly efficient reproductive structures (internal fertilization, hermaphroditism, brood protection),
– abbreviated generation and life time, often combined with neotenic trends, e.g. in cladocerans, ostracodes, syncarids.

The limited dispersive potential of meiofauna should reduce the chances for allopatric speciation by geographic isolation and support sympatric speciation by ecological, physiological or behavioural isolation. Minute deviations in the ecological or behavioral inventory open up new niches and evolutionary chances (e.g. in harpacticoids, see IVESTER and COULL 1977; MARCOTTE 1984). Slight differences in physiological and ecological reactions result in divergences in the ecological specialization within an identical morphological frame.

As a result, in meiofauna the percentage of morphologically similar or even identical, but ecologically divergent and specialized species within closely related flocks should be high. This was recently corroborated by WESTHEIDE (1991) and is probably responsible for many identification problems. The flock of cryptic species in the harpacticoid genus *Tisbe* (BATTAGLIA et al. 1978; GABRICH et al. 1991) or *Tigriopus*, the species complex of *Canuella perplexa* (Harpacticoida), *Paracanthonchus caecus* (Nematoda) and *Pseudomonocelis* spp. (Turbellaria, Proseriata) differ in their ecological range, but hardly in structural details. Differences in substrate preference and biometry characterize also the ostracode forms of *Cobanocythere labiata* (WESTHEIDE 1991).

However cryptic species and morphologically identical species flocks are also brought about by sexual isolation through polyploidy. In the marine oligochaete *Lumbricillus lineatus*, a local 3n population has developed which must be pseudo-fertilized by the normal 2n stock (CHRISTENSEN and O'CONNOR 1958) for production of offspring.

Also diverging reproductive strategies are a means of sympatric speciation in meiobenthic animals. Sibling species with gonochoristic or hermaphroditic lines occur in the polychaete genus *Ophryotrocha*. It contains simultaneous or consecutive hermaphrodites, some of them being oviparous, some viviparous (ÅKESSON 1973).

In these examples where differences are not to be assessed by morphological characters, structural discrimination is mostly substituted by physiological methods, i.e. enzyme electrophoresis, and more recently, by immunological and RNA-fingerprinting methods (WESTHEIDE and SCHMINKE 1991). It was also suggested to prove the genetic relationship by antiserum tests which had already been successfully applied for predator-prey relationships (FELLER et al. 1979). In some rare cases a natural relationship can even be evidenced by a successful, positive or negative infestation with parasites (the "rule of FAHRENHOLZ"). ÅKESSON (1977) succeeded in transmitting sporozoan parasites from *Ophryotrocha* species to Dinophilidae and suggested a close relationship between the groups.

Compared to macrofauna, sympatric speciation is for meiofauna probably a more important mechanism for the fairly specialized, often K-selective animals. This mode of speciation is contrasting to the classical allopatric speciation based mainly on geographical isolation. As a consequence, the distributional patterns of meiobenthic animals with their restricted dispersive capacity do not always allow for easy explanations considering mainly the conventional lines of allopatric speciation.

Discussing the amazing structural uniformity of many geographically disjunct meiofauna there must be made two caveats:

1. Close morphological similarity in geographically separated forms does not necessarily correspond to identical physiology and ecology (see above). In cryptic species, uniformity is restricted to structure only, not to the genetical background. A population exposed to an unusual physiographic milieu may initially cope with this altered environment by its wide adaptive capacity (non-genetic adaptation); but stepwise its genetical range may become shifted as an energetically less costly solution. Eventually, this will result in a genetical interpopulation variance which better answers the new adaptive demands (BATTAGLIA and BEARDMORE 1978). Heterozygosity offers adaptive advantages and selective changes particularly in stressed areas of steep environmental gradients (e.g., salinity, pollution). Before this differentiation becomes apparent in structural details, it may be expressed and genetically fixed in divergences such as community and life history parameters, physiological reactions and ecological preferences. Growth, reproductive effects, metabolic rates, or adaptational ranges may differ when experimentally tested in disjunct populations of an identical morphospecies (e.g. *Ophryotrocha*; LEVINTON 1983; *Aktedrilus*; GIERE 1980). Establishment of these cryptic differences by modern methods, preferably by RNA and protein fingerprinting, will probably reveal a much less developed specific uniformity in meiofauna than assumed so far. A differentiation becoming gradually fixed in the genome will be particularly effective in species with a low distributional capacity, which is characteristic of meiobenthos.
2. Many cases of morphological "identity" are possibly due to superficial morphological analysis. Rigorous scrutinization by experienced taxonomists, application of electron microscopy, will probably reveal many more differences than reported so far. As a result, the global distribution of many a former "species" will have to be differentiated and split into various restricted areas of better identified natural new species.

More detailed reading: STERRER and AX (1977); BOTOSANEANU and HOLSINGER (1991)

CHAPTER 7

The Distribution of Meiofauna

The general zoogeographical pattern of meiofauna, at least in the marine field, seems still a "meiofauna paradox". Why are there so many identical meiofauna taxa, even at lower systematic levels, in completely divergent areas? Why is there a close relationship despite apparent adaptations which prevent dispersal? How can identical forms bridge oceans, occupy completely disjunct shores (Fig. 71; GERLACH 1977b; WESTHEIDE 1987b; but see previous Chap.)?

General conclusions on the distributional pattern of meiofauna are still problematical, since the information available often mirrors the local conditions only, and in many areas investigations lack completely. Nevertheless, a cosmopolitan distributional pattern as shown in Fig. 71 occurs remarkably often in soft-bodied marine meiobenthos such as interstitial polychaetes (e.g. *Hesionides* spp., *Microphthalmus*-complex), but also in ciliates (e.g. *Remanella rugosa*), turbellarians, gnathostomulids, gastrotrichs (*Urodasys viviparus* and other species). Apparently, it is less common in "hard meiofauna" like harpacticoids. The presumed amphiatlantic distribution of *Paraleptastacus macronyx* has been revealed to be a composite pattern of several separate species, the representative along the American shores being *P. coulli* (HUYS, pers. comm.). Within harpacticoids in general, WELLS (1986) states that "interstitial species tend to show a higher degree of endemism (76%) than primarily epibenthic or phytal species (63 and 68% respectively). None are truly cosmopolitan, though a few have a rather wide distribution."

7.1 Mechanisms of Dispersal

7.1.1 The Plate Tectonics Theorem

STERRER (1973) concluded from the frequent uniformity of amphiatlantic meiofauna that their close similarity on the genus or even species level would be maintained by an extremely slow speciation despite their separation by continental drift since about 200 million years. Evolution to proceed rather slowly is also required explaining the "relict refuge model", which underlines the stability of habitat conditions resulting in the relictary character of many meiofauna. Maintenance of habitat and climatic conditions combined with low

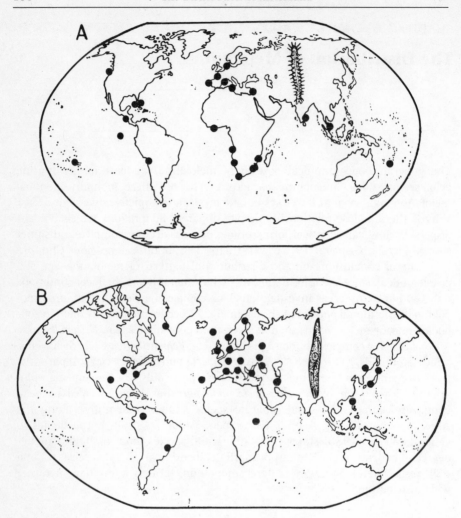

Fig. 71A, B. Examples of cosmopolitan distribution of meiofauna species. A *Hesionides arenaria* (Polychaeta). B *Gyratrix hermaphroditus* (Turbellaria). (A WESTHEIDE 1977b; B STERRER 1973)

number of offspring and restricted means of dispersal would result in a low speciation pressure.

However, this extremely slow speed of speciation contradicts the experience that the recent species rarely developed before the Eocene period. This would refute explanations by plate tectonics between Europe and America and a Gondwanda-based common origin for similarities on the species level (HARTMANN 1986, 1988). Recent findings on the rich meiofauna of the Galápagos Islands also contradict the idea of a particularly slow speciation in meiobenthos, since within only 3 million years (geological age of the islands) the pristine beaches of the Galápagos archipelago have been colonized by a rich and radiating meiofauna (WESTHEIDE 1991). In many cases a more detailed knowledge of land-bridges (e.g. the Scotland-Iceland Ridge) and their periods of existence would reveal the existence of shallow-water connections later than assumed so far.

7.1.2 Water Column Transportation

If the "geological vehicle" of plate tectonics is questionable, which other means of dispersal are there to explain the present-day's distribution pattern of meio-fauna? Recruitment and colonization mediated by "water-column processes" is the other common explanation of meiofauna distribution. Recent investigations revealed a partially "hyperbenthic" occurrence of meiofauna which was regularly caught in sediment traps high above the ground (PALMER 1988; ARMONIES 1988a). This underlines the necessity to take into account the "drift" of meiofauna after suspension and to differentiate this drift in several natural, often interacting pathways relevant for distribution: erosion/suspension, emergence/suspension and rafting.

7.1.2.1 Erosion/Suspension

This transport mechanism is most relevant in hydrodynamically more rigid areas where tidal currents and breaking waves scour the sediment and suspend preferably the meiofauna of the upper sediment layers. Sediment disturbance by larger fish and crustaceans, and aboveground structures such as boulders and plant culms can modify the hydrodynamic turbulences and may cause additional suspension of meiofauna.

Interstitial species, despite their well-developed adaptations against suspension (see Chap. 4.1), have been found to account for 10–30% of the drifting meiofauna (HAGERMANN and RIEGER 1981). In flume experiments, about 5% of all meiofauna became suspended and occurred drifting in the water column.

PALMER and GUST (1985) compared in their calculations the suspended meiofauna organisms with quartz grains of 40 µm diameter, thus inferring a completely passive transport of the animals once suspended in the water column. Indeed, on the basis of detailed flume experiments and calculations, BUTMAN (1989) confirmed that the degree to which larval (and meiobenthic) organisms can actively influence their settling and resuspension behaviour is very limited. By passive transport, meiofauna can easily become dispersed with velocities of 10 km per day. Breakers in gales have been shown to erode the bottom down to 25 m water depth and to carry particles over distances of 50 km; but also regular tidal currents are able to alter the small-scale distributional pattern by suspension and resettlement, especially in the epibenthic harpacticoids, less so in nematodes (BELL and SHERMAN 1980; DECHO and FLEEGER 1988; ARMONIES 1990).

Following continental coastlines and drifting along large ocean current systems, erosion by hydrodynamic forces and subsequent passive drift may be a relevant means of meiobenthos dispersal resulting in short-term and local fluctuations. Moreover, as a means of recolonization, these hydrodynamic forces will contribute to compensating for destruction of meiofauna due to local devastation.

In streams and creeks, erosion and passive water-column transport are the most important means of meiofauna dispersal (PALMER and GUST 1985; PALMER 1990a). Recruitment of meiobenthic populations in the sediment from the water column and streambed surface is even more relevant than immigration from adjacent areas and can explain the rapid recolonization within some weeks after severe disturbances (e.g. spring floods). Up to 2500 meiofauna organisms per 1 m³ (mainly rotifers, chironomid larvae, oligochaetes and copepods) have been found drifting in the waters of a stream (PALMER 1992). Although clearly flow-dependent, even here an active control over drift entry seemed to operate, resulting in an increase of drift and significantly increased drift rates at night. Interestingly, the threshold velocity for a drastic increase in meiofauna drift was significantly lower than that for sediment erosion. Beside these pathways, the refuge into deeper, hyporheic sediment layers after lotic disturbances and following re-entry into the streambed has been tested and found to be of significance for recruitment of rotifers and copepods (PALMER et al. 1992).

7.1.2.2 Emergence/Suspension

An analysis of the suspended meiofauna revealed that some groups such as harpacticoids, ostracodes and turbellarians are particularly subject to aquatic drift while nematodes and oligochaetes are only rarely found in the water column. This observation does not support the contention that animal drift result only from passive forces and meiofauna individuals are comparable to sand grain transport (FEGLEY 1988). It rather indicates a biologically influenced, more active way of meiofauna suspension where behavioral and structural differentiations gain importance. Emergence and subsequent dispersal seem particularly important factors for phytal meiofauna (see Chap. 8.4; HICKS 1986, 1988; WALTERS and BELL 1986; KURDZIEL and BELL 1992). From seagrass meadows more than 50% of all harpacticoids has been found to ascend nightly into the water column. There is apparently a direct coupling to the overall hydro-dynamic situation: the most active ascent was in calm weather (ARLT 1988). But this active emergence is not restricted to the phytal. ARMONIES (1988a,b) showed that in the sandy flats of the Island of Sylt (North Sea) 87% of all harpacticoids, 67% of ostracodes and 42% of turbellarians leave the sediment at night. This emergence considerably increases the chance to be drifted away and become dispersed (see also WALTERS 1991). Emergence seems sometimes also to be related to reproductive processes. Its activity may vary depending on sex distribution and developmental stage (BELL et al. 1989).

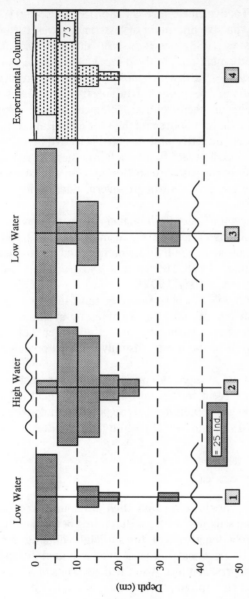

Fig. 72. Migratory reaction of meiofauna to the tidal wave. Graphs *1–3* field samples taken during one 18-h tidal cycle beginning with low water; graph *4* experimental core simulating high water conditions but omitting wave action. (After MEINEKE and WESTHEIDE 1979)

The counterpart of active emergence is the process of re-entry. The original contention of a merely passive and more or less random resettlement from the suspended stock (ECKMAN 1983) has now been differentiated. It is rather a mixture of passive resettlement and, in the direct vicinity of the bottom, a fairly selective re-entry (BUTMAN 1989) which not only depends on physical details like structural complexity and roughness of the substrate (PALMER 1988; ECKMAN 1990), but certainly also on chemical and biological signals. Apparently, it is a fairly quick process (periods of hours and days for colonization) and also quite an active (selective) one. An active and repeated selective behaviour would be an explanation for the puzzling fact that meiofauna, despite tidal currents and suspension, can maintain a microscale distribution in an almost identical pattern with certain centres of patchiness through several tidal cycles (FLEEGER et al. 1990).

A "biological conception" controlling meiofauna occurrence would correspond to the observations of active vertical migrations of meiofauna in the sediment column influenced by the hydrodynamic conditions and by vibration stimuli (Fig. 72; BOADEN 1968; BOVÉE and SOYER 1974; MEINEKE and WESTHEIDE 1979; FOY and THISTLE 1991).

Emergence of meiofauna was observed to be most intense in the dark. This nightly behaviour of meiofauna may be explained by avoidance of the strong benthic predation pressure. Ascending into the water column at night would diminish predation by the mostly visually oriented predators (ARMONIES 1989b).

Altogether, distributional capacities by emergence, suspension and reentry/ settlement of meiofauna seem to depend on (1) anatomical and behavioural features of animals, (2) light conditions, (3) the population density (SERVICE and BELL 1987; WALTERS 1991), and (4) the hydrodynamic patterns of the sediment surface.

7.1.2.3 Rafting

Microalgae (diatoms), cyanobacteria and their mucous excretions can from dense mats covering the sediment surface. Here, they become easily suspended, forming rafts which carry, together with the surficial sediments, also epibenthic meiofauna (especially harpacticoids) at least over a limited distance of the sea (HICKS 1988; HARTWIG, pers. comm.). Also the dense rafts of *Sargassum* floating in the ocean contain a rich (phytal) meiofauna. Recently, ARMONIES (1989c) found numerous meiobenthic organisms in the "sea foam" excreted by algal cells after blooming periods (*Phaeocystic*). Since these foam mats drift regularly in the warmer season on the surface of the sea and accumulate on beaches, these relatively long-lasting rafts may contribute considerably to meiofauna dispersal over wide areas. Even the thick fibrous cover of coconuts which float for long distances over the oceans has been found to harbour numerous meiobenthic organisms (GERLACH 1977b). Whether this is a relevant distributional vehicle for meiofauna remains open.

There is a fairly novel, and, in its biological significance for particle transport, probably not to be underestimated marine phenomenon: marine snow. Apparently, these aggregates of detritus, bacteria and mucus have considerable relevance for general particle transport in the oceans (RIEBESELL 1992), but also for meiofauna dispersal. They throw a new light on meiofauna suspension and drift in the water column. It seems that many meiobenthic organisms drifting in the ocean are bound to aggregates of marine snow. Equipped with various adaptations for adhesion and clinging, they could even maintain their nutritional preference for a substrate rich in bacteria. In collections from oceanic sediment traps high above the sea bottom, 80% of all nematodes, polychaetes and larvae, plus 20% of the harpacticoids found where observed to firmly cling to marine snow particles (SHANKS and EDMONSON 1990).

7.1.2.4 Conclusion

The contention seems rather far-fetched that all those meiofauna groups which share a high amphiatlantic similarity have rates of speciation so slow or even stagnant that their present distribution pattern on the species level could be explained by plate tectonics. Instead, various interacting mechanisms of water-column transportation seem more plausible tools to explain migration and genetic

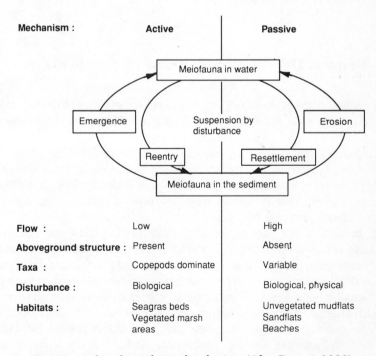

Fig. 73. Modes of meiofauna distribution. (After PALMER 1988)

exchange of meiofauna. They alone could account for colonization of remote islands without any previous contact to continents such as the Galápagos or New Zealand.

Particularly the recent reports on marine snow in the open ocean with adhering meiofauna, and on the existence of a distinctly demersal zooplankton are likely to overcome the traditional border-lines between plankton and benthos and render their classical definitions somewhat arbitrary. Admittedly, these more recently discovered distributional pathways do not yet allow evaluation of their general relevance. However, it is conceivable that an intensive benthic-pelagic coupling could attain considerable relevance for meiofaunal dispersal and trans-oceanic similarity (Fig. 73).

Discussing trans-oceanic pathways of meiofauna transportation, even the possibility of a man-made dispersal cannot be excluded, at least locally and for some meiofauna forms. It was suggested by GERLACH (1977b) that, among other means of meiofauna dispersal, also wet ballast sand could be a possible source of meiofaunal transport across the seas. Sand from the sea shore was taken regularly and in great amounts as load for empty sailing vessels during the last century. At least between the more frequented ports of the tall-ship period this may have caused some continental exchange of interstitial animals.

More detailed reading: PALMER (1988a); HICKS (1988); BUTMAN (1989); ARMONIES (1990); ECKMAN (1990)

7.2   General Distribution Patterns of Meiobenthos

Before dealing with the meiofauna in selected biotopes (Chap. 8), some considerations on the general distribution of meiofauna, and on typical horizontal and vertical profiles should serve as a basic orientation.

Large-scale meiobenthic distribution is mainly related to physical and chemical parameters with some sedimentary and biogenic heterogeneities as modifying factors. Meiofauna in general has a high capacity to realize new ecological niches. This results in relatively high meiofauna abundance and diversity even in extreme biotopes.

In the oceans, the tides represent the determining factor for zonation and abundance of meiobenthos (HULINGS and GRAY 1976), although the distributional pattern in singular groups is influenced by a more complex factorial combination. On non-tidal shores, biotic interactions become more apparent, resulting in a rather small-scale meiofaunal patchiness. Aggregations in the range of a few centimetres are common in meiofauna and often even persistent through tidal cycles. Recent studies support the view that most of this patchy distribution is generated by selective feeding preferences and direct or indirect trophic interactions (FINDLAY 1981; FLEEGER et al. 1990; BLANCHARD 1991).

7.2.1 Large-Scale Horizontal Distribution

Meiofauna occurs in all aquatic biotopes and climatic zones. However, a latitudinal comparison of meiofauna diversity between the global climatic zones still suffers from the inadequate meiobenthological knowledge mainly in the polar and tropical regions. Recent studies revealed in Antarctic beaches a rich and highly endemic meiofauna (HARTMANN 1990 – ostracodes, HERMAN and DAHMS 1992) which contrasts to the poorer Arctic meiofauna. There is apparently a close relation of the Antarctic to the adjacent abyssal fauna which is, as yet, little known.

Even the drifting sea ice seems to contain considerable populations (up to 9000 individuals per $1000 \, cm^3$ ice!) of well-adapted turbellarians, harpacticoids, cyclopoids and nematodes feeding mainly on the rich diatom stocks in the lower levels of the ice layer (DAHMS et al. 1990; GRAINGER 1991; CAREY 1992). Sometimes the Arctic sea ice contained $> 20\%$ rotifers, and while the Antarctic ice was devoid of nematodes, acoel turbellarians dominated by $> 60\%$ (GRADINGER et al. 1993). Also subpolar intertidal beaches (Iceland) contain a rich meiofauna ('OLAFSSON 1991). The numerous turbellarian species in polar areas are closely related, if not identical with those in boreal latitudes, but they contrast distinctly to the subtropical (Mediterranean) and tropical turbellarian fauna.

A depression of diversity and abundance in tropical meiobenthos compared to boreal latitudes, as indicated in COULL (1970) for the Bermuda reef platform, cannot be generally confirmed. Today there are numerous studies from various warm-water areas which show that both abundance and diversity of meiobenthos in all subtropical/tropical zones of the globe is in fact very high (McINTYRE 1968; SALVAT and RENAUD-MORNANT 1969; RENAUD-MORNANT et al. 1971; FAUBEL 1984; GRELET 1985; ST. JOHN et al. 1989; ALONGI 1990b; WESTHEIDE 1991). Particularly the biogenic calcareous sands of the tropics with their rich structural heterogeneity and often high content of detritus favour colonization by meiofauna. In many tropical regions of the Pacific Ocean, annelids attain a major role in the abundance of meiobenthic groups, often second to nematodes. Contrastingly, in New Zealand beaches oligochaetes seem to be scarce which might also explain the low number of otoplanid turbellarians, their main predators (RISER 1984).

The organic content of the sediments as a decisive nutritional factor seems to play a key role for meiofauna density. Along the coasts of South Africa, meiofauna abundance was positively correlated with detritus content of the sediment (McLACHLAN et al. 1981). Consequently, the Mediterranean beaches with little organic matter harbour only a relatively poor meiofauna.

The eulittoral zone of temperate to boreal regions often contains the richest meiofauna. Here, the finer sediments are more densely populated than coarse sand: up to 16×10^6 ind. m^{-2} are not unrealistic (McINTYRE 1968). Differentiated for sediment types of a North Sea shore, muddy sediments tend to have the maximum of meiofauna abundance near the low tide level, while in sandy shores the maxima are closer to the mid-tidal and high-tide line. The same trend is recognizable also for the species richness. In the exposed upper areas of the eulittoral and supralittoral, meiofauna abundance and diversity decreases. Only some oligochaetes have their preferred habitat in this zone (Fig. 48). Along (perpendicular to) the water line, only little differences in meiofauna numbers can be expected, at least along fairly homogeneous shores. In sublittoral bottoms despite the more stable physiographic conditions, meiofauna abundance is generally three to four times less than in eulittoral bottoms. This accounts probably for the lower concentration of food. Recent data from a North Sea survey indicate, however, that the general decrease both in density and diversity of meiofauna from the south to the north cannot simply be attributed to grain size conditions and/or organic content alone (HUYS et al. 1992). Other factors such as feeding biology apparently interact with these abiotic parameters.

Since the general decrease in abundance and biomass towards greater depths is even more drastic in the macrobenthos, meiobenthos often gains in relative importance both with respect to abundance and biomass. Locally, e.g. in the deep-sea and in muddy fjords, the quantity of meiobenthos can equal that of macrobenthos, attaining even a competitive significance. In greater depths, fluctuations of meiobenthic stock are attenuated, but there is still a close coupling of (meio-) benthic dynamics with the primary production in the surface layers (MARCOTTE 1984; FAUBEL 1984); this even refers to deep-sea bottoms (PFANNKUCHE 1992; see Chap. 8.3).

7.2.2   A Typical Vertical Distribution Profile of Meiobenthos

In a vertical sediment profile, the upper few centimetres usually harbour more meiofauna than the deeper horizons. This relates to the richer supply with oxygen and food particles. In finer sediments with their steeper physiographic gradients, this general pattern is particularly clearly revealed. In a core of silty sediment, YINGST (1978) counted in the uppermost two centimetres 71% of all meiofauna present. Mud flats contained in the uppermost single centimetre about twice as much meiofauna as in sandy bottoms in a 10-cm column (SMITH and COULL 1987). However, there are exceptions from this general pattern. Thiobiotic meiofauna prefer the chemocline at the oxic/sulfidic interface and are only exceptionally encountered in the surficial layer (GIERE et al. 1991; OTT et al. 1991; see Chap. 8.5).

A differentiated vertical distribution entails preference and avoidance reactions with migrations along the gradient system. Tidal currents with their powerful "tidal pump" have a massive influence on fluctuations of the vertical distribution (JOINT et al. 1982). Some meiobenthos has been shown to negatively react to the tidally changing swash zone. During low tide, many meiobenthic animals avoid the reduced water content and the changing temperature and salinity by downward migrations (Figs. 4, 71; BOADEN 1968; BOADEN and PLATT 1971; MEINEKE and WESTHEIDE 1979). In tidal beaches nematodes and harpacticoids have been shown to migrate up to 25 cm through the sand during one tidal cycle (HARRIS 1972). Compared to nematodes, in populations of harpacticoids, ostracodes and turbellarians short-term fluctuations (within the range of some weeks) are even more pronounced (Fig. 74).

Diurnal and particularly seasonal variations in the vertical occurrence of meiofauna often correspond to the temperature pattern. In the summer, most animals of boreal shores live closer to the surface, in winter they migrate deeper down and tend to live in closer aggregates (Fig. 75).

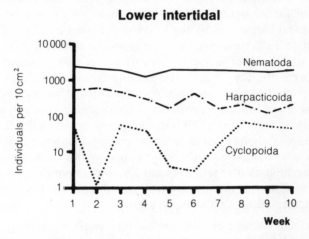

Fig. 74. Short-term changes (within weeks) in the vertical distribution of meiofauna. (After ARMONIES 1990)

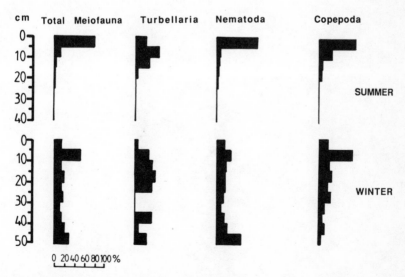

Fig. 75. Vertical occurrence of eulittoral meiofauna at summer and winter conditions. (After HARRIS 1972)

The schematic outline of meiofauna occurrence described above may vary considerably when local distributional patterns are considered. The existence of meiobenthic "communities" or consistent "associations" characterizing a particular zone of habitat is rather improbable, taking the patchiness and high variability in meiofauna occurrence into account. The establishment of interstitial coenoses as a basis for an ecological grouping (REMANE 1933, 1952a) was influenced by corresponding conceptions in the early macrobenthos research relating depth and sediment type to communities. The "*Halammohydra* coenosis" in coarse sublittoral sands (*H. octopodites*, some stenoecious archiannelids, turbellarians and harpacticoids), the "*Turbanella hyalina* coenosis" in sublittoral fine sediment, or the "*Otoplana* coenosis" in exposed eulittoral beaches (otoplanids, some gastrotrichs) rather denote the occurrence of some conspicuous species than a stable community. The structure of meiofauna assemblages seems to be regulated by other factors than depth and sediment type only and probably biotic factors are of decisive importance. This became obvious in the comprehensive study by HUYS et al. (1992) on the harpacticoids of the North Sea. At the same sample stations, their distribution pattern contrasted to that of nematodes. The higher spatial stability of nematodes seems to be exceptional and enables a discrimination of relatively monotonous communities depending on sediment structure and organic content (see Chap. 5.6; BONGERS and HAAR 1990; VANREUSEL 1990; VINCX et al. 1990).

More detailed reading: general review on meiofauna ecology – McINTYRE (1969); meiofauna in tropical regions – ALONGI (1990b)

Meiofauna in Selected Biotopes

The variety of meiofauna biotopes is amazing: Meiofauna has been found in void systems of Antarctic sea ice (see Chap. 7.2.1), in the gas passages of Spartina culms (see Chap. 8.4) and in the interstices of gill chambers in amphibious crabs (RIEMANN 1970). This biotopical diversity, and the variability in meiofauna distribution and composition renders it problematical to describe singular biomes with their "typical" meiofauna. In a compilatory work, this difficulty is enhanced by the multitude of studies with differing scientific perspectives and methods. Nevertheless, the following chapters try to characterize the meiobenthos in a selection of relevant biotopes emphasizing the constraints by and the adaptations to the pertinent ecological conditions. Necessarily, this overview cannot represent an exhaustive or a complete survey.

8.1 Meiofauna in Brackish Waters

The brackish-water fringe between marine and freshwater sediments has always been an important area for meiofauna settlement and migration. Brackish sediments extend well into the upper littoral of the sea shore and into the limnetic biome. Surprisingly many subterranean habitats (see Chap. 8.2.2) have slightly brackish water conditions. The gradual transition from marine to limnetic salinities combined with a fairly stable abiotic regime supported during long geological periods evolutionary trends in meiofauna to adapt to reduced salinities. This could explain the relatively high percentage of typical brackish-water species among meiobenthos (FENCHEL 1978). The brackish coastal groundwater attains a considerable relevance as an evolutionary pathway for immigration of marine meiofauna into the continental groundwater system (see Chap. 6.2).

In contrast to macrofauna, many marine or freshwater meiofauna species have a high tolerance capacity for brackish water and its ecological stress factors. This results in a surprisingly high meiofauna diversity in brackish-water areas. The overall diversity of the meiofauna in the brackish Baltic Sea exceeds that of the macrofauna, but it decreases from the western Belt Sea to the eastern Bothnian Gulf where mainly nematodes represent the meiofauna (ARLT et al. 1982). High tolerance for brackish-water conditions contributes also to the wide progression of marine species into brackish waters: Among the Baltic Sea turbellarians, no endemic fauna could be reported. All the species occurred also in the supralittoral of the adjacent North Sea where they represent a true brackish-water fauna (ARMONIES 1988d). The surprising degree of faunal identity in Alaskan, Canadian and European coastal turbellarians can also be referred to the well-developed brackish water tolerance capacity in meiofauna, but also to the uniformity of brackish water biotopes. It results in a community of Plathelminthes with a circumpolar distribution (Ax and ARMONIES 1990). Lack of indigenous Baltic Sea species is also known from halacarids (BARTSCH 1974).

Even in estuaries with their highly astatic hydrographic regime, thalassogenic meiofauna populates in considerable density even the higher limnic reaches (ALONGI 1990a). In the brackish-water zone of an estuary, the sediments are found to harbour many euryhaline marine species of nematodes and turbellarians co-occurring with freshwater and brackish-water forms (RIEMANN 1966; SOPOTT-EHLERS 1989). On the other hand, the meiobenthic rotifers colonizing the brackish region are limnogenic species, while the percentage of marine species is very low. A considerable share of faunal groups other than the usual nematodes and harpacticoids was also underlined by CASTEL (1992) in a review paper on brackish-water lagoonal systems. While the species number may become reduced by the astatic conditions, CASTEL stressed the high abundance of meiofauna in the often nutrient-rich lagoons where meiofauna densities often range from $3-5 \times 10^6$ individuals m^{-2} and, thus, exceed the average values for deeper marine sites. In the shallow waters, meiofauna may be an important nutritional resource for juvenile fish.

The high diversification of meiobenthos documented in all haline regimes might be the reason why, at least in more stable brackish regions, the "brackish-water species minimum" around salinities of 8–10‰ S (REMANE 1940) seems less characteristically developed in meiobenthos than it is both for macrobenthos and plankton. There is, of course, a depletion in species richness in this transition area, as pointed out by GERLACH (1954) for nematodes, but it is not comparably distinct when all meiobenthic species are pooled.

8.2 Meiofauna in Freshwater Biotopes

There are numerous types of freshwater biotopes, each with a differentiated meiofauna. Groundwater aquifers, river shores, cave pools and lakes, all harbour a particular meiobenthos. Ecologically, this meiofauna often attains considerable importance since, together with bacteria and protozoa, meiobenthic animals are involved in the regeneration of waste water and in the supply with drinking water. Compared to the marine biota, the meiofauna studies of the varying and often remote freshwater biotopes are relatively few and often still in their pioneer stages. Within the framework of this book, freshwater meiobenthology will not be considered with equal comprehensiveness as its marine counterpart. Detailed faunistic and zoogeographical information on the subterranean or stygofauna is given in the compendium, edited by BOTOSANEANU, *Stygofauna Mundi*, and

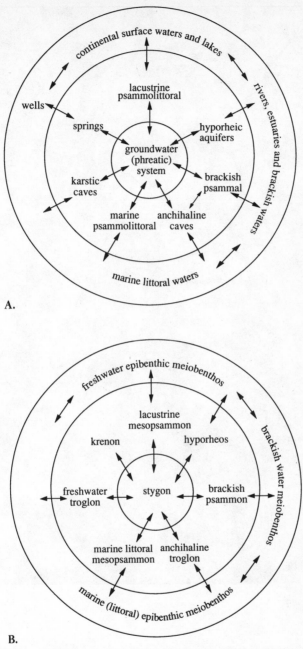

Fig. 76. Schematic diagrams of freshwater habitats (**A**) and their meiobenthic assemblages (**B**) indicating their numerous interactions. (After WARD and PALMER 1993)

Fig. 77. Artist's impression of the freshwater interstitial environment and its fauna

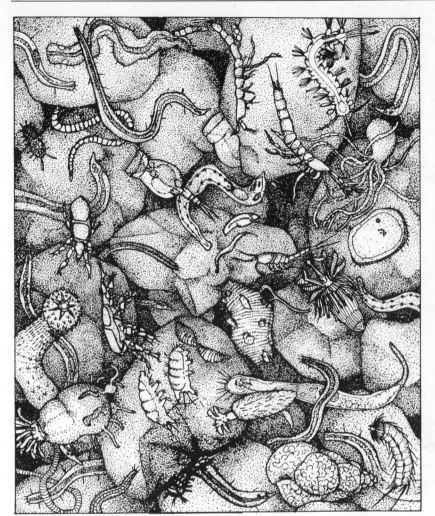

Fig. 78. Artist's impression of the marine interstitial environment and its fauna

many important papers are published in a scientific journal specializing in freshwater meiobenthos, *Stygologia* (STOCK and BOTOSANEANU, eds.).

The biological nomenclature in freshwater biotopes is rather inconsistent and confusing. Different and complicated terms relate the fauna to the sediment, the water movement etc. Particularly the groundwater ("stygobiotic") and cave-dwelling ("troglobitic") meiofauna have attracted much attention, since many archaic relict groups, especially among crustaceans, have been found here. Although the first ecological studies on freshwater meiobenthos were performed in river beds, this "hyporheic" and (deeper down) "phreatic" meiobenthos is less well known than the limnetic meiobenthos in lakes or the "stygofauna" in the subterranean aquifers and continental groundwater. Figure 76 indicates the close connections and interactions between the various biotopes and their coenoses, between the ground water system and the surface waters, but also the marine habitats. The interconnecting pathways may be regarded both as routes of dispersal and evolutionary pathways for the meiobenthos (see Chap. 6).

The biotopical constraints and the typical faunistic composition of the freshwater meiofauna differ much from that in marine realm, although also here nematodes frequently dominate in abundance and biomass. But in many freshwater biotopes, rotifers are often of equal importance, followed by copepods (cyclopoids and harpacticoids), tardigrades (*Macrobiotus*), cladocerans (Chydoridae), naidid oligochaetes, chironomid larvae, collembolan insects and hydracarid mites. An artist's view of the biotope may illustrate this low accordance with the typical marine meiobenthos (Figs. 77 and 78). However, this generalized picture often varies unaccountably in the different freshwater biotopes (SCHWOERBEL 1961a, 1967; PENNAK 1988).

8.2.1   River Beds and River Shores

Many studies (SCHWOERBEL 1961b; WILLIAMS 1989) differentiate between the benthos of the river bed and that of the river shore (similarly with creeks and streams). The "hyporheic" habitat of the river bed is mainly characterized by the direct vicinity of the flowing water while the "phreatic" sediments deeper down and at the sides of the water merge continuously to the groundwater horizons (Fig. 79; PENNAK and WARD 1986).

However, the numerous physiographic connections and faunal interdependences render this delineation of biotopical subunits equivocal, and indicate a close similarly (DANIELOPOL 1989, 1991; LAFON and DURBEC 1990).

The abiotic factor system of hyporheic habitats is characterized by astatic fluctuations. The extremely inhomogenous sediments of the river bed (MD = 0.5–6.0 mm), the varying current system (depending on the degree of exposure), the seasonal changes in temperature and water level, together with a variable chemical milieu, often contribute to the instability. Frequently, river pollution to which the hyporheic fauna is directly exposed, is an additionally aggravating factor for colonization with fauna. In this environment, characteristic features

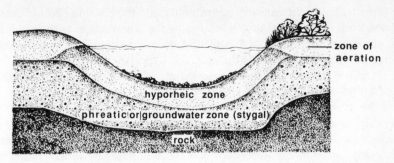

Fig. 79. The habitat zonation of a river bed. (After PENNAK and WARD 1986)

of the pore water compared to the overlying river water are pH values lowered by 1–2 units, an increasing content in CO_2, often paralleled by an increase in the concentration of some other chemical parameters such as silica and iron due to enhanced adsorptive forces in the sediment. The accumulation of detritus in the interstices of the sand combined with intense bacterial growth accounts for a typical decrease in oxygen concentration (Fig. 80), sulphate and nitrate content.

Among the biogenic factors, the load of detritus is a most important trophic source in hyporheic biotopes. Its rich bacterial and phytobenthic epigrowth serves as a food basis for most meiobenthos. Beside a good permeability in coarser sediments, it is a rich supply with organic matter that seems to be a key factor for settlement of riverine meiofauna, while the animals seem rather indifferent to a reduced content of oxygen in the pore water. Particularly sediments based on crystalline primary rocks with their "open" pore system are effective traps for detritus, while calcareous rubble and sand tend to become clogged

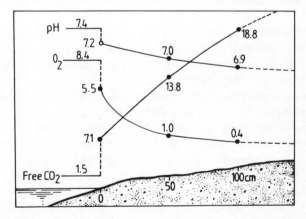

Fig. 80. Oxygen decrease and other abiotic parameters in a typical hyporheic biotope. (PENNAK 1940)

with fine silt. This reduces water permeability and the filtering effect of detritus particles, and, in turn, negatively influences the colonization with meiofauna. On a general scale, the hydrological and geomorphological regime seems to override all other factors. Thus the epigean, hyporheic and groundwater zone of a river are connected, their meiofaunal communities regularly interact (DOLE-OLIVIER and MARMONIER 1992).

In the hyporheic fauna, beside the frequently dominating nematodes, also rotifers and copepods (both cyclopoids and harpacticoids), juvenile oligochaetes and insect larvae are of relevance. PALMER (1990a) documented that the meiofauna composition in the hyporheos is often dominated by other groups than nematodes. She found in the bed of a low-gradient stream below $10 \, cm^2$ up to 6000 meiobenthic organisms with 35–85% rotifers as the prevailing group, followed by 20% juvenile oligochaetes, then chironomid larvae, and, of subordinate rank, nematodes and copepods. In another stream, WARD and VOELZ (1990) found small crustaceans and chironomid larvae to account for 2/3 of all meiofauna. A group also characteristic for the hyporheos are chydorid cladocerans. Small isopods, rather frequent in European river beds, seem to be absent in North America (WILLIAMS 1989). STRAYER (unpubl.) reported in a survey of numerous unpolluted North American streams a particularly high ratio of oligochaetes (naidids, enchytraeids, juvenile tubificids) and cyclopoids beside nematodes and harpacticoids. Most of this fauna is detritivorous, while predators such as mites are rather rare.

Particularly the river bed and less so the shore is densely populated by meiofauna, often with abundances of 20,000 to 30,000 ind. m^{-2}. SCHWOERBEL (1967) even reported 60 000 to 100 000 ind. m^{-2}. These figures underline the relevance of the hyporheic community for the biological production and self-purification of the river. Comparing the fauna of a river bed with the plankton of a nearby mountain lake, PENNAK and WARD (1986) found substantially more interstitial crustaceans in the river. Since meiobenthic grazing on bacteria has been found to stimulate microbial growth (see Chap. 9.3.1), it indirectly contributes to the regenerative potential of organically enriched waste water, a process of considerable relevance for our supply with drinking water.

In general, the upper course of a river with its bed of pebbles and gravel, its huge internal surface area and high structural complexity harbours a more diverse hyporheic meiofauna than the sandy to muddy lower reaches. Here, with increasing organic load, the riverine meiobenthos are predominated by juvenile oligochaetes and chironomids, while the soft sediments of lowland streams, often with occasional formation of hydrogen sulphide, are mainly populated by nematodes and specialized rotifers. The deeper the sampling extends into the underlying phreatic habitat and the groundwater aquifer, the more homogenous becomes the fauna with little or no altitudinal differentiation. In contrast to the epigean fauna, it is not the elevation, but rather the site-specific physiography which structures the stygofauna (see Chap. 8.2.2, below) of rivers. Comparing hydrologically different arms in the bed of the river Rhône (France), a good relation was found between disturbance and meiofauna structure: the more undisturbed and static the riverine habitats, the higher the portion of groundwater species in the meiobenthic community (increasing from 8 to 47%, see WARD and PALMER 1993).

In gravel bars of rivers, areas with prevailing influx of upwelling groundwater into the interstitial system seem to cause an increase of stygofauna related to the epigean fauna. Contrastingly, in areas of sediment aggradation, the epigean fauna tended to enter and accumulate in the sediment with the infiltration of surface water. At the upstream end of gravel bars, epigean fauna clearly dominates, whereas the contribution of groundwater fauna progressively increases along the length of the bar. Hydrology apparently structures the composition of riverine meiofauna also temporally: under low water conditions (often during the summer months), the epigean fauna is confined to a thin surface horizon while the stygobionts dominate throughout all the deeper layers. The situation reverses at times of high water discharges. This pattern of hydrological and geomorphological basic determinants may become modified by spatial discontinuities in sediment water flow, but also by the local biotic factors (DOLE-OLIVIER and MARMONIER 1992; WARD and PALMER 1993).

Although surface films of bacteria, diatoms and other organisms stabilize the sediment against erosive forces, disturbance of the hyporheic community by sudden floods with subsequent heavy erosion and faunal suspension is a common phenomenon not only in montane rivers. This faunal reduction is usually compensated in a few weeks. This quick recovery is possible through passive transport from regions higher up the river, but originates probably also from the refuge areas represented by the debris accumulating in lentic niches along the shores. A seasonality in hyporheic abundance patterns is well developed with minima in the spring following the spring highwaters. In the summer, after a phase of low stream velocities, the fauna of deeper horizons often suffers from stagnant pore water conditions which favour development of hydrogen sulphide. In cold winters, the surface layers become easily exposed to frost with the concomitant noxious impact on the inhabitants. In the cold season, nematodes, copepods and oligochaetes were found deeper down in the sediment (PALMER 1990a). Also hydracarid mites survived frost by evading deeper down into the hyporheic interstitial (SCHWOERBEL 1967).

In his pioneering experimental work on the hyporheos of a mountain stream, SCHWOERBEL (1967) found the highest abundance of meiofauna close to the water's edge in sand of 0.5 to 1 mm grain size. Among experimental tubes, nematodes preferred those filled with the finest sand (0.25 mm), harpacticoids were most frequent in gravel of 4–6 mm, and the third group of importance, chironomid larvae, dominated in sand of 1–4 mm grain size.

Our knowledge of hyporheic (and phreatic) meiofauna, particularly of their quantitative composition, is limited. The reason for this is often inherent technical and methodological problems. Only in the sandy bottoms is quantitative work with regular station profiles possible. In the pebbles and gravel of most river beds, sample holes can be driven only with massive corers and the entering pore water is pumped up for faunal analysis.

The typical method for hyporheic studies is pumping with the "Bou-Rouch pump" (Bou 1974). A small pump is mounted on a perforated metal tube which has been hammered into the river bed and remains in position for repeated sampling of interstitial water. This pore water and its inhabitants can enter the tube through the perforating holes. It is arguable whether this method will yield reliable reproducibility of consistent area-related quantification. Most of the methods based on pumping do not allow exact reference to a distinct depth or area. Here, the corers designed by DANIELOPOL and NIEDERREITER (1987) and

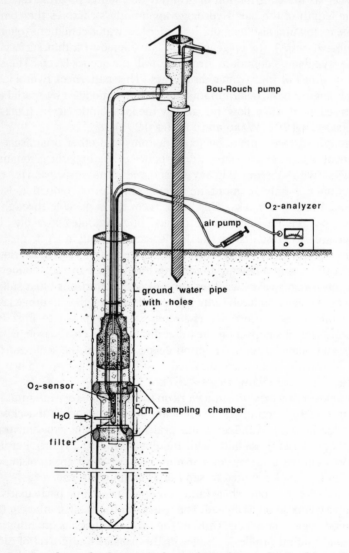

Fig. 81. A coring device for fractionated sampling of groundwater fauna. For details see text. (After DANIELOPOL and NIEDERREITER 1987)

TABACCHI (1990) provide better results. They allow for fractionated vertical sub-sampling of interstitial water and fauna by subdivision of the coring tube in small chambers which are evaluated separately (Fig. 81).

Quantitative results are obtained with a more sophisticated, but expensive method. A sediment core obtained in a metal tube is in situ shock-frozen by liquid nitrogen and then retrieved undisturbed after previous anaesthetization of the fauna by electro-positioning (BRETSCHKO and KLEMENS 1986).

The limitations of quantitative sampling indicated above render calculations of population density per area or volume problematical. Hence, for hyporheic fauna, estimates of production rates are extremely rare. As indicated by KOWARC (1990), the meiofauna production in mountain streams is rather low with a P/B-ratio of 3–6, probably caused by the low temperatures and general oligo-trophic conditions.

Compared to the bottom of the stream, the (deeper) hyporheic biotope can be considered a refuge area for fauna avoiding currents and drift. It attains considerable importance as a sheltered recruitment zone for insect larvae. These hyporheic biotopes have a transitional position between the exposed river bed and the stygobiotic groundwater sediments.

8.2.2   The Groundwater System

Groundwater represents a freshwater reservoir of eminent importance. About 40% of all freshwater on earth is stored in the continental groundwater system (DANIELOPOL 1989). The homogeneity of its sediments, constancy of its physio-graphic conditions and the protective capacity through long geological periods are the main biotopical features of the groundwater system. Its abiotic milieu is characterized by constantly low temperatures, a somewhat lowered pH, a slightly undersaturated oxygen content, a high amount of free CO_2, and very oligotrophic conditions (drinking water, wells).

The fauna: the stygobiotic and/or deep phreatic fauna is specifically adapted to this factorial combination: low metabolic activity, slow locomotion and growth, long generation and life times, hardly any diurnal rhythms. Typical stygobiotic species are mostly cold-stenothermal forms with low productivity (few, large eggs), i.e. they are specialized K-strategists. Nevertheless, they are often widely distributed in the continental groundwater net and are consistent enough to form a characteristic community (*Bathynella-Parastenocaris* com-munity), often related to the fauna of caves and springs. Good examples for this distributional relation are the freshwater polychaete *Troglochaetus beranecki*, and the amphipod genus *Bogidiella*. Other typical representatives of the groundwater fauna are the harpacticoids *Chappuisius* spp., several isopod genera, mostly from karstic sediments (*Microcharon, Microparasellus*), haplotaxid and rhyacodrilid oligochaetes, the amphipods *Microniphargus, Salentinella* and *Stygobromus* (the latter so far found only in North America), and the crustacean order Thermosbaenacea (Pancarida).

There are several hypotheses about the origin of the present-day's stygofauna and the pathways of groundwater colonization. According to the "regression model" (STOCK 1980), many marine animals, living originally in the interstitial of the sea shores, invaded during a regression period brackish and freshwater habitats and finally adapted to conditions in the continental groundwater. Other driving forces of a "stygobization" might be large-scale desiccation of the epigean soil layers. This model is related to the contention that the numerous endemic and archaic animal groups of the groundwater passively persisted in the static refuges of an environment which conserved many life conditions of former periods ("relict refuge model"; BOTOSANEANU and HOLSINGER 1991). Contrastingly, DANIELOPOL and ROUCH (1991) incline to the view that the groundwater fauna is generally represented by animals which keep actively invading the stygobiotic freshwater milieu from other, often marine biotopes. BOUTIN and COINEAU (1991) point out that the colonization process is a combination of both suggestions. Their "two-phase model" is characterized by several successive steps: at first active dispersion and vertical transition from surface waters to the interstitial, later passive persistence combined with gradual adaptations which further separates the interstitial fauna of the continental groundwater from that of the marine habitats.

Meiofauna from the sea shores, apparently preadapted by their frequent exposure to brackishwater conditions (see Chap. 8.1), have been particularly successful in slowly adjusting to freshwater conditions, provided they were exempted in their new localities from the stress in surficial habitats. Their small size adapted the interstitial fauna to become competent colonizers of the ground-water system and caves. Interestingly, many interstitial troglobites from anchiha-line and freshwater cave systems are typically highly adapted to low oxic conditions (STOCK 1986) which would favour colonization of the groundwater. Originally of a wider occurrence, the animals became increasingly isolated through geological sea or land level changes (transgressions, regressions, emergence, isostasis). One possible evolutionary consequence would be a subsequent intense radiation/speciation within small "founder" populations during longer periods of times. The often isolated location of the "evolutionary ports into the groundwater system" many have contributed to the high allopatric speciation in the stygobios. This would explain both the rich number of zoological peculiarities and also the high diversity combined with low abundance of stygofauna. Another possible consequence would be slow speciation in the stable refuges with little competition. This would favour structural stasis and lead to the often relictary character of stygobiotic species. Morphologically primitive, often neotenic crustacean groups like Bathynellacea and Pancarida (see Fig. 60) are good examples. As a result, many stygobiotic animals from the continental groundwater are considered relicts of a primitive fauna mirroring their old marine "plesiotope"

In North America, the glacial edge of the last ice age seems to represent an important distributional borderline for the hypogean fauna. A typical ground-water fauna, speciose and rich in K-selected endemists, e.g. Bathynellacea, was found only in the formerly unglaciated areas. Both the upland plains, influenced by the ice cover, and the coastal plains under the impact of the marine regime have been found poorer in groundwater fauna (STRAYER, pers. comm.).

Probably all the different and debated "models" presented above have been and still are effective in an interactive way, creating the confusing mosaic of archaic and derived forms typical for the stygobios. In any case, most stygobiotic animals are doubtless of marine origin, and the most frequently used pathway for colonization of the stygal is certainly the seashore with its river mouths, brackish lagoons and marine caves. Once adapted to the physiological constraints of freshwater, subterranean species seem rather confined to the continents, not able to bridge the oceans (SCHWOERBEL 1967). Yet, the surprisingly frequent circum-mundane distribution of stygobiotic species cannot be conclusively explained. A realistic distributional picture of these character forms is only in its initial steps, not to speak of quantitative estimates. It can be expected that many more zoological "preciosa" will be discovered. This contributes to the zoological and zoogeographical peculiarity and relevance of this biotope (DANIELOPOL 1990b; BOTOSANEANU and HOLSINGER 1991).

8.2.3 Stagnant Waters, Lakes

The meiofauna of lake sediments is much more easily accessible for quantitative sampling and thus more thoroughly investigated than that of other freshwater biotopes. However, even from lakes there are not many quantitative data on meiofauna abundance, turnover and production available, although their high ecological potential becomes increasingly evident (see below). There is a need for comprehensive, interdisciplinary projects in lacustrine (meio-) benthology such as the recently started *"Cytherissa* project" in the Austrian Mondsee (DANIELOPOL 1990a), focussing on a dominant ostracode genus and its environment.

The permanently submersed bottoms of lakes and ponds (the "hydropsammal", WISZNIEWSKI 1934) mostly consist of fine sand rich in organic particles and silt, often covered with plants. The concentration of dissolved inorganic and organic substances in this sediment is often 40–50% higher than in the overlying lake water. During aestival warm-water conditions, occurrence of oxygen deficiency and formation of hydrogen sulphide in the subsurface layers is frequent. Here, mostly burrowing macrofauna and epifauna prevails, the meiofauna is restricted to the uppermost centimetres. Nematodes usually dominate, followed by copepods, ostracodes, juvenile oligochaetes and chironomid larvae. Tardigrades occur irregularly, but can locally reach high abundance (NEEL 1948; HOLOPAINEN and PAASIVIRTA 1977). All other groups are considered insignificant.

In the damp shore sediments above the water level ("hygropsammal", WISZNIEWSKI 1934), the more exposed sites with medium sand (Md > 250 μm) usually comprise a belt of 1 to 3 m width, depending on exposure and slope of the shore. This zone is well supplied with oxygen and rich in detrital food due to the debris washed ashore. PENNAK (1940) found in this zone a relatively rich psammofauna with rotifers sometimes dominating in extreme densities (>10000 ind. $10\,cm^{-3}$). Next to them, harpacticoids with just a few species (e.g. *Parastenocaris*; *Phyllognathopus*), but in fairly high numbers, populated the hygropsammal. In addition, nematodes, oligochaetes (including aeolosomatid annelids) and tardigrades belonged to the ecologically dominant groups of lacustrine shores; all others have been stated as less important.

However, literature data show much variation: STRAYER (1985) reported from the shore sediments of Lake Mirror (USA) that 70% of all meiobenthic animals were nematodes, accompanied by turbellarians (often found only occasionally), gastrotrichs, cladocerans, copepods, just a few rotifers and tardigrades and hardly any harpacticoids. Beside the usual divergences in methodology, these variations of meiobenthic population sizes seem to depend on the geographical location and climatic situation of the water body studied. Marked seasonal fluctuations often determine the population dynamics and render quantitative comparisons from climatically different lakes and seasonally varying sampling periods problematical. Moreover, in many freshwater studies, the nematodes seem neglected or inappropriately retrieved and probably grossly underestimated, reports of only $10-50\,cm^{-3}$ are certainly biased. Also, the rich ciliate fauna is usually not taken into account. PENNAK (1939) counted about 10000 protozoans compared to only about 500 metazoans per $10\,cm^2$!

One of the first studies which reliably quantified the whole spectrum of lacustrine meiofauna and related it to macrofauna has been the work of STRAYER (1985). He described that the average square metre of lake bottom was inhabited by 1.2 million meiobenthic individuals, 60 times the number and 1/3 of the biomass of the corresponding macrobenthos. Nutritionally, most of this fauna was based on diatoms, even in food selective species like in rotifers. The large majority (80%) of the meiobenthos was grazed down by the predaceous larvae of Tanypodida (Diptera) and other insect larvae.

The ecological importance of the limnopsammon becomes even more underlined when the threefold higher turn-over rate of meiofauna versus macrofauna is considered. This can result in a double to fourfold meiobenthic production. Also indirectly can the limnetic meiobenthos favour production by the general activation of bacterial growth due to grazing and bioturbation (see Chap. 9.3.1).

STRAYER (1986) analyzed, on the basis of his comprehensive data set from Lake Mirror, the limnetic benthic size spectra comparing them with corresponding figures by SCHWINGHAMER (1981a) for the marine benthos. The limnobenthos cannot be grouped into natural units of micro-, meio- and macrobenthos, separated from each other by clearly indicated "troughs" which make the curve of the size spectrum tripartite. Explanations for this remarkable difference between marine and freshwater benthos will be discussed later (Chap. 9.1.1)

More detailed reading: hyporheic fauna – SCHWOERBEL (1961a); PENNAK and WARD (1986); PALMER (1990b, 1992); WARD and PALMER (1992); BRETSCHKO and KLEMENS (1986) (methods); groundwater meiobenthos – DELAMARE

DEBOUTTEVILLE (1960) (monograph); evolutionary aspects – DANIELOPOL (1990b); BOTOSANEANU and HOLSINGER (1991); lake meiobenthos – WISZNIEWSKI (1934); PENNAK (1940); STRAYER (1985)

8.3 Meiofauna in the Deep-Sea

The deep-sea is the largest, but probably the biologically least explored biotope. New results from numerous expeditions, often obtained with sophisticated instruments (e.g. the "multiple Aberdeen corer", BARNETT et al. 1984) and application of newly developed methods (e.g. analysis of chloroplastic pigments, of adenosine nucleotide content, and electron transport system activity) have changed our view on deep-sea meiofauna considerably. Compared to the shallow benthic zones the abyssal bottoms are rather static and monotonous, but local areas of sandy sediments with water currents of $5-10 \, \mathrm{cm \, s^{-1}}$, interspersed into the widely prevailing mud, cause a certain structural heterogeneity. THISTLE (1988) even reported "benthic storms" strong enough to suspend the deep-sea sediment, and thus negatively affect meiofauna populations. In less exposed areas, a fluffy layer of sedimented phytodetritus with an extremely high water content can form a surface floc of several centimetres depth. While the temperature of the deep bottoms is widely $1-2 \, ^\circ \mathrm{C}$, it is considerably elevated in the Red Sea ($21 \, ^\circ \mathrm{C}$), in the Mediterranean (about $10-12 \, ^\circ \mathrm{C}$), and in areas of hydrothermal activity. Today, there is evidence of a far more direct interaction between the surface and the ocean bottoms than earlier conceived (see below).

Normally, abundances of 100 to 1000 meiobenthic organisms per $10 \, \mathrm{cm^2}$ (without foraminiferans) are quite typical for the deep-sea (Table 14). In deep-sea bottoms of polar regions, they seem to be clearly higher responding to the rich supply with phytodetritus from the surface (PFANNKUCHE 1992). In hadal depths they can decrease to $10-100$ per $\mathrm{cm^2}$. Compared to surface values, these figures underline the often limiting role of the organic particle flux (measured in

Table 14. Meiofauna composition (%) and abundance in samples from increasing water depth. (COULL et al. 1977)

Taxon	400 m	800 m	4000 m
Foraminifera	30.8	33.1	65.2
Nematoda	45.1	59.7	30.2
Harpacticoida	10.7	2.4	2.0
indetermined	5.1	1.5	1.1
Polychaeta	2.8	1.6	0.5
Organisms per $10 \, \mathrm{cm^2}$ (average values)	442	892	74

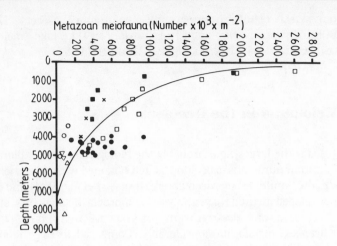

Fig. 82. The decrease in meiofauna abundance with ocean depth. (TIETJEN 1992)

sediment traps or in the bottom sediments as organic carbon or total nitrogen) for the existence of a deep-sea meiobenthos (TIETJEN 1989).

At least for the better studied Atlantic Ocean, there is a clear relation between abundance and depth (Fig. 82). Meiofauna biomass from the abyssal plains is often only 50–100 mg fresh wt m^{-2} (corresponding to approximately 4–8 mg C m^{-2}). These low values underline the oligotrophic character of wide abyssal regions, although in the areas with a rich import of organic matter values around 1 g wet wt m^{-2} and more have been recorded. The few data from the Pacific and Indian Ocean are somewhat inconsistent.

To date, biochemical parameters related to biomass render often more reliable results than direct counting and weighing. Determination of proteins, carbohydrates and adenosine nucleotides are frequently used in modern deep-sea research and their correlation with animal numbers is well established (PFANNKUCHE 1992).

The restrictive trophic impact probably also accounts for the distribution of deep-sea meiofauna in a vertical bottom profile: about 90% of all meiofauna are concentrated in the upper 2 cm where the detritus accumulates. It emerges that in the surface horizons of the deep-sea sandy bottoms nutrient supply directly determines the abundance and distribution of the meiofauna. In the deeper layers, particularly of finer sediments, oxygenation tends to become an additional key factor for meiofauna density. In areas with a high input of food particles (seasonally from the plankton blooms in boreal latitudes; permanently at the foot of the continental slope or in sites where currents accumulate debris) population densities beyond 1000 ind. 10 cm^{-2} have been recorded. Also in upwelling areas and along the ridges of continental plates, a richer nutrient supply gives rise to a generally higher meiofauna density (ALONGI 1990b).

Consumption rates equivalent to $7.9 \, mg \, C \, m^{-2} d^{-1}$ have been measured from the Mediterranean deep sea (GUIDI, pers. comm.).

Patchiness of phytodetritus accumulations and sediment disturbance seem to correspond with a non-random meiofauna distribution in the deep-sea. A higher diversity was recorded in the northern latitudes where the amount of phytodetritus is rich. In areas of little sedimentation of both phytodetritus and terrigenous matter, the meiofauna is more evenly distributed. Increasing sediment heterogeneity due to the seasonal input of phytodetritus also alters the species composition of deep-sea meiobenthos (LAMBSHEAD and GOODAY 1990).

The typical deep-sea meiobenthos organism is highly adapted in its biology and ecology to scarcity of food. Favoured by the prevailing low temperatures, a slow growth and a long life span result in a low maintenance expenditure. Also the predominant mode of nutrition, a rather passive suspension and deposit feeding, is energetically more favourable than a more active picking of selected food particles. The metabolically costly reproduction must also be subject to the imperative of energy conservation and results in low egg numbers often combined with brooding. A high capacity of asexual multiplication, as frequently realized in protozoans, is another character interpreted as energetically conservative.

What is the composition of typical deep-sea meiobenthos? Only a few years ago it became known that Foraminifera (Protozoa) play the dominant role among the abyssal meiobenthos (Fig. 83). Usually 50% of all individuals (maximally 90%) and about 30% of their biomass are made up of foraminiferans (SHIRAYAMA and HORIKOSHI 1989). Hence, this group alone matches in abundance all the remaining metazoan meiofauna. Often there occur 100–1000 individual foraminifera per $10 \, cm^2$ (maximally 2000 ind. $10 \, cm^{-2}$) or 150–200 ind. per ml sediment, and species number can attain $40 \, cm^{-2}$. Beside the foraminiferans, other rhizopods such as Amoebina and the large Xenophyophoria have also been found to richly populate the deep-sea bottoms (LEVIN 1991). It seems that there is no square centimetre of deep-sea bottom which is not interwoven with rhizopod pseudopodia. Although some of them exceed meiobenthic size, they have a considerable impact on the abyssal meiofauna, since they have a structuring and predatory effect on the metazoan populations.

Among the metazoans, nematodes (often Desmoscolecidae) prevail and, in lower numbers, harpacticoids and polychaetes, often followed by juvenile bivalves. But there are variations: in studies from the northern Atlantic, polychaetes ranked second after nematodes. Even in the hadal troughs beyond 10000 m depth, meiofauna does exist (e.g. harpacticoids). Interestingly enough, even representatives of oligochaetes, a group thought to be of limnogenic/terrigenic descent and lacking propagatory stages, have been found in > 7000 m depth.

The nutrition of the deep-sea meiobenthos is mostly derived from the surface production:

– Marine detrital particles (decaying plankton) sink as phytodetritus down to the deep-sea bottom: In the deep-sea, phytodetritus can still make up to 1–3%,

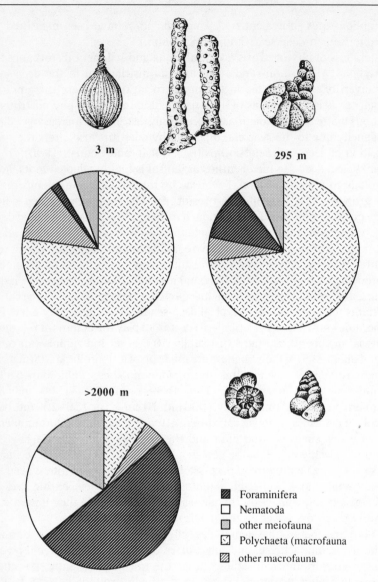

Fig. 83. The relative increase in Foraminifera with ocean depth among the benthos. (After SHIRAYAMA and HORIKOSHI 1989)

in other estimates 5–10% of the surficial primary production. This rich, but seasonally varying input of organic matter has been measured as concentration of chloroplastic pigments in the sediment (THIEL et al. 1988/89; GOODAY and TURLEY 1990; LAMBSHEAD and GOODAY 1990).
– The bacterial film colonizing mucus strands, and excretions of the ubiquitous benthic protozoans (see above) have an important trophic role.

– Terrigenous detritus often accumulates off the mouth of big rivers and at the foot of the continental slopes.
– Remains of dead large animals and plants represent large food packages.

On a geographic scale, the horizontal distribution pattern of deep-sea meiobenthos is rather monotonous. The ubiquitous muds harbour an association of deposit feeders of high generic conformity, but where sediment structure changes and sandy areas prevail, suspension feeders with mucus filtration become more frequent. Geological heterogeneity of the bottom may contribute to variations and increased diversity in the colonization pattern of meiobenthos (ALONGI 1990b). In particular this seems valid for the meiobenthos of hydrothermal vent areas. Although, so far, the meiofauna in these "oases of life" in the deep-sea have been little investigated, some studies indicate an enhanced abundance of meiobenthos benefiting more or less directly from the enormous biomass and production of sulphur bacteria (VAN HARTEN 1992). On the other hand, the species spectrum wil be restricted to those few forms which can tolerate the hostile sulphidic conditions (DINET et al. 1988).

Also biogenic structures like tubes, rhizopod tests and shells, which persist for long periods under the stable deep-sea conditions, have been shown to enhance meiofauna diversity. With increased sediment complexity, epistrate feeders are attracted and contribute to local variations. Around mud balls of the polychaete *Tharyx* in 1000 m depth, meiofauna had an increased density. The ameliorating effect can be attributed to a complex interaction of favourable factors such as enhanced growth of bacterial stocks, increased vertical transport of solutes, better protection from predators and establishment of hydrodynamically favourable sheltered zones with accumulation of debris (THISTLE and ECKMANN 1990; ECKMANN and THISTLE 1991).

Beside local and seasonal fluctuations in sediment structure, it cannot be ruled out, however, that meiofauna variations are also attributable to inadequacies in sampling methods and differing gear. The meiofauna content even between the cores of one multiple corer deployed in the deep-sea bottom can vary considerably.

The impoverished food supply in the deep sea contributes to an overall decline in body size of the benthos. Thus, the faunal size spectrum and abundance in the deep sea favours the meiobenthos and can result in a biomass relation between macro- and meiobenthos of 1 : 1 (THIEL 1972a, 1975; TIETJEN 1992).

Considering productivity rates, the relevance of deep-sea meiobenthos becomes even more impressive: 80% of all metabolic processes refer to meiobenthic organisms (SHIRAYAMA and HORIKOSHI 1989). Within meiobenthos, per definitionem a small-sized group, the nematodes were found to further decrease in body size with increasing depth, parallel to the diminishing chlorophyll content of the bottom (SOETAERT and HEIP 1989). The same seems to hold also for harpacticoids. In a deep-sea area of the South Pacific, more than 50% of the specimens were less than 200 µm in length (SCHRIEVER, pers. comm.).

In general, productivity rates in deep-sea bottoms are two to three orders of magnitude lower than in shallow water sediments. The specific role of the meiobenthos for the biological productivity of the deep-sea is still rather difficult to assess, since the problems of measuring production (see Chap. 9.2) are aggravated in the deep-sea. Promising methods are:

- Recording of community respiration in a limited sediment area under a bell-jar (PFANNKUCHE and LOCHTE 1990; PFANNKUCHE 1993).
- Quantitative analysis of the most important nutrient input, the chloroplastic pigments. This parameter, depending directly on the production of phytoplankton in the euphotic zone, evidences the close coupling of the meiobenthos in the deep-sea and particularly the continental slopes with the planktonic production at the ocean's surface (THIEL et al. 1988/89; Fig. 84). Measurement of chloroplastic pigments (see GREISER and FAUBEL 1988) thus contributes significantly to a reliable assessment of meiofauna production in the deep-sea.
- Measurement of particulate ATP content as well as the activity of the electron-transport system reflect the (meio-)benthic metabolic activity.

Altogether, the meiobenthos of the deep-sea is strongly dominated by biotic factors such as food competition and predator-prey relations, and less so by physiographic constraints. It is a community of K-strategists with a high diversity, but rather low abundance limited by food scarcity.

Why has the composition and abundance of the deep sea meiofauna with its dominance of protozoans not been recognized earlier?

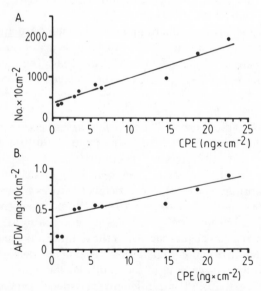

Fig. 84A, B. The correlation between phytodetritus (chloroplastic pigments) and meiofauna abundance (**A**) or biomass (**B**) in the deep-sea. (PFANNKUCHE 1985)

– In general the deep-sea bottom is still widely unexplored.

– Benthic foraminiferans in their irregular shells, sometimes even hidden in sedimented *Globigerina* tests and covered by agglutinated detritus, have often been overlooked.

– The bow wave of the large deep-sea corers and grabs, often stirs up the top mm-layer consisting of flocculent detritus particularly rich in meiobenthos.

In this context, the possible role of meiobenthos for the formation of manganese nodules in the Pacific Ocean should be briefly mentioned. Although the chemical processes involved in the massive accretion of valuable heavy metals are not yet understood, each nodule is densely covered and in its internal interstices colonized by a diverse meiobenthic community of deep-sea protozoans and also metazoans. It may be of relevance that Foraminifera have been found to selectively excrete manganese, iron and other metals as xenobiotic particles (xanthosomes). Perhaps these excretions serve as the initial granules for the formation of new nodules, since in all of them a mineral centre has been found (RIEMANN 1985). Thus, it is conceivable that meiobenthic organisms in one or another way influence the growth of these structures of high economical potential (SHIRAYAMA and SWINBANKS 1986). Their expected large-scale exploitation by deep-sea mining would certainly massively threaten the slow-growing deep-sea meiobenthos.

> *More detailed reading*: phytodetritus – THIEL et al. (1988/89); manganese nodules – MULLINEAUX (1987); Foraminifera – GOODAY et al. (1992); abundance and biomass of meiofauna – TIETJEN (1992); monograph on deep-sea – GAGE and TYLER (1991)

8.4 Meiofauna in Phytal Habitats

REMANE (1933, 1940) characterized the phytal as a biotope populated by an abundant and diverse fauna. Algal belts on rocky substrates offer a range of complex habitats richly exploited by meiofauna (CRISP and MWAISEJE 1989). Even less structured sea grass beds harbour twice as many meiofauna species as the adjacent sediments (HICKS 1986). A phytal meiofauna of 10^6 individuals/ m^2 of algal cover is not uncommon, and may correspond to 10% of the macrofauna biomass in these biotopes.

The phytal occupies a primary role on hard bottoms, since it is the predominant suitable substrate for vagile benthic animals. For the meiobenthos, a phytal belt with finely ramifying substructures offers a rich microhabitat in a complex hydrodynamic spectrum and a great range of protection for its inhabitants. Habitat complexity, as traditionally defined, i.e. the "relation of surface area to volume per plant weight unit" is only partially applicable, since it is not the absolute area which is relevant for meiofauna colonization but the realizable area depending on the exposure, protective effect, cover with sediment etc. These are the significant variables determining both the structural complexity of algae and the range of colonization with meiofauna and explaining why smaller, filamentous plants usually harbour a richer meiofauna than larger ones (GIBBONS 1991). As for seagrass, the density of plants per area as well as the shape of the vegetated area have been found to influence the meiofauna density.

Along boreal shores, there are three general phytal biotopes distinguished by their decreasing structural complexity: the *Laminaria-Delesseria* zone, the *Fucus* zone and the *Zostera* zone (REMANE 1940). The corresponding algae in warm-water areas (HALL and BELL 1988), *Cladophora, Corallina* and *Delesseria,* harboured a much richer "meiophyton" than *Laminaria, Fucus* and sea grass blades. Also the complex pelagic *Sargassum* rafts are regularly inhabited by a rich meiobenthic fauna.

Macroalgae (e.g. kelp) provide three subhabitats for meiofauna (HICKS 1985): the surfaces of the thallus fronds, the sediment and detritus deposits accumulating at the bases of the fronds and the interstices of holdfasts (rhizoids). The phytal harpacticoid *Porcellidium* has been found to migrate between these subhabitats along a vertical plant gradient to avoid tidal stress (GIBBONS 1991). In Mediterranean *Posidonia* beds a structurally complex and meiofaunally rich "stem stratum" has been distinguished from the more monotonous "upper leaf stratum" and "lower leaf stratum" (NOVAK 1989). A biotope typical for the American tidal flats are the extensive *Spartina* salt marshes. Both the root system, the culms and leaves of the plants are the basis for a richly structured, well-protected habitat. With its abundant supply of organic matter, *Spartina* marshes harbour rich and genuinely adapted meiofauna populations of ecological importance (RUTLEDGE and FLEEGER, in press). A peculiar subhabitat whose general relevance for meiofauna remains to be clarified is the rich system of "gas passages" (aerenchyme) in the stem of *Spartina alterniflora* (HEALY and WALTERS 1993). They are regularly inhabited by an amazingly rich and diverse meiofauna dominated by mostly specialized oligochaetes. Together with the meiofauna found under the leaf sheaths, inclusion of these biotopes increased the overall abundance of salt marsh meiofauna by an order of magnitude!

Despite divergences in the climatic regime and species composition, algae of comparable complexity and structural differentiation, but from geographically disjunct zones, are colonized by meiofauna of "taxo-ecological similarly".

Representatives of these meiobenthic "isocommunities" (HICKS 1985) are enoplid nematodes which dominate both North Sea and Chilean phytal habitats. Many oncholaimid nematodes are also characteristic inhabitants of the phytal, epsilonematids populate the stems of sea grasses. A similar parallelism can be found in families of harpacticoids, ostracodes and halacarid mites. Characteristic harpacticoids in algae are *Parathalestris, Scutellidium* and *Harpacticus,* typical ostracodes are *Xestoleberis* spp., *Paradoxostoma* sp. and *Loxoconcha* sp. and among the halacarids the Rhombognathidae. Another common and widely distributed component of phytal meiofauna are the small species of tanaidaceans (Crustacea, Peracarida).

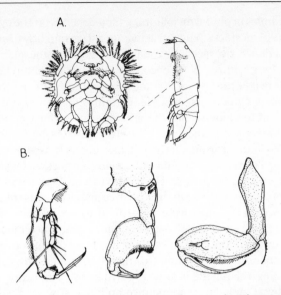

Fig. 85A, B. Structural adaptations of harpacticoids to a phytal environment. **A** The flattened harpacticoid *Porcellidium* in lateral view, and its "sucker disk" (mouth parts) in ventral view. **B** Some grasping legs from various phytal harpacticoids. (Combined from TIEMANN 1975; HICKS 1985)

What are the adaptive features that members of the phytal meiofauna have in common? The rather elevated position of some plants on the shoreline, frequently exposed to waves, requires a highly developed attachment capability of the animals (Fig. 85): flattening of the body, development of "haptic" organs such as sucker-like structures, adhesion by mucus secretion, prehensile legs and hairy spines are of high adaptive value (e.g. *Porcellidium*, *Ectinosoma* and *Thalestris* in harpacticoids, Paradoxostomatidae, Xestoleberidae, Bairdiidae in ostracodes, Rhombognathidae in Halacaroidea).

Only a few members of the phytal meiofauna utilize their substratum, the plants, directly by piercing the cell and sucking their cytoplasm. Some tardigrades, halacarids (Rhombognathidae), siphonostomous Cyclopoida and ostracodes (some Xestoleberidae, Paradoxostomatidae) have adopted this specialized mode of living. Other more sessile members of the phytal meiofauna regularly utilize the plants as prominent substrates for filter feeding (e.g. rotifers, cladocerans). Rotifers are particularly frequent in the phytal fringe of brackish waters. The nauplii of some harpacticoids (Thalestridae) use their phytal substratum in a peculiar way by eating the medullary tissues and producing algal galls. Many phytal nematodes, harpacticoids and ostracodes grasp and crack diatoms with specialized mouth parts.

Because of close structural and trophic ties of phytal meiofauna to the plants and their surface films, there exists usually a marked seasonal variation in population dynamics of phytal meiofauna which depends on the cycles of growth and decay of the algal stocks. The attractiveness of certain algae for meiobenthos seems to be mediated by specific exudates secreted by the plants, and fauna responding to these will rarely be encountered outside the phytal habitat.

For phytal meiofauna, by far the most important nutritive response is the surface film formed by detritus, bacteria, diatoms and protozoans. This organic film is the basis of the food chain, the grazing ground for the meiofauna. The decaying frond ends are particularly densely populated by bacteria which, in turn, attract meiofauna grazing on them. Experiments have also demonstrated a close relationship between meiofauna settlement (harpacticoids) and the detritus layer deposited on the blades of seagrass (MEYER and BELL 1989). Mobilization and reduction of this layer through the nutritive activities of meiofauna enhances the plants' growth. Fine, divaricate algae tend to accumulate more sediment and detritus than less complex, larger ones. Since coarse sediments, trapped on algae, seem to increase the overall structural complexity and trophic potential, they favour meiofauna colonization. On the other hand, rich accumulations of fine deposits reduce the habitable area and structural complexity with a parallel reduction in phytal meiofauna abundance and diversity.

Under certain conditions, a good part of the meiofauna is less tightly linked to the phytal proper, but shows close relations to the surrounding sedimentary fauna in the environment. The meiofauna in the basal "pockets" of the fronds filled with sediment and detritus, and of the algal rhizoids, are frequently derived from local soft bottom sediments. Some are even derived from the organic layer covering the rocky substrate and are found similarly associated with the detrital film on the surfaces of the fronds. In colonization experiments using artificial seagrass blades, the settling meiofauna originated both from the ambient natural phytal and from the sediment (BELL and HICKS 1991). Furthermore, a number of phytal meiofauna species are known to periodically leave their substrate for (nightly) excursions into the overlying water column and can subsequently be caught in suspension traps (KURDZIEL and BELL 1992). Thus, the phytal has an important function as a biotope coupling benthic and pelagic fauna. On rocky shores, the meiofauna of the phytal zone represents an additional food source usually only available in soft bottom areas: it is particularly accessible to small predatory fish. In both systems meiofauna are therefore of potential importance as a link to higher trophic levels (see Chap. 9.3).

Studying phytal meiofauna poses some problems of quantitative sampling. The plants have to be carefully placed in sampling bags underwater so that any loss of fauna is avoided. Using an open cylindrical jar with a thick and softly attaching rim of flexible silicone sealant has been proven to be another simple and useful tool (GIBBONS and GRIFFITHS 1988). After injection of some formalin in order to release the attached meiofauna, the water volume of the sample is syphoned off. For better reproducibility, this procedure should be repeated several times.

More detailed reading: *Posidonia* meadows – NOVAK (1989); reviews – HICKS (1985); GIBBONS (1991); monograph – REMANE (1940)

8.5 Meiofauna in Anoxic Environments – the Thiobios Problem

8.5.1 Historical Aspects

In 1969, FENCHEL described a characteristic group of animals from oxygen-free, reducing sediments rich in H_2S, consisting mainly of ciliates, but also of some meiobenthic metazoans. The various species of the "living system of the sulphide biome" (FENCHEL and RIEDL 1970) seemed to depend in different ways "on the reducing conditions" of the habitat. Among the taxa present, the "most conservative" bilaterian phyla, Plathyhelminthes and Aschelminthes, predominated while crustaceans were absent. The above authors were not the first to find animals under low-oxygen and high-sulphide conditions believed to be intolerable for free-living fauna. Specialized animals had long been known within many groups to thrive in hypoxic or even completely anoxic conditions exposed to free sulphide. But it was FENCHEL and RIEDL (1970) who realized that there is a distinct, definable community of eukaryotic life specifically adapted to reducing conditions. OTT (1972) underlined that "the sulphide system has a homogeneous and stable nematode fauna of its own right". This sulphide community was termed by BOADEN and PLATT (1971) "thiobios". Suggestions "that the sulphide biome would contain at least some primary {faunal} elements ... of the oldest biosystem on earth" (FENCHEL and RIEDL 1970) were later vehemently objected to (REISE and AX 1979). The latter authors, without having performed measurements of the oxygen content, could not find "a specific meiofauna confined to oxygen-deficient horizons of the sediment".

Most papers on thiobiotic animals underline the relevance of the redox potential discontinuity for interpretation of the vertical distribution. Today, this threshold cannot be envisaged any longer as a horizontal border separating the oxic surface from anoxic depths. Numerous geochemical cycles and microbial processes together with the bioturbative impact of sediment fauna create a complicated three-dimensional "landscape of the redox potential" (OTT and NOVAK 1989) with oxic/anoxic areas often surrounding a verticalized redox threshold.

Today, we have a more detailed knowledge of the ecological and physiological diversity of the "sulphide fauna". The sulphidic ecosystem, dominated by the presence of reduced substances such as dissolved sulphide, methane and ammonium, is much too complex to allow for simple right or wrong answers to the early debate about the existence of a thiobios. The following section will explain this in some detail.

8.5.2 Composition of the Fauna

What is the composition of the typical meiofauna thiobios, which groups are regularly encountered in the sulphide biome?

Ciliata: there is a rich ciliate fauna found mainly around the RPD-layer (e.g. *Kentrophoros*), and many of these thiobiotic ciliates maintain a veritable "kitchen garden" of prokaryotic symbionts (FENCHEL and FINLAY 1989). Other ciliate species (*Metopus, Plagiopyla*) harbouring methanogenic bacteria as symbionts live as true anoxybionts deep down in the black layers (FENCHEL et al. 1977; FENCHEL and FINLAY 1991). They lack normal mitochondria but possess hydrogenosomes which perhaps originated from them. The hydrogen excreted by the ciliates is coupled to the reduction of CO_2 by the methanogens and the resulting methane is released.

Plathelminthes: among the turbellarians numerous representatives of the Acoela (e.g. *Solenofilomorpha, Oligofilomorpha, Parahaploposthia*), the Catenulida (Retronectidae) and of the Kalyptorhynchia are found in the deeper horizons around or underneath the oxic/sulphidic interface (see Fig. 86; STERRER and RIEGER 1974; BOADEN 1975; CREZÉE 1976; POWELL 1989). Their preference for hypoxia was observed rather early, and has now been experimentally proven by MEYERS et al. (1987, 1988). SCHERER (1985) found most thiobiotic turbellarians in the close vicinity of macrofauna burrows where they probably take advantage of an enriched food supply (Fig. 86).

Gnathostomulida: while in most meiobenthic groups the thiobiotic species are exceptional specialists, it seems that all the gnathostomulids prefer mild sulphidic and low oxic to anoxic biotopes (MÜLLER and AX 1971). They are regularly encountered along the tube walls of endobenthic burrowers, they were found in considerable diversity confined to the reduced fine sand underneath cyanobacterial mats or the roots of surf grass (WESTPHALEN, in press; FARRIS, unpubl.), they even dominated all other meiofauna in the permanently anoxic sediments underneath deep-sea brine seeps (POWELL and BRIGHT 1981; POWELL et al. 1983). This phylum of primitive Bilateria is thus a most typical representative of thiobiotic meiofauna.

Gastrotricha: the close relation of some gastrotrich genera to the reduced milieu of the thiobiota is documented by their scientific names: *Thiodasys, Turbanella thiophila, T. reducta* (BOADEN 1974, 1975). Frequently, these and other gastrotrichs can be found in the black layers of sand underneath the chemocline.

Nematoda: the most abundant animals in reduced sediments are various groups of nematodes. Most of them are unusually slender to threadlike forms (JENSEN 1987b) belonging to Siphonolaimidae and Linhomoeidae (Monhysterida), but also among the enoplid family Oncholaimidae (*Pontonema*), the chromadorid families Comesomatidae (*Sabatieria*), and Desmodoridae (subfamily Stilbonematinae: *Leptonemella, Eubostrichus*) are typical inhabitants of sulphidic habitats. Their counterparts in the limnic thiobios are members of the genus *Tobrilus* and some dorylaimids such as *Eudorylaimus andrassy*. Corresponding to their taxonomic divergence is the wealth of structural peculiarities of which only a few cases have been closer investigated: reduction or absence of the gut (*Astomonema*), accumulation of large globular

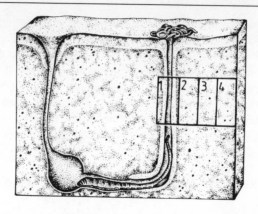

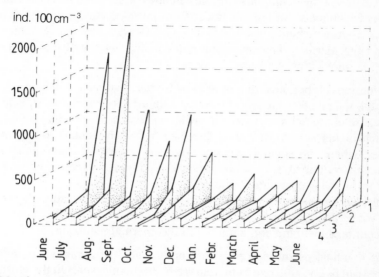

Fig. 86. The preferred occurrence of meiofauna (here mainly turbellarians) along the tube walls of biogenic structures. Diagram showing the distribution and annual fluctuation around the tube of the lugworm, *Arenicola marina*. *Numerals in the sediment block* stand for equidistant subsamples away from the tube into the surrounding sediment. They correspond to those indicated in the *lower distributional graph* (After SCHERER 1984)

granules in the intestinal cells (*Siphonolaimus*, *Sphaerolaimus*, *Sabatieria*, *Terschellingia*), cover with a "fur" of epibacteria (Stilbonematinae), inclusion of crystals in the muscle cells (*Tobrilus*, *Sabatieria*). A relation of these structures to a thiobiotic life, assumed by various authors, has been proven in the Stilbonematinae (OTT et al. 1991; Polz et al. 1992). In *Tobrilus*, formation of crystals could also be observed in species not occurring underneath the chemocline. Thus, their function as sulphide-precipitating structures is still ambiguous (SCHMIDT 1989).

Oligochaeta: various isolated macrobenthic species, mainly tubificids, are tolerant to hydrogen sulphide (particularly *Euilyodrilus heuscheri*), but in the meiobenthos there are two marine genera, *Inanidrilus* and *Olavius* (Tubificidae) with numerous species which all have a typically thiobiotic occurrence (GIERE 1981; GIERE and LANGHELD 1987; ERSÉUS 1984a, 1990b). Occurring mostly in calcareous sands of warm water regions, all of these species (about 50 described so far) are gutless and without nephridia and, as far as studied, have incorporated subcuticular bacteria (see below).

8.5.3   The Reduced Habitat and Its Impact on the Meiobenthos

The structure of the ubiquitous reduced "sulphide areas" and their relevance as biotopes for animals can only be understood under geochemical and microbiological aspects which, in turn, lead to a more comprehensive and dynamic view of their biota. Marine sediments become reduced and contaminated with dissolved hydrogen sulphide in two general modes:

a) In marine soft bottoms, rich in organic matter, hydrogen sulphide develops by the microbial reduction of oxidized sulphur species, mainly of sulphate. Because of the rich concentrations of sulphate in seawater, marine habitats reach higher concentrations of dissolved hydrogen sulphide than limnetic ones, where the sulphide mainly originates from the degradation of the proteins contained in organic matter. If produced in excess, free hydrogen sulphide will accumulate in the pore water.

b) Volcanic, geothermal and other geological activities often lead to the venting of water and gases rich in hydrogen sulphide, often combined with rich amounts of methane and ammonium (geothermal reduction).

As manifold as the processes and origins are the areas where sulphide biotopes can be encountered: from deep-sea hydrothermal vents to the subsurface layers of tidal mudflats and mangroves, from marine basins like the depths of the Baltic Sea and the Black Sea to meromictic lakes, from local gas seeps connected to submarine oil fields to coastal muds polluted by sewage. All of these diverse biotopes have been found to harbour, at least temporarily, meiobenthic animals. The complicated interaction of numerous chemical and microbial processes has not yet been completely clarified, but recent insights throw a new light on the sulphide biome and on its conditions for sustaining animal life (e.g. JØRGENSEN and BAK 1991; WATLING 1991): previous views of the thiobiome as a static two-dimensional world with an oxic horizon superimposed on the sulphidic depth were much too simple. Intensive research during the last few years showed the complexity of the sulphide system. Some important aspects to be considered in this scenario are:

– Microbial sulphate reduction which produces sulphide is possible both under anoxic and, as recently discovered, oxic conditions (see JØRGENSEN 1977; JØRGENSEN and BAK 1991).

- In natural marine sediments, thiosulphate is the predominating sulphur species and it disproportionates to a large extent into sulphide and sulphate (JØRGENSEN 1990; FOSSING and JØRGENSEN 1990).
- Microchambers (often only 50–200 μmφ) with reduced conditions amid an oxic sediment exist (JØRGENSEN 1977; WILSON 1978; GOWING and SILVER 1983; CARY et al. 1989; FENCHEL 1992) and enable co-occurring aerobic and anaerobic processes on a mm scale.
- The narrow light-coloured haloes around tube structures in the sulphidic depths contain chemically fixed oxygen, i.e. nitrate, but probably no dissolved free oxygen (JØRGENSEN and REVSBECH 1985; WATLING 1991).
- We must differentiate between oxidized sediment layers which can have highly positive redox values, but no free oxygen (see Chap. 2.1.1.3), and the oxic layers. Only in the latter is dissolved free oxygen available for the respiration of animals (SIKORA and SIKORA 1982; JØRGENSEN and REVSBECH 1985; WATLING 1991).
- We have to realize the complicated dynamic fluxes between the oxic, oxidized and anoxic/sulphidic strata, as well as their continuous mixture and alteration by bioturbation.

All these details, still not fully understood, are part of a complicated and changing system. It is characterized by intermingled processes and differentiated gradients in the concentration of O_2 and H_2S. The few general features existing in this complex web are (a) the sharp and narrow boarderlines (sometimes in the range of mm or even just cell diameters between oxic/anoxic and anoxic/ sulphidic microenvironments (see Fig. 7; REVSBECH et al. 1980; BOCK et al. 1988), and (b) their general interdependence based on microbial metabolism (FREITAG et al. 1987).

Areas of steep gradients between oxic (oxidized) and sulphidic layers are preferred habitats of rich bacterial stocks (JØRGENSEN 1977; ALLER and YINGST 1978), but the ecological potential of sulphide-exposed habitats (see Chap. 8.5.4) can only be utilized by a thiobios physiologically adapted to the hostile conditions. This entails resistance not only to longer lasting hypoxia or even anoxia, but also to the toxic and highly permeable hydrogen sulphide (GIERE and LANGHELD 1987; OTT and NOVAK 1989; POWELL 1989; SCHIEMER et al. 1990). For meio-benthos, the adaptive mechanisms involved are probably manifold, but largely unknown.

Today, there is evidence that not only protozoans, but also some metazoan meiobenthos can survive for long periods of time in complete anoxia without any apparent access to oxic layers. Nematodes of the genera *Desmoscolex*, *Tricoma* and *Cobbionema*, and some tubificid oligochaetes (*Tubificoides* sp.) have been reported from the permanently anoxic depths (> 300 m) of the Black Sea (ZAJCEV et al. 1987; own unpubl. data) and from Lake Tiberias (the nematode *Eudorylaimus* and the tubificid *Euilyodrilus heuscheri*; POR and MASRY 1968). The worms from Lake Tiberias could be kept for several months in a sealed jar under complete anoxia. Even in the symbiotic oligochaete *Inanidrilus leukodermatus*, a survival in a sealed jar with original sediment and rich development of H_2S (smell) could be found for 5 months despite the need for bacterial oxidation of reduced sulphur species. Gnathostomulids and nematodes occurred in reduced sediments cut off from oxic seawater by a thick-layered seep of brine (POWELL and BRIGHT 1981; JENSEN 1986). Various other species of nematodes, turbellarians and gastrotrichs have been reported from the black depths of various tidal flats. Occurrence in anoxic sediment below the chemocline was supported in ecophysiological experiments by WIESER et al. (1974). They found for *Paramonhystera wieseri* that oxygen deteriorated the eco-physiological capacity and viability of this nematode, existing solely in the black sediment depths. Similar results were obtained by FOX and POWELL (1986, 1987) for the turbellarian *Parahaploposthia*, which was found to be CN- and H_2S-insensitive. SCHIEMER and DUNCAN (1974) reported that the nematode *Tobrilus gracilis* stays metabolically largely anaerobic even in the presence of oxygen, an unusual statement in the light of new insights into animal anaerobiosis.

Many animals with a high tolerance for anoxia can still maintain an oxic metabolism at extremely low residual oxygen concentrations (POWELL 1989; GNAIGER 1991). Therefore, any statements about animal life under anoxic conditions should be carefully substantiated by microelectrometric (see Chap. 2.1.3) and physiological measurements.

All thiobiotic animals described take up oxygen, if available, but the rate of its utilization can be considerably reduced compared to typical aerobic species (Fox and POWELL 1987; SCHIEMER et al. 1990). In many of the aposymbiotic species, the role of mitochondria seems to be of importance. Either the amount of these organelles in the tissues is increased or their structure differs considerably from their normal appearance (DUFFY and TYLER 1984; GIERE et al. 1988b; JENNINGS and HICK 1990). In macrobenthic species, the mitochondria have been identified as the site for sulphide oxidation (POWELL and SOMERO 1986). Parallel to conditions in the macrobenthic thiobios (POWELL and ARP 1989), also in some meiobenthic thiobios the properties of a haemoglobine seem adapted to scavenging the slightest traces of oxygen (e.g. COLACINO and KRAUS 1984, for the gastrotrich *Neodasys*). BOADEN (1975, 1977) discussed the role of haem proteins as efficient oxygen scavengers in various red-coloured species of Gnathostomulida and Turbellaria. In some typical thiobiotic meiobenthos the activity of oxygen-metabolizing, sulphide-insensitive enzymes was recorded to be higher than in oxybiotic and macrobenthic fauna, indicating the potential toxicity of oxygen radicals (MORILL et al. 1988). It is conceivable that in an environment free of dissolved oxygen, oxidized substances such as nitrate or thiophosphate can become enzymatically activated to serve as oxygen donors. This pathway has been proven for some prokaryotes and the primitive ciliate *Loxodes* (FINLAY et al. 1983).

Notwithstanding further analysis for symbiotic bacteria, it is evident that a good part of thiobiotic animals can tolerate H_2S in the extreme sulphidic environment even without "bacterial metabolic help". This corresponds to numerous macrobenthic animals (for reviews see SOMERO et al. 1989; FISHER 1990; VISMANN 1991). Production of sulphur-containing granules or crystals or protection from sulphide as insoluble external precipitates (GIERE et al. 1988 for *Tubificoides benedii*) may be of importance. Yet the quantitative role of these processes still has to be assessed. In various tissues of sulphide nematodes, iron has been found and suggested to bind reduced sulphur (NUSS and TRIMKOWSKI 1984; NICHOLAS et al. 1987; GIERE 1992).

8.5.4   The Food Spectrum of the Thiobios

Oxic/anoxic and sulphidic chemoclines offer a favourable food source for meiofauna in an environment with reduced competition and predation (GIERE et al. 1991), if the problem of sulphide toxicity has been solved.

It has frequently been concluded that thiobiotic meiofauna directly or indirectly utilize the rich stock of "sulphur bacteria" which develops preferably in the oxic/sulphidic chemocline (FENCHEL 1969; FENCHEL et al. 1977; YINGST and RHOADS 1980; GROSSMANN and REICHARDT 1991). Many groups probably graze on the bacterial mats of their environment. Other thiobiotic species evolved symbioses with bacteria as an adaptation to the thiobiotic environment. All these animals take advantage of the microbial potential to metabolize reduced substances such as sulphide, thiosulphate and possibly methane.

Studies on *Inanidrilus* (Oligochaeta), on *Kentrophoros* (Ciliata) and Stilbonematinae (Nematoda) indicate that these forms use their microbes as a convenient food source. Grazing on their symbiotic epibacterial fur has been shown for stilbonematine nematodes (OTT and NOVAK 1989; OTT et al. 1991; POLZ et al. 1992). Within the gutless oligochaetes and ciliates, at least some species derive their food directly from their internal symbionts by digesting their subcuticular bacteria via phagocytosis (GIERE and LANGHELD 1987; FENCHEL and FINLAY 1989). The food spectrum of the thiobiotic turbellarians in the Solenofilomorphidae and Kalyptorhynchia has not yet been ascertained.

Dissolved Organic Matter.   As a characteristic of the meiobenthic thiobios, the length/width ratio of their body is extremely large (BOADEN and PLATT 1971; GIERE 1981; JENSEN 1986, 1987b). This favours transepidermal uptake of dissolved organic substances by the thread-like animals. Dissolved organics in considerable amounts have been shown to be present in the anoxic sediment horizons (LIEBEZEIT et al. 1983; Chap. 2.2.1). Acetate, an important substrate for sulphate reducing bacteria (JØRGENSEN 1977; GIBSON et al. 1989; MICHELSON et al. 1989), is utilized by nematodes and cannot be ruled out as an important organic food source for the thiobios. BOADEN (1977) even postulated absorptive feeding on dissolved organics as a general feature of the (primaeval) thiobios.

Endosymbiosis with bacteria and transepidermal nutritional pathways, occurring frequently in the thiobiotic (meio-)benthos, would render possible the frequent reduction of the intestinal tract. The degeneration of mouth and anus has created generic names like *Astomonema* (Nematoda), *Inanidrilus* (Oligochaeta) and *Astomus*, recently described as *Parenterodrilus* (JOUIN 1992) (Polychaeta).

8.5.5   Distributional Aspects

Attraction of thiobiotic species to the rich bacterial stocks in the deeper, sulphidic sediment layers often leads to patterns of vertical distribution which differ characteristically from the usual concentration of oxibiotic meiofauna (BOADEN 1977; GIERE et al. 1982; OTT and NOVAK 1989). FENCHEL (1969) emphasized the divergent vertical occurrence of ciliates (see Fig. 27), JENSEN (1981) showed a very heterogeneous vertical pattern within the genus *Sabatieria* (Nematoda) with *S. pulchra* as a typical thiobiotic species living deep down in the anoxic sediment and *S. ornata* occurring close to the surface. Compiling the vertical distribution of various nematodes from the Kattegat (Baltic Sea), an assemblage of oxibiotic species was clearly separated from the thiobiotic species (Fig. 87).

Among limnic nematodes, a similarly differentiated pattern is developed in the genus *Tobrilus* (SCHMIDT 1989). POWELL (1989) summarized experimental studies on various turbellarian species with a different preference reaction to oxic and sulphidic layers respectively (Fig. 88).

In general, a preference for layers around the oxic/sulphidic chemocline could be experimentally proven for different meiofauna like thiobiotic ciliates,

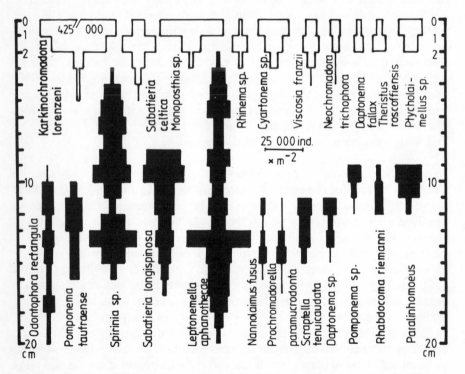

Fig. 87. The vertical distribution of various nematodes in sediments of the Kattegat. A separate oxibiotic and thiobiotic fauna assemblage is apparent. (After JENSEN 1987b)

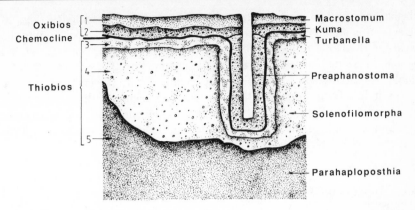

Fig. 88. The occurrence of various oxibiotic and thiobiotic turbellarians in a depth profile of a tidal flat. *1*, Normoxic; *2*, microoxiphilic; *3*, anoxic, no sulphide; *4*, low sulphidic; *5*, high sulphidic. (After POWELL 1989)

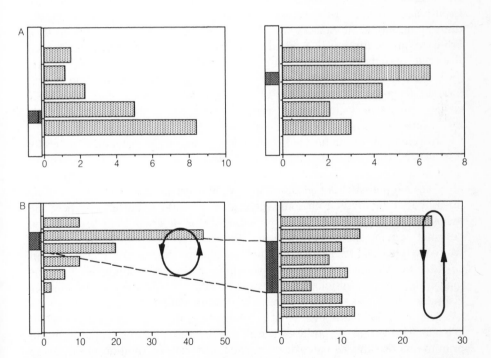

Fig. 89A, B. Migration experiments with thiobiotic oligochaetes (**A**) and nematodes (**B**) preferring the chemocline (*dark field in vertical bar*). The more diffuse this transitional layer, the wider the migratory radius. (**A** after GIERE et al. 1991; **B** after OTT et al. 1991)

turbellarians, gutless oligochaetes and stilbonematine nematodes (Fig. 89). However, this preference seems to result from a very dynamic positioning. It is typical for thiobiotic animals that they frequently migrate between oxic and sulphidic horizons regularly traversing the chemocline. They do not maintain a stable position linked consistently to the chemocline. In the bacteria-symbiotic thiobios it has been suggested that the bacteria recharge their energetically valuable reduced sulphur store in the sulphidic layers and gain energy oxidizing the acquired sulphide/thiosulphate in the oxic layers (SCHIEMER et al. 1990; GIERE et al. 1991; OTT et al. 1991).

8.5.6   Aspects of Diversity and Evolution of Thiobios

Compared to its oxic counterparts, the thiobiotic meiobenthos is impoverished, at least in diversity, while its abundance and biomass can be rather high. This becomes particularly clear in shallow sulphidic areas. In the polluted Baltic Sea basins and bights, the viable benthos in the anoxic and sulphidic bottoms is mainly represented by dense masses of the nematode *Pontonema vulgare* which feeds on dead macrofauna (LORENZEN et al. 1987). Under similar circumstances, the nematode *Terschellingia communis* can also develop huge populations. Occurrence of nematodes is probably not so much influenced by oxygen deficiency than by ammonia (BONGERS and HARR 1990). Deep-sea sediments beneath a brine stream, continuously exposed to sulphide and frequent anoxia, harboured a specialized fauna of predominantly meiobenthic forms (POWELL and BRIGHT 1981; POWELL et al. 1983). Compared to the rather scarce meiofauna typical for oxic deep-sea sediments (see Chap. 8.3), hydrothermal vents with their copious supply of bacterial food seem to harbour a relatively rich thiobiotic meiofauna (DINET et al. 1988; VAN HARTEN 1992).

It seems that in the hostile sulphidic environments, meiofauna is generally less affected than macrofauna, of which often only very few species remain alive (JØRGENSEN and WIDBOM 1988; AUSTEN and WIDBOM 1991). The high eco-physiological potential of meiofauna in oxygen-deficient environments is under-lined by investigations from some (geologically young?) vent areas where the meiofauna did not contain any specialized forms differing from those in the "normal" environment (FRICKE et al. 1989; SHIRAYAMA and OHTA 1990). It was concluded that in a restricted sulphidic biotope the fauna is recruited from colonizers of the ambient oxic fauna. Conditions in recent sulphidic locations may initiate an ecophysiological adjustment, which is only later, through longer evolutionary periods of time, followed by anatomical adaptations.

The evolutionary relevance of a thiobiotic fauna has always been much debated. The various hypotheses centre around the question on the origin of primitive metazoan life in an oxic or anoxic Proterozoic earth atmosphere. There are various conflicting conceptions both about oxygen conditions and the metabolic nature of the most primitive fauna in particular (JENKINS 1991;

MANGUM 1991; RUNNEGAR 1991). Originally, restriction of thiobiotic animals to complete anoxia was not postulated. FENCHEL and RIEDL (1970) included in their reasoning for an old thiobiotic fauna "the necessary assumption of a low-oxygen atmosphere in the pre-Cambrian age". Also BOADEN (1977) considered low-oxygen concentrations as important for a thiobios when he described their "metabolism adapted to very low levels of dissolved oxygen". On the other hand, BOADEN (1975, 1977, 1989b) described a scenario of an anaerobic, interstitial and holobenthic primitive "thiozoon".

An oxic fauna prevailed since the Cambrian, probably even since the late Ediacaran. Absence of fossil remains of minute meiobenthic animals from the pre-Ediacaran geological period, crucial for this problem, in no way precludes the existence of a primitive animal life. Although geological strata of an earlier, pre-Ediacaran period such as "red beds" are indicative of a chemically oxidized status, this does not necessarily mean the presence of free oxygen molecules dissolved in the water, which alone would be relevant for primitive animals with an aerobic metabolism. Even if during long Proterozoic periods the overlying water contained low concentrations of free oxygen, the concomitant meiobenthic animals with an endo-/mesobenthic life style presumably still lived in complete anoxia (REVSBECH et al. 1980; WATLING 1991, GIERE 1992). This has to be concluded from modern insights into sedimentary ecology (see Chap. 2.1.3).

Why should an endobenthic life exist under these hostile conditions? Compared to recent geological periods, Proterozoic life conditions will be dominated by scarcity of organic matter as a nutritive basis. This ecological reason, but also structural constraints, may have favoured development of animals of meiobenthic size. Organic matter will have accumulated in the upper sediment layers probably leading to the local enrichment of microbial life as compared to the water column. This better trophic supply plus the shelter from erosion (and UV radiation?) will have favoured the existence of meiobenthic life underneath the surface in the upper sediment layers.

As much as these arguments allow for the possibility of a continuous existence of some relict meiobenthic thiobios in a "plesiomorphic biotope" (BOADEN 1975), this hypothesis is rejected by other authors (cf. MANGUM 1991). Conclusive evidence not only on the existence of meiofauna in these early periods, but also on the ecophysiological availability of free oxygen will hardly be achieved. Although there is no doubt that the "lower animal groups" (Protista, Gnathostomulida, Turbellaria and Nematoda) hold a dominating position among thiobiotic fauna, today almost all systematists confirm a derived phylogenetic position of most thiobiotic species.

This has to be taken into account when formulating a differentiated and modern definition of a thiobios (e.g. POWELL et al. 1983; MEYERS et al. 1987, 1988). Meiobenthic species live in the complicated system around the chemocline where they are adapted to a set of chemical and ecological microniches. They continuously move around through hypoxic, oxidized and sulphidic gradients. They are able to take advantage of the potentials and to cope with the risks of an environment dominated by reduced sulphur. They are probably not primitive representatives of an archaic low-oxic or anoxic biotope, but rather specialists differing from the "normal", oxic fauna.

In an arbitrary situation based mainly on hypotheses, one should not link the term thiobios cogently with its uncertain Proterozoic origin and possible anaerobic metabolism. Instead, the term thiobios should be based on its ethymological derivation, reflecting hydrogen sulphide (concomitant with other reduced substances) as the dominating factor:

> "Thiobios represents a diverse community of organisms characteristic for biotopes where hydrogen sulphide and other reduced substances are regularly dominating ecofactors. The thiobios is directly or indirectly linked to sulphidic habitats" (GIERE 1992).

This would leave the possibility of oxygen being present or not to the local biotopical conditions and to new geological insights. At present, the existence of a meiobenthic thiobios can neither be denied nor should it be ecologically underestimated. Among the thiobiotic soft-bottom fauna, meiobenthic forms play a substantial role. They are mostly specialized animals with differentiated metabolic and trophic pathways populating extraordinary "sulphur environments". One may conceive a strictly anaerobic life (although today physiologically not yet proven) and a completely aerobic life of free-living benthos as extremes in the wide continuum of varying oxygen supply. The fact that aerobic animals today are much more frequent than anaerobes would only reflect the dominance of present-day oxic environments. Further research will discover many more and highly interesting meiobenthic thiobios, possibly including free-living metazoans able to thrive permanently in anoxia, with a relevance reaching far beyond the meiobenthos.

> *More detailed reading*: evolutionary aspects – FENCHEL and RIEDL (1970); BOADEN (1977, 1989b); physiological aspects – POWELL (1989); MEYERS et al. (1987); ecological aspects – JENSEN (1987); OTT et al. (1991); GIERE (1992); geochemical aspects – JØRGENSEN and BAK (1991); anoxic ciliates – FENCHEL and FINLAY (1991); review on metazoan anaerobiosis – BRYANT (1991)

8.6   Meiofauna in Polluted Areas

8.6.1   General Aspects

The multitude of pollutants and the diversity of meiofauna investigations from polluted habitats do not allow considering all the singular studies. Hence, the following section will rather concentrate on principal aspects and problems of more general relevance. When mentioning various methods of evaluation, the frame of this book does not allow for discussion of underlying mathematical principles and explanation of calculatory procedures. For this, the reader is referred to the literature cited.

Compiling the more than 200 meiofauna studies which deal with the impact of pollution (COULL and CHANDLER 1992), it turned out that a slight majority is performed in the field while mesocosm experiments and field experiments are rather rare (only 15%). The multitude of the pollution problems studied can be grouped in six categories (percentage of papers in brackets): petroleum hydro-carbons (28), metals (23), organic waste (18), mixed pollutants, e.g. industrial waste (16), pesticides (5), others (10).

The relevance of many laboratory assays testing meiofauna and toxicants in the aqueous phase may be doubted. Although many pollutants are sorbed to the sediment particles and often recorded in higher concentrations in the sediment than in the supernatant, the sediment tests proved often less toxic than those in the unnatural aqueous conditions. It is the bioavailability and not the absolute concentration of pollutants in the water which determines the noxious effect on meiofauna. Due to synergistic effects, mixed pollutants usually act more detrimentally than the toxicants alone.

Even with meiofauna and its rapid turnover, it is important to conduct a pollution study for a sufficiently long period of time in order to obtain reliable results. It has often been shown that shortly after the pollution event the overall abundance of meiofauna tends to increase (depending on the species, the nature and concentration of the pollutant etc.), superficially pretending little short-term impact. This could easily lead to problematical and erroneous conclusions about a non-detrimental effect. Only if the often long-lasting erratic fluctuations in species composition after the pollutional event, and the more indirect conse-quences will be analyzed, may negative effects revealed. The reduction of diversity, decrease of egg production and hatching rate of juveniles, perhaps even the increase in intersexuality (MOORE and STEVENSON 1991), are particularly impor-tant parameters with severe consequences for the community development. It is generally more important and also promising to concentrate on the highly sensitive reproductive and larval stages than to focus on the overall abundance and biomass (GIERE and HAUSCHILDT 1979, for oil pollution; CHANDLER 1990, for insecticide contamination). On the other hand, it might be deceptive that egg-producing females often are less sensitive to PCBs and other lipophilic toxicants than males. This may be referred to the sequestering and deposition of noxious substances into the lipid-rich yolk of the eggs.

Thorough research on the impact of pollutants generally requires a close combination of chemical analysis with field studies and laboratory experiments. This is, of course, valid also in meiofauna studies where the field data have often been found to diverge considerably from unifactorial laboratory bioassays; but the above claimed "triade" in pollution research (CHAPMAN 1986) can be realized only very rarely.

For reliable results from field studies, the general lack of data on the local status prior to the polluting event usually is the decisive problem. Base line studies which include meiofauna are extremely rare, generalizations from "similar" biotopes in other places do not represent a reliable reference regarding the notorious patchiness and variability in meiofauna. One of the few long-term studies on meiofauna are the investigations along the Belgian coast by the group of C. HEIP, which focussed mainly on nematodes, but considered also harpacticoids and other taxa (HERMAN et al. 1985).

The above reasons would rather speak against meiobenthos as a favourable object to document pollution hazards. However, as HEIP (1980b), HICKS (1991) and recently WARWICK (1993) have pointed out, there is a suite of undisputable advantages in meiofauna over macrofauna in pollution studies:

- High abundance of animals in small areas, even in "poor" biotopes like estuaries and exposed beaches which often become polluted, render the resulting data statistically well treatable.
- High diversity allows for conclusions on changes in species composition, even in "poor" biotopes (see above) where macrofauna is often very scarce.
- A still relatively rich meiofauna in eutrophicated areas with subsequent hypoxia and formation of hydrogen sulphide where macrofauna is often exterminated (e.g. JOSEFSON and WIDBOM 1988, AUSTEN and WIDBOM 1991, in the depths of the Gullmarsfjord, Sweden).
- Meiobenthic animals generally lack a planktonic phase. Hence, the members of the populations are more consistently exposed to a local pollutant accumulated in the sediment.
- Short generation cycles and rapid growth allow for easier answers on a possibly noxious impact, especially since the usually more sensitive reproductive and juvenile stages can be regularly included in the assessment, even in short-term investigations.
- Concerning one of the most important sources of pollution, namely heavy metals, it seems that meiofauna in general reacts more promptly and drastically than macrofauna (MCINTYRE 1977; VAN DAMME et al. 1984).
- In contrast, meiofauna is relatively insensitive to mechanical disturbance and destabilization of the sediment. In many cases this offers the chance to differentiate between impairment by mechanical and chemical actions (AUSTEN et al. 1989; WARWICK et al. 1990a).
- Within the spectrum of meiobenthic groups, there are some sensitive taxa of particularly high indicative value such as harpacticoid and ostracod crustaceans. They are often widely distributed and, thus, relevant in many areas.
- Small samples and convenient, inexpensive sampling gear allow for simple sampling logistics including the necessary replicate samples (finance and reliability aspect). The possibility of easily transporting the whole, unextracted sample back to the lab reduces the time in the field; bulk fixation of the complete sample allows convenient timing of evaluation (time aspect).
- The rather two-dimensional distribution of meiobenthos as compared to plankton simplifies not only sampling but also the interpretation of results.

But the problems of pollution studies with meiofauna should not be concealed:

- Basing a pollution study on microscopic animals not readily visible to a layman is not an easy task. It is difficult to convince the public that their role may be of equal importance in the ecosystem as the often familiar macrofauna (the psychological problem). Beside the personal skills helping to present the results in a convincing manner, one should concentrate on those species for pollution assessments that are wide-spread and allow for direct comparisons. They should be really common, directly linked to a food chain of apparent relevance even to laymen (e.g. harpacticoids → fish) and easily to be demonstrated, allowing for convincing arguments about their ecological role. In this respect, the use of rare and cryptic species such as interstitial gastropods for pollution studies (POIZAT 1985) seems rather problematical and limited.

– In order to cope with the high temporal and spatial heterogeneity, frequent sampling along a dense station grid is required for reliable documentation of trends (see Chap. 7.2.2; the problem of working capacity).

– The rather complicated identification of meiofauna requires a high expertise and high-standard optical equipment (the identification problem). Does the high ecological variability in meiofauna which corresponds to the multitude of micro-niches allow for a summative identification of higher meiobenthic taxa? Large-scale and comparative meiofauna studies have shown that data, based on taxonomic ranks as high as families, still allow a fairly clear separation of stations according to their degree of disturbance/pollution (see Chap. 8.6.2; WARWICK 1988; HERMAN and HEIP 1988). While in harpacticoids discrimination between sites seems better at the species level than in nematodes, the latter yielded fairly consistent results with data aggregated to higher taxonomic ranks. For nematodes, it is only beyond the family level that a major loss of information occurs (HEIP et al. 1988). By and large, it becomes increasingly clear that, if carefully used, relatively little information is lost in multivariate analyses if the taxa are determined only to higher taxonomic ranks. This work is sufficiently simple to be feasible by non-specialists even, and reasonably time-consuming to be of practical value (HEIP et al. 1988; WARWICK 1988).

Production of easy-to-use modern (illustrated) keys, e.g. the nematode keys by PLATT and WARWICK (1983, 1988, in prep.) must continue and, for many purposes, even a pictorial key restricted to the identification of families in all major meiofauna groups would be of invaluable help in assessing the pollutional impact (see below). This cognition could greatly alleviate the notorious identification problem for applied meiofauna studies. Simultaneously, it could be a strategy to create a new reputation for meiofauna studies bringing its values to the attention not only of the scientific community, but also of decision making agencies.

8.6.2   Methods for the Assessment of Pollution Using Meiofauna Data

Many meiofauna investigations on pollution effects are based on a divergent reaction of the two main taxa, nematodes and harpacticoids. This general notion has been referred to the different trophic niche occupied by the two groups

(MONTAGNA et al. 1989). Most nematodes are linked to a short, detrital/bacterial-based food chain. In the case of organic enrichment their number will increase rapidly. In contrast, harpacticoids are mainly microalgal-based and oxygen-sensitive members of the food web. They will react negatively to an increase in the organic load (compare RUDNICK 1989; VINCX et al. 1990). Basing on a paper of PARKER (1975), RAFFAELI and MASON suggested already in 1981 use of this divergent ecological character of nematodes and copepods for assessing the impact of pollution by simply calculating the nematode/copepod ratio. This easy and general indicator of pollution would abolish the difficult and tedious species identification. Investigating a Scottish west coast beach, RAFFAELI and MASON showed that the overall ratio of nematodes/harpacticoid copepods increased with increasing degree of pollution due to the reduction of the more sensitive harpacticoids. Although they realized that this N/C-ratio would shift depending on variations in the grain size distribution, they contended that if this trend were to be generalized for all sandy eulittoral coasts: a value > 100 would indicate pollution.

This attractively simple parameter immediately arose much debate (COULL et al. 1981). Not always did scrutiny of areas with an obvious pollution gradient result in the expected increase of this ratio (BOUCHER 1980; HUYS et al. 1992), even a re-examination of the very beach in Scotland which RAFFAELI and MASON (1981) had sampled showed that the original data could not be generalized (LAMBSHEAD 1984). Apparently the ecological spectra of the individual species within nematodes and harpacticoids are too diverse and are influenced independently by a complex set of environmental parameters to stand for a universal validity. Too many other factors besides pollution play a decisive role, as indicated already in the relation to grain size distribution. This led WARWICK (1981) to a refinement of the N/C-ratio taking into account the various feeding types involved. Based on the examination of the mouth parts and the armature of the buccal cavity he could show that epigrowth feeders react different from predators and deposit feeders. Their separation would certainly add precision to the nematode/copepod ratio. However, this would also eliminate its attractive simplicity and, again, the specialists's expertise would be required. In a later paper, RAFFAELI (1987) specified and partly corrected his earlier oversimplified contention: mainly for harpacticoids, the simple N/C-ratio does not account for the varying habitat adaptations of the species and, consequently, not the different ecological reactions of the two major harpacticoid groups which live in two distinctly separate ways. The species in coarser sand are mesobenthic (interstitial), and thus depend much on grain size. In fine sand and mud the mostly epibenthic or endobenthic-burrowing species mainly depend on the water content. Therefore, they live predominantly in water-saturated, sublittoral substrates. Comparatively, nematodes seem much more independent of habitat structure. They react to an increased content of organic matter (generally typical for finer sediments) with improved food conditions, their populations usually increase. This would explain the divergent reactions of these meiofaunal groups in many cases of pollution. As long as the physico-chemical conditions are not drastically impaired, will

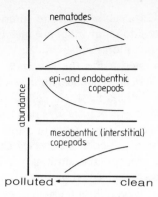

Fig. 90. The divergent response of nematodes and harpacticoid copepods to a pollution gradient. For further explanation see text. (RAFFAELI 1987)

the nematode number increase with a rising load of organic pollution regardless whether fine or coarse sediments prevail. An exception are gradients of chemical toxic pollutants. Here, the nematode populations will decrease already rather early (Fig. 90). In cases where mixed wastes are released, an intermediate reaction would be the response. In harpacticoids the reaction is different: the strongly structure-related interstitial species will rapidly disappear with increasing concentration of sewage and detritus clogging the void system. Contrastingly, the endo-/epibenthic species will at first not significantly change their abundance, will perhaps even grow in population size with increased food supply. But this trend will terminate at a point where oxygen depletion and formation of hydrogen sulphide mark a drastic metabolic limitation for all harpacticoids.

The different reaction of the two meiofauna groups have been corroborated in various studies (e.g. SANDULLI and DE NICOLA 1991), also in experimental work. The suggestion by SHIELLS and ANDERSON (1985) to restrict calculations of the N/C-ratio just to the mesobenthic forms in order to get reliable results does essentially confirm the refined version of the N/C-ratio. Hence, although not for universal application, this refined version seems still useful in cases, where in a limited area the temporal effects of a polluting incident and its recovery phase shall be monitored (RAFFAELI 1987).

Another method to ascertain an impact of pollution, applicable for the interstitial fauna of sand, has been proposed by HENNIG et al. (1983a). Normally, the authors found in South African beaches a close correlation between population density and sediment structure. This correlation resulted in an "optimal" grain size for each group which was altered in cases of a pollutional event: a positive deviation would indicate organic enrichment by domestic sewage, a negative deviation would point to industrial (chemical) inlets. Also from this study emerged a differed reactive spectrum of nematodes and harpacticoids. However, also this simple method is probably restricted in its applicability and valid only in very homogeneous and rather unpolluted sands. They will hardly correspond to an average European beach. Here, interaction of too many factors other than grain size composition may influence meiofauna abundance and will render a clear relation to pollution problematical.

In an interesting approach using the rationale and methods of cladistic systematics and applying them for ecological analyses, LAMBSHEAD and PATERSON (1986) compared sample stations for their pollution stress. A plesiomorphic (primitive) feature in phylogenetic studies was paralleled in ecological analyses with the absence of a species at a given station, and an apomorphic (derived) feature was paralleled with the presence of a species. Consequently, the resulting "cladogram" of pollution grouped the stations according to their degree of pollution. It could be shown, indeed, that this grouping coincided with the geographical situation in a given pollution gradient separating the polluted sites from those without pollutional stress. It seems that the transfer of this fairly objective method, so successful in systematics, to applied ecology is a promising tool which should be further scrutinized for its general applicability.

From most meiofauna studies emerges that pollution entails marked changes in diversity (Fig. 91). Hence, diversity indices are calculated in many pertinent meiofauna studies, although the validity and meaning of diversity as an indicator of ecological disturbance are much debated (see Chap. 9.1.2; for restrictions and calculation see HEIP et al. 1988, and ecological textbooks). Beside the problem of correctly interpreting a diversity index, there remains the difficulty of meiofaunal species identification (see Chap. 8.6.1).

Investigating areas with an apparent pollution gradient, GRAY (1979, 1981) related the decline in (meio-)benthic diversity to the impact of pollution. He

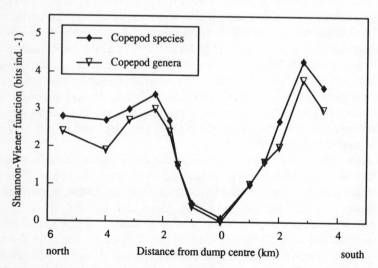

Fig. 91. The decrease in harpacticoid copepod diversity related to the distance from a polluted site in the Firth of Clyde, Scotland. Note corresponding curves from evaluation of species and genera. (MOORE and BETT 1989)

allocated to the number of species, plotted in cumulative percentages, their corresponding abundance, arranged in geometrical grouping. The result is in cases of a non-disturbed environment a log-normal distribution, a straight line on a probability scale. In the case of a disturbance (by pollution), there would be a break in the line indicating deviation from an equilibrium community. Thus, the impact of pollution could be evidenced in its spatial or temporal progress directly from the changes in the log-normal plots.

After publication of this method, its mathematical basis (LAMBSHEAD et al. 1983) as well as its ecological assumptions have been criticized (PLATT et al. 1984; WARWICK 1986). From general ecology it is known that diversity is not always diminished by disturbance. Minor disturbances often enhance diversity (HUSTON 1979). Hence, every kind of disturbance, not only pollution, may cause the break of the straight log-normal line; it will respond also to changes in the nature of the habitat and even to the methodology. This would render an interpretation of pollution as the only causative factor highly problematical. As a consequence of these theoretical and ecological shortcomings, today this method, at least in its application for pollution studies, has been largely abandoned.

LAMBSHEAD et al. (1983) introduced into ecology an apparently easily applicable method to assess stress through pollution, the "k-dominance method". Its alternative name ABC-method (WARWICK 1986) indicates that it relates abundance to biomass in two comparative curves. It is based on the contention that in undisturbed biotopes the more K-selected specialists (persisters) among species account for a high individual biomass, although they only attain a low population abundance and, thus, only a low numerical rank. In contrast, communities of r-selected generalists (colonizers), dominating in disturbed areas, are typically characterized by just a few species of low individual biomass, but with mostly large populations which attain a high numerical rank (Fig. 92). (The general meaning of the character complexes r- and K-strategists has been surveyed in PARRY 1981).

If now, on the abscissa the species are plotted in a log-scale according to their rank, and, on the ordinate their corresponding cumulative percentage-dominance, the resulting course is highly indicative when comparing curves for abundance and biomass: a disturbance can be postulated (which is often based on pollution), if the curve for abundance is elevated above that for biomass. If the biomass curve exceeds that of the abundance, the region in scrutiny can be considered undisturbed (Fig. 92). Essentially, the method is a graphic comparison of the two components of diversity, species richness and evenness (PLATT et al. 1984)

Although this convenient method allows only for a rather coarse dicrimination of a few levels of pollution, it does not require control samples, the two curves representing an "internal control". Regrettably, this simple method is of only limited value for small animals, e.g. for meiofauna. Their changes of biomass are minute regardless of their nature as r- and K-strategists (BEUKEMA 1988; HEIP et al. 1988). However, a recent study by WARWICK et al. (1990)

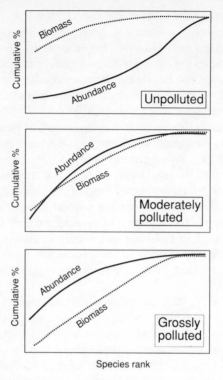

Fig. 92. Hypothetical k-dominance curves for variously polluted areas. (WARWICK 1993)

proved the use of the k-dominance method also for meiofauna as a valuable indicator of disturbances. On the other hand, this study also revealed the problems of interpretation involved: each kind of disturbance will have its impact on the course of the curves; the major groups react differently to sediment disturbance (bioturbation), nematodes being affected, harpacticoids less so.

Recently, BONGERS and HAAR (1990) and BONGERS et al. (1991) in studies on the "quality" of freshwater or marine habitats, based the characterization of an aquatic biotope as polluted or not on the character of its nematode community. They assigned to each of the more frequent taxa (species, genera or even families) a "c-p-value" according to their nature as a colonizer or persister. "1" means an extreme colonizer and "5" an extreme persister. Using these values, they calculated a "maturity index",

$$MI = \overset{n}{\underset{i=1}{S}} \, v(i) \cdot f(i),$$

where $v(i)$ is the c-p-value of taxon i and $f(i)$ its frequency. Thus, in principle, it is again a comparison of the ecological nature of the species with their abundance.

Applying the calculation of this maturity-index to various areas stressed by an oil spill, an overload of organic matter, heavy metal contamination etc., they could show that the index is widely applicable for discrimination of stressed from non-stressed biotopes, freshwater as well as marine habitats, deep-sea as tidal flat areas. In disturbed sites, the MI is low (about 2.1), in areas with little or no disturbance the values are high (about 2.6). With increasing reliability of the classificatory c-p values of the species/taxa considered, this method could provide a relatively simple means of assessing biotope disturbance. However, it requires a taxonomic differentiation feasible only by the (nematode) specialist. Moreover, the source of stress, whether a mere sediment turbation or a case of pollution, can hardly be identified.

Perhaps the biologically most meaningful method of linking community data to an environmental variable such as pollutional impact is Multi-Dimensional Scaling (MDS) (FIELD et al. 1982). Although its computational basis is fairly complex (KRUSKAL and WISH 1978), it yields an easy-to-conceive graphic document, a "map" of sample similarities influenced by environmental variables, e.g. pollution. Its sensitivity and its general applicability have been proven in numerous examples both for macro- and meiobenthos, and both for abundance and biomass values (WARWICK 1993). Moreover, MDS configurations based on taxonomically higher ranks than species resulted in an essentially similar

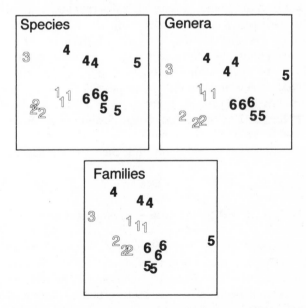

Fig. 93. Multi-dimensional scaling ordinations (root transformed abundance) for copepod data from the Firth of Clyde, Scotland. Samples from unpolluted sites are *stippled*, those from polluted sites in *black*. Use of different taxonomic ranks results in principally the same configuration of samples. (WARWICK 1988)

pattern (Fig. 93). Hence, this method represents a valuable tool revealing the impact of an environmental variable, e.g. pollution, even in cases where mere diversity assessment could not give a significant answer (WARWICK 1988).

In summarizing articles, WARWICK and CLARKE (1991), CLARKE (1993) and WARWICK (1993) evaluated methods for assessing changes in community structure (and community stress). Applying these methods to a range of studies, they used meiobenthos data in direct comparison with macrobenthos studies and discriminated between:

1. Univariate methods where relative abundances of the various species are reduced to a single index. The appropriate statistical test is the classical ANOVA; the most frequently used method within this category is the Shannon Wiener diversity index H' (SHANNON 1949).
2. Graphic/distributional methods where relative abundance (or biomass) of species is plotted as a curve. The typical example is the k-dominance curve. This group of methods comprises more information about the distribution of the fauna than the single index in (1).
3. Multivariate methods of classification and ordination which compare faunal communities considering both their specific identity and their (relative) quantitative importance. These latter methods are exemplified in techniques such as the multi-dimensional scaling ordination (MDS).

Since the two first groups of methods are not species-specific, they can yield identical results for communities of completely different taxonomical composition. Although they are less sensitive in detecting community changes than the multivariate methods, they allow a fairly clear judgement on the presence of detrimental (e.g. pollution) effects. This is not as easy in the last group of methods. These document faunal changes with precision and in favourable generality, but give few indications of the possible reasons. Nevertheless, to disclose community level stress, both naturally or pollution-induced, the future seems to be in the more sensitive and more information-retaining multivariate analyses. Probably the option is the comparative use of various methods compensating their shortcoming and optimizing their specific advantages (CLARKE 1993).

8.6.3 Oil Pollution and Meiofauna – a Selected Case

So far, the more general aspects of pollution studies in meiobenthology have been considered. Now, some case studies should reflect the real scenario. Oil spills as a source of frequent chemical pollution will be selected as examples because their threatening impact to shore life is particularly spectacular. Therefore, the effect of petroleum hydrocarbons on these shore sites, often rich in meiobenthos, is most frequently considered in meiobenthic studies. Moreover, crude oil and its derivatives are typical pollutants characterized by a mixture of noxious and instabile compounds interacting in numerous ways with the physical and chemical environment. The detrimental consequences exemplify well the difficulties to ascertain and identify any causative agents, a situation as frustrating as common in the daily cases of environmental pollution.

Crude oil is a natural product to which many meiofauna species can adapt to a certain degree. Natural oil seeps in the sea are comparable with organic enrichment (Montagna et al. 1989) to which many specialized bacteria react positively with extreme multiplication. Around natural oil seeps the stocks of nematodes are particularly rich, while harpacticoids, in the absence of the microphytobenthos, are underrepresented. Some of the negative results recorded from oil-contaminated areas probably refer rather to the impoverishment of oxygen induced by the degradation of excess organic matter than to the toxicity of the oil (Bodin 1988).

Chemically, crude oil cannot be defined precisely, its detailed composition may differ with each oil well and change continuously ("ageing of oil"). Depending on the addition of dispersants after a spill, on the local physiographic and climatic conditions, the impact on the fauna can vary considerably. Tolerance experiments with oil do not yield reliable results which can be extrapolated to the field situation since, in the laboratory, the unnatural stress situation and the artificial absence of multifactorial combinations will bias the data.

The resulting variance in meiofauna field studies on the impact of oil spilled on the shores will render the different case studies rather incomparable and may explain the divergence of data obtained.

In the large oil spill off La Coruña (Northern Spain) in 1976, all eulittoral meiofauna on beaches adjacent to the oil outflow became exterminated and

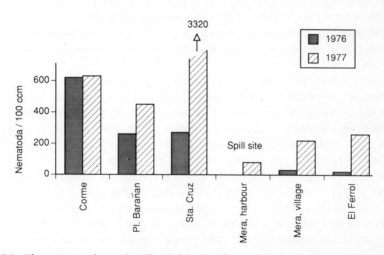

Fig. 94. The impact of an oil spill on the abundance of nematodes in beaches off La Coruña (northern Spain). Comparison of two data sets: 6 weeks and 1 year after the spill. (After Giere 1979)

only a few opportunistic species could be recorded 1 year after the spill (Fig. 94). A similar massive destruction was reported by WORMALD from a Hong Kong oil spill (1976) and by ELMGREN et al. (1983) from a spill in the brackish water of the Baltic Sea.

In less destructive cases of oil spills, the immediate meiofauna response is usually a strong reduction of harpacticoids, ostracodes and turbellarians, and a less severe impact on annelids. The decrease in nematode populations is often only subordinate and difficult to separate from natural fluctuations. As a long-term reaction, the common pattern is a slow recovery accompanied by erratic fluctuations in abundance. In exposed, sandy shores complete recovery may be achieved in 1 or 2 years, in muddy bights and estuaries damage can persist for a decade.

In the complex natural ecosystem, this generalized pattern of detrimental effects is, however, hardly predictable. There may even occur a strong short-term population growth in some species/groups due to the destruction of more sensitive competitors and predators, mostly followed by sudden population breakdowns. In this situation, small-scale experiments without the natural species interactions will give unnatural results. More reliable studies can be expected from mesocosm experiments, which attempt to maintain the natural ecosystem (ELMGREN and FRITHSEN 1982).

In other cases, e.g. after the oil disaster of the Amoco Cadiz in 1978, a grave impact on the meiofauna could not immediately be found, although some-times sandy beaches buried under a thick layer of oil slick were sampled. Judging from meiofauna composition and abundance also 1 year later the situation did not reveal any major damage, even at the species or genus level (GIERE, unpubl. data). These surprising results referred to the N/C-ratio (BODIN 1988), and were confirmed by the long-term study of BOUCHER (1980) (see also RENAUD-MORNANT and GOURBAULT 1980). Although the Amoco Cadiz spill represented a rare case of "reference" in which the nematode fauna, monitored for years prior to the oil spill, could be compared with the situation after the spill, no significant differences were evident.

On the other hand, there was a change in species composition after the spill, which can be interpreted as a disequilibrium; but compared to the reference, the population fluctuations were too inconsistent to be clearly explicable. Often just those species considered sensitive increased in density after the spill. The complexity of field conditions did not allow for a straightforward interpretation (BODIN and BOUCHER 1983). The situation in meiobenthos was controversial to that in macrobenthos where a detrimental effect could clearly be shown (see CLARKE 1993). Perhaps the response to the pollution would have been more clear-cut if the data were considered and evaluated at a higher taxonomic level, a notion made for macrofauna by WARWICK (1988). Then the numerous and inevitable small-scale changes in the natural environment (both spatial and temporal ones) affecting the species composition and confounding the pattern caused by the pollutant, would not have altered the proportions of higher taxonomic ranks and not become evident, blurring the overall picture.

It was not until the impact studies of the Amoco Cadiz oil spill focussed attention to the sensitive harpacticoids that a drastic decline in their abundance and diversity right after the spill was found, followed by some erratic fluctuations with occasional population "explosions" in the summer (BODIN 1988, 1991). After 2 or 3 years, the "degradation phase" ended, but it took almost the same time for the harpacticoids to recover and establish the status previous to the spill. Studying specifically the more susceptible reproductive and juvenile stages of harpacticoids, BODIN (1991) noticed that reproduction was either severely reduced or delayed. This caused massive disorders in the population dynamics accompanied by deficiencies in copepodite stages.

8.6.4 Conclusions

Many incidental phenomena, such as exposure of the polluted site, the seasonal conditions at the moment of the pollution accident and the type of pollutant, contribute to the course of events, the deletion and recovery of the fauna. The scientifically fairly well covered case of the Amoco Cadiz showed in its divergent results that even on the basis of longer-term and careful investigations, general statements about the impact of an oil spill for eulittoral meiofauna have to be taken with caution. However, some conclusions on the impact of oil can be drawn:

a) After an oil spill, the meiofauna of "high energy" sites will suffer from immediate and drastic population variations, but these will last for a relatively short time (in the order of months to 1 year) as compared to more sheltered areas. Here, the fluctuations will be dampened and retarded, but will persist much longer (in the order of several years or even one decade).
b) The embryonic development will be retarded in cases of light oil contamination and stopped in severe pollution (GIERE and HAUSCHILDT 1979; BODIN 1991).
c) Juveniles are more sensitive to oil pollution than eggs or adults.
d) Only the most recent formulations of oil dispersants do not enhance the toxicity of the oil.

For pollution studies in general, some principal rules can be gathered from the oil spill studies:

– Simple field studies after the polluting event, especially if performed on a short-term basis only, will yield biased data not accessible to an unequivocal interpretation.
– Without information on the status quo ante and without analysis of the species composition, the biological status of the area and its phase of recovery can hardly be ascertained. Using the methods of k-dominance and multi-dimensional scaling (see Chap. 8.6.2) could somewhat compensate for this lack of information.

– When focussing on the reproductive and juvenile stages, particularly in experimental work, meiofauna investigations can be a valuable tool.
– In a stressed field situation, characteristics of population dynamics and behaviour gain an important role
– Results derived from simple tolerance experiments of "indicator species" hardly allow for predictions about mode and speed of recovery in the field. Instead, preference reactions would give a more "natural" and reliable answer.
– Ecophysiological parameters such as respiration measurements could indicate the status of physiological resistance ('OLAFSSON et al. 1990).

If carefully performed, meiofauna studies on pollution are not only a valuable supplement to the common macrofauna bioassays. With modern and case-specific methods, the inherent advantages of meiofauna studies listed above could render conclusions on anthropogenic pollution in a more rapid and reliable range than possible with macrofauna.

More detailed reading: PLATT et al. (1984); HEIP et al. (1988); WARWICK et al. (1990); WARWICK and CLARKE (1991); COULL and CHANDLER (1992); CLARKE (1993) WARWICK (1993)

CHAPTER 9

Synecological Perspectives in Meiobenthology

9.1 Structural Aspects of Meiofauna Populations

9.1.1 The Size Spectrum

In the marine realm, meiobenthos can be delineated from macrobenthos not only arbitrarily by the mesh size of sieves used for processing, but by the size spectra of the fauna which separate meiobenthos both from micro- and macrobenthos (Fig. 95; SCHWINGHAMER 1981a, 1983). Scrutinization of data from many investigation areas has confirmed the wide applicability of this benthic grouping. It emerged from calculations based on body size, biomass, even assimilation effectivity (GERLACH et al. 1985; WARWICK et al. 1986a).

What are the reasons for this consistent phenomenon which seems to define meiobenthos biologically and not just technically by arbitrary limits of separation? SCHWINGHAMER assumed the relation of body size to grain size to be responsible and argued that the animals have either to be small enough to live interstitially or large and powerful enough to displace the sediment particles. However, the separated grouping of meiofauna was found valid both in muddy bottoms and in sand. Also salinity gradients are irrelevant since also for fauna from estuarine bottoms this scheme turned out to hold true.

According to WARWICK (1989), the underlying factors are of a more principal biological and evolutionary nature (see WARWICK et al. 1986a). Following the principle of competitive displacement, competition with the other benthic elements causes a delineation not only in size and biomass, but probably also in a whole array of biological characteristics such as metabolic efficiency. The demarcation towards the Protista (microbenthos) may be based on the fact that these are mostly hapto-sessile organisms adhering to sediment particles while most meiobenthos is vagile. The numerous differentiating features towards the macrobenthos are summarized in Table 15.

It seems that the diverging size spectra result from mutually structuring effects evolved during geological times from biological interactions between larger and smaller benthos. As a consequence, today, the different benthic groups occupy different ecological niches thus avoiding competition. This principle becomes evident in larvae: as long as larvae of macrobenthic species are in the size range of meiofauna, they prevent direct competition and predation by evasion as meroplankton into the pelagial. They do not settle prior to reaching

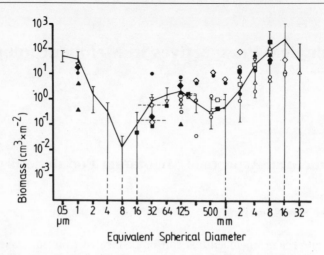

Fig. 95. Size spectra of benthic marine fauna. (SCHWINGHAMER 1981a)

Table 15. A biological delineation of meiobenthos vs. macrobenthos. (WARWICK 1984)

Animals	Weight <45 µg	Weight >45µg
Development	Direct, all benthic	With planktonic stages
Dispersal	Mainly as adults	As planktonic larvae
Generation time	Less than 1 year	More than 1 year
Reproduction	Mostly semelparous	Mostly iteroparous
Growth	Attain an asymptotic final size	Life-long permanent growth
Trophic type	Often selective particle feeders	Often non-selective particle feeders
Mobility	Motile	Also sedentary

a size beyond the main prey size of the meiobenthos. According to WARWICK, it is mainly this principle which is responsible for the size separation of the benthic fauna. Consequently, in a diagram of size structure, the size range of meroplanktonic larvae of the macrobenthos will exactly fit in the trough between the peaks of benthic meio- and macrofauna (Fig. 96).

In polar seas or in freshwater biotopes, where macrofauna has predominantly holobenthic larvae, and planktonic stages are rare, it would be of predictive value to ascertain whether the characteristic bi- or trimodal curve of the size spectrum became unimodal without separating troughs.

In fact, the graph of size spectra of the limnetic benthos is a unimodal curve without any seperation between meio- and macrofauna. According to STRAYER (1986), this is due to the numerous oligochaetes and chironomid larvae, characteristic members of the freshwater meiobenthos, which in their size spectrum exactly fill the trough between typical meio- and macrobenthos. But also WARWICK's (1989) arguments are corroborated: while in freshwater

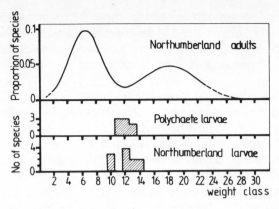

Fig. 96. The size spectrum of benthic fauna related to that of planktonic larvae. (WARWICK 1989)

habitats many factors are comparable to the marine situation, the reproductive biology of the benthos is essentially different since planktonic larvae are largely lacking. As an additionally divergent factor, in freshwater a good part of the temporary meiofauna, the insect larvae, never reach a competitive size with macrofauna because after metamorphosis they leave the aquatic biota. Thus, the competitive situation between meiofauna and macrofauna in freshwater is principally different from the marine realm: Contrasting reproductive modes and, subsequently, size spectra do not evolve, and this is evidenced by the unimodal graph for the size spectra.

It could be inferred that through the size separation which is based on ecological and evolutionary differences, today there is little interaction and competition between meio- and macrobenthos; but the evidence rather contradicts this argumentation (COULL and BELL 1979; BELL and COULL 1980; REISE 1985; WATZIN 1986). Particularly between temporary and permanent meiofauna of the same size class there is a fairly high degree of interactive effects (see Chap. 2.2.4) which probably have a continuously structuring impact on the relations between macro- and meiobenthos.

9.1.2 Factors Structuring Meiofauna Assemblages

Which are the predominant factors structuring a meiobenthic assemblage and determining its diversity? There is no generally valid answer. One contention infers that biotic factors such as food supply, predation, competition and reproductive strategies are decisive, while others emphasize the structuring impact of abiotic parameters such as exposure, temperature and salinity. Of course, there are good examples for both aspects in meiobenthic ecology, mainly depending on the area investigated, the animal group studied, and the methods used. Biotic, particularly nutritive interactions constrain the ecological niches of meiofauna, they become narrow, and in these micro-niches specialization of meiobenthos is fairly high (GRAY 1978). This would lead to smaller populations and increased diversity. The rich diversity, typical of meiofauna, may, therefore, be related to the dominance of highly discriminative deposit feeding (WARWICK et al. 1990b).

From cage experiments in the North Sea tidal flats, REISE (1985) concluded that it is mainly predation pressure which structures the populations of meiobenthos while physical disturbances such as low temperatures and storms act more marginally and irregularly only. But in many habitats meiofauna is also subject to strong biotopical constraints resulting in similar biological adaptations (e.g. interstitial fauna, see Chap. 4.1). Here, functional similarity and competition, not so much predative interaction, seem to exert a strongly structuring effect (RHOADS and YOUNG 1970; WOODIN and JACKSON 1979). This non-predative interaction would lead to displacement, cause a less stable and mostly reduced diversity, and ultimately to "amensalism," a suppression and spatial partitioning of species (see Chap. 2.2.4).

Biotic interactions will gain importance in physically more stable environments such as atidal seas, groundwater systems, or deep-sea bottoms. Here, competition and predation will enhance the number of more specialized species, usually grouped as K-strategists, in a highly interactive and patchily distributed meiofauna community of clearly hierarchial structure. Contrastingly, the eulittoral shores represent physiographically more extreme habitats. They harbour a meiofauna exposed to a system of rigid abiotic factors. Here, numerous irregular disturbances, seasonal variations, and tidal fluctuations become the structurally dominant factors (HULINGS and GRAY 1976). They favour a meiofauna assemblage with a more uniform species composition and a high percentage of opportunistic r-strategists, with reduced hierarchial structure and little biological interaction. The distribution in this type of assemblage, mediated by the physical regime, is more homogeneous and dominated by a community of high abundance and low diversity.

Naturally, there are all transitional phases and "compromises" between these rather extreme structural sketches of a meiofauna assemblage. As was pointed out for macrobenthos, also populations of opportunists can become regulated in their colonization sequence by biological interactions such as facilitation, inhibition or depletion (WHITLATCH and ZAJAC 1975; CONNELL and SLATYER 1977).

Studies of natural disturbance use diversity as a sensitive and highly reactive meiofaunal community parameter. In this respect they correspond to investigations on disturbance caused by pollution (Chap. 8.6). Diversity integrates various interactive factors such as intensity of disturbance, nature of the biotope and structure of the community. But it is this interdependence from numerous factors relevant also for a meiofauna assemblage which renders the interpretation of diversity indices so difficult. "Up to now it has ultimately not conclusively been explained which factors or factorial combination is responsible for the high diversity of meiofauna" (HERMAN and HEIP 1988).

General benthic ecology, but also meiobenthology, could show during the last years that disturbance and diversity are not necessarily negatively correlated. Inversely, stability of a biotope does not in each case enhance diversity (WARWICK et al. 1986b) as has been earlier generally assumed. A natural ecosystem with its incessant cases of disturbance is never completely stable (HUSTON 1979), because disturbances interfere with the established ecological balance and create new niches. Smaller disturbances (biotic and abiotic ones) continuously offer new chances for colonization and adaptation. As much as catastrophical disturbances may devastate the benthic communities, so will smaller events open new niches, and thus reduce the intensity of competitive displacement. This opening of new niches will support those species which, under a higher competitive pressure, would have been removed by selection. It can be inferred that in extremely stable ecosystems diversity would probably be lower than in slightly disturbed ones. Consequently, even in eulittoral areas with disturbances of predominantly abiotic origin, a considerable meiofauna diversity can be maintained, provided the interferences are not too drastic. This could explain the high turbellarian diversity in tidal flats of the Isle of Sylt, North Sea (ARMONIES 1986; REISE 1988), which in a merely stability-controlled system should be populated by a more uniform assemblage.

After a severe devastation of meiofauna, the first phases of recolonization proceed very rapidly, as evidenced in recolonization experiments (HOCKIN 1982a). The more the set of ecological niches is occupied, the more this process gradually slows down. Since the width of niches becomes continuously narrower, a higher specialization is necessary to successfully colonize in a later phase. Thus, the structure of the meiobenthos assemblage will become altered from a mainly r-selected to a more K-selected one. The more isolated a "defaunated island" is, and the larger the devastated area, the slower the process of recolonization from the surrounding undisturbed biotopes.

As a result, the recolonizing meiofauna assemblage will at first be relatively poor and low in diversity. Within the variety of differently reacting meiofauna groups, harpacticoids are, in spite of their general sensitivity, the most commonly and rapidly recovering taxon due to their active emergence (see Chap. 7.1.2). In contrast, the usually more "sediment-bound" nematodes have a lower potential of recolonization. But these general features which corroborate the zoogeographical "island theory", also depend on the degree of maturity and complexity of the neighbouring "donor" associations, whether they are dominated by opportunistic generalists or highly adapted specialists (AZOVSKY 1988). Thus, disturbance creates a mosaic of differently composed patches depending on the state of the previous fauna and its capacity of recovery. In this context isolation in small ponds or coastal water bodies etc. may be of structuring relevance for meiofauna communities (CASTEL 1992). In the long term, this can lead to genetic drift of the populations.

However, meiofauna recolonization after defaunation of an area is to a considerable degree also a matter of chance, influenced by unpredictable events such as storms, irregular exposure, specific reproductive patterns prevailing in the area, and seasonal and trophic conditions. These often accidental factors render the time scale of recovery after a disturbing event fairly unaccountable. The composition of the establishing meiofauna will never be exactly identical to the status quo ante, albeit the general structural traits may become predictable (RHOADS and YOUNG 1970).

This textbook will not detail the characteristics, the strong and weak points of the various existing diversity indices (for reference see HEIP et al. 1988). Provided the taxa have been keyed out to a sufficiently indicative and homogenous level (see Chap. 8.6), the ease of calculating diversity indices contrasts to the difficulty of a correct interpretation. Especially if not applied correctly, can "diversity indices hide more than they reveal" (PLATT et al. 1984). Along this line the drastic characterization of a diversity index by CRISP and MWAISEJE (1989) is understandable, who stated that "it is most frequently used to describe examples of impoverished data collecting." Particularly often it is overlooked that the mostly used SHANNON-WIENER index (SHANNON 1949) is strongly distribution-dependent. For a distribution-independent data matrix, other indices, e.g. the simple BERGER-PARKER index $(d = N_{max}\text{-}N_T)$, should be applied $(N_{max} = \text{abun-}$ dance of the most frequent taxon, and $N_T = $ overall abundance of meiofauna in the ·sample).

9.2   Meiofauna Energetics: Abundance, Biomass and Production

9.2.1   General Considerations

The interactive relations of meiofauna to other faunal elements and the contribution of meiobenthos to the energy flux through the benthic ecosystem can be assessed by measuring numerical parameters such as population density (abundance, biomass), production and, for better comparison, turnover rate of meiofauna. A detailed report on the inherent problems, with description of the relevant methods and a critical evaluation of their applicability is given by FELLER and WARWICK (1988). Also the competent accounts by GRAY (1981) and CRISP (1984) should be thoroughly studied. Hence, methods will not be considered here in detail except for some general aspects.

Basing on abundance data, the biomass at a given time and for a given area ("standing stock") will be determined. However, simple numeric abundance parameters remain of very limited ecological value if life history data are not considered. GERLACH (1971) pointed out that meiofauna in terms of weight attains just 3% of the overall biomass; however, their nutritional share within the food web approximates 15%. It is only when life history parameters are combined with abundance data that meaningful estimates of energy flow can be made. This is achieved by calculating production, "the gain in organic substance per unit of time" (REMMERT 1992). Since this value does not directly depend on biomass, it is a more realistic basis for comparisons between meio- and macrofauna.

Production is, however, difficult to calculate. The relevant biological parameters such as fecundity, natural and predative mortality, generation time, and other data on population dynamics are little known and highly variable. In many common species our ignorance of these simple life history data is even far more misleading in quantitative calculations than the notoriously imperfect methodological approaches. Especially for meiobenthos, life history parameters mostly originate from laboratory studies with their unnatural conditions. Moreover, life history cannot be considered a static process, the parameters vary intraspecifically with environmental conditions and food supply (see VRANKEN et al. 1988, for nematodes). Interspecifically, this variability is even enhanced. Particularly for meiofauna with their short life cycles, the problem measuring production in adequate time intervals remains a difficulty. An annual time span is often too long to yield reasonable values. The annual number of generation varies considerably (Table 17), and in many populations, growth of distinct generations (= cohorts) is not identifiable since reproduction is continuous and/or generations overlap. It is important to assess life-history data for the meiofauna group in question. Only then can calculation of production rates be based on adequate time intervals and both realistic and reliable values be obtained.

A simple and often used estimate of production is the P/B ratio. Relating population production (over a time period, mostly annual) to the mean biomass, it expresses the flux of matter, its turnover rate, and thus, compensates for the differing biomass and life time in the various taxa (Table 18).

The high variability in biological parameters entails a high inaccuracy of data when generalized for a large taxon like nematodes, harpacticoids, or even meiofauna. The more detailed data on single species we have, the more reliable become our compilations, and the more useful they are for testable ecological hypotheses and predictions. However, assessment of the animals' life history is extremely tedious, and at present, often not considered as a modern science worth being financially supported.

It is mainly among the harpacticoid copepods that we find careful and detailed autecological analyses of growth and production. For *Canuella perplexa*, CECCHERELLI and MISTRI (1991) showed that direct measurements yielded results considerably different from indirect calculations which greatly underestimated production. Another harpacticoid species whose production has been assessed in detail and whose population development followed over a whole decade is *Parastenhelia megarostrum* from New Zealand (HICKS 1985, 1991). The detailed studies by FLEEGER and PALMER (1982) and MORRIS and COULL (1992) focussed on the common harpacticoid *Microarthridion littorale*. The latter authors found a remarkable dominance of the naupliar stock whose large interannual variations (due to predation and natural mortality) determined the overall population size of the species. Various *Tisbe* species have been studied in detail revealing differences of growth and of reproduction in the morphologically closely related species flock (BATTAGLIA 1957; GAUDY and GUERIN 1977). The remarkable harpacticoid *Drescheriella glacialis* from Antarctic sea-ice showed extensive growth and short generation times despite the low ambient temperature (BERGMANS et al. 1991). The concluding remark by CECCHERELLI and MISTRY (1991) should be emphasized here: "a great deal of more data is needed before it becomes possible to refine the relationships and to establish... equations according to taxonomical and ecological groups."

In order to circumvent the laborious direct measurements, generalizing approaches have been developed which summarize existing single data. For instance, total biomass of meiofauna has been related in this compilatory way to total annual production. Regarding the different methods applied in the various studies and the different faunal properties existing in the various areas, results and formulae obtained from these reductive and generalized compilations should not be overestimated. They are certainly mathematically not very meaningful if they produce figures calculated to the fourth place after the point. Most energy-flux diagrams existing in literature which consider meiobenthos and contain quantitative figures do not as yet stand up to the exactness that their figures insinuate. A factual scrutinization would prove them to be $\pm$ rough or fictious estimates with a rather arbitrary relation to reality. (In a general energetic flux diagram these sources of error do not only refer to animal groups. This is pointed out for the production of diatoms by PINCKNEY and ZINGMARK 1991). "Generalizations... remain useful but also dangerous tools" (VRANKEN et al. 1988). The value of the existing models based on meiofauna is more in their qualitative character than in a representation of realistic quantities of energy metabolized.

9.2.2 Assessment of Energetic Parameters

Following these general outlines some examples should indicate how to assess the various parameters required for estimates of production. They simultaneously illustrate the extreme variability and some of the pitfalls emphasized above. Since the (marine) meiofauna is in almost all habitats dominated by nematodes, the abundance obtained for nematodes is often representative for the whole community and thus allows for some first generalizations (see Fig. 35; Chap. 5.6). This refers also to biomass: in eulittoral salt marshes nematodes represented >90% of all living biomass and still almost 80% in sublittoral reaches (SIKORA et al. 1977).

9.2.2.1 Abundance, Density

As evidenced in Table 16, the figures for meiofauna density vary greatly and authors giving average values should be aware of this difficulty. The review by MCINTYRE (1969) reports a span between 30 and 30,000 ind. $10\,\text{cm}^{-2}$. If 1000

Table 16. Meiofauna abundances from various habitats (compiled from McIntyre 1969 and some other authors)

Habitat and locality	Meiofauna abundance (ind. $10\,cm^{-2}$)
Sandy beaches, tidal:	
West coast of Scotland	Between 1000 and 4000
West coast of Denmark	Between 750 and 1900
Shores of Indian Ocean	Between 1000 and 10,000
East coast of Australia	About 470
Sandy beaches, atidal:	
East coast of Sweden	Between 200–1000
Kattegat (low tides)	About 500
Fine sandy to muddy tidal flats:	
Lynher estuary; England	About 12,500
Coast of Netherlands	About 2500
Vellar estuary, India	Between 420 and 3800
Shallow subtidal sediments:	
Sand, east coast of Denmark	Between 600 and 1300
Fine sand, Belgium	About 2200
Fine silty sand, Helgoland (North Sea)	About 4000
Mud, Fladen Ground, North Sea	Between 900 and 3200
Mud, English Channel	Between 90 and 200
Mud, Buzzards Bay, east coast of USA	Between 280 and 1860
Mud, Finnish coast, low brackish water	Between 11 and 100

to 2000 ind. $10\,cm^{-2}$ can be assumed as an average value (Coull and Bell 1979), meiofauna would exceed macrofauna in abundance two to three times. Particularly high values for meiofauna, sometimes ten times exceeding the above figures, have been recorded from silty mud and fine sand in eulittoral lagoons and tidal flats. Towards sublittoral depths these extreme densities become reduced and in greater depths do not exceed 2000 ind. $10\,cm^{-2}$.

9.2.2.2 Biomass

With a generalized value of about $1-2\,g$ dry wt. m^{-2} and peaks around $5\,g$ (Coull and Bell 1979), the meiofauna of shallow littoral bottoms usually attains only less than 10% of the corresponding macrofauna values. However, in more extreme biotopes such as sandy beaches, brackish water regions and deep-sea bottoms (Thiel 1972b; Tietjen 1992, see Chap. 8.3), the macrofauna is much more scarce, resulting in a relative increase of meiofauna (Fig. 97). Here, sometimes the biomass-relation between the two faunal groups can become 1:1.

Some methodological comments on biomass values: if stated as fresh weight, one should indicate whether based on unfixed or formaline-fixed material. Fixation will enhance the weight values considerably (Wiederholm and Erikson 1977; Widbom 1984) and also alter other weight parameters like carbon content.

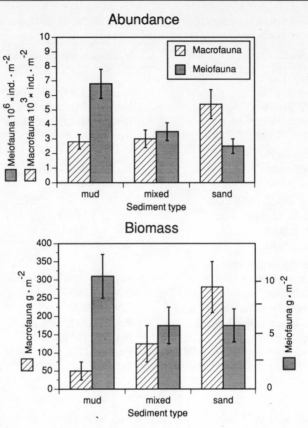

Fig. 97. Abundance and biomass (wet weight) of meiofauna vs. macrofauna in different sediment types of the eastern Baltic Sea. (After ANKAR and ELMGREN 1976)

In order to reduce sources of error due to different water content, assessment of dry weight is required. If direct weighing is not possible, the compilation of typical dry weight values for different meiofauna groups (FAUBEL 1982) or single species (HUYS et al. 1992) may serve as a useful basis.

For conversion of wet weight to dry weight a factor of 0.2 to 0.25 (GERLACH 1971; FELLER and WARWICK 1988), or also 0.15 (SCHIEMER 1982) is frequently adopted. The most reliable weight parameter is the carbon-content (ash-free dry weight). As a rough approximation, a factor of 0.4 can be used for conversion of C-content from dry weight (FELLER and WARWICK 1988). Frequently derived from the C-content, the caloric value serves as a basis for further calculation of production. SIKORA et al. (1977) could demonstrate for meiobenthos the small variance and superiority of caloric values compared to the far more demanding direct calorimetry. They calculated caloric values by conversion from C-values (WINBERG 1971).

Parameters often measured to quantify the overall amount of meiofauna present are the ATP content, the dehydrogenase activity, and activity of the electron transport system (ETS). The value of these parameters is their restriction to processes in living cells. The methods involved are well established in physiological studies based on isolated organisms under controlled conditions. However, referring them to the field situation and applying them to complex multi-species communities, there remain severe problems of interpretation due to of non-reproducible fluctuations and inconsistent variations. Enzyme activity as well as electron transport are dynamic processes depending on the physiological status, age and other individual variables. This could explain the high inter-variability often recorded in replicates. It renders interpretation of these data problematic. If used for whole sediment cores, the necessary separation of animal-related values from bacteria-related ones is of particular difficulty. Expressing the ATP content in terms of the more illustrative wet weight, GOERKE and ERNST (1975) used a factor of 740 based on experiments with nematodes.

The methodological procedures to measure these parameters are relatively simple and the results very exact. As with many of the "modern" quantitative methods, it is not so much a problem of assessing accurate values, but it requires good experience and a solid understanding of the often complicated underlying physical and (bio-) chemical processes to succeed in a reasonable interpretation (see Chap. 2.2.2).

9.2.2.3 P/B-Ratio, Generation Time

P/B-ratios are mostly calculated for the annual cycle, but considering the high number of generations in meiofauna populations and the seasonal fluctuations in standing stock, the P/B per cohort or generation is probably more meaningful and also less variable (CECCHERELLI and MISTRI 1991). The problems in generalizing an average figure of annual generations for meiobenthos have been stressed by HEIP et al. (1985a), who showed that in nematode species alone it fluctuates between 1 and 40 (on average 5–15; see Table 17). GERLACH (1971) considered an overall annual generation number of 3 to be realistic; according to VINCX (pers. comm.), the figure was 20 for populations from fine North Sea sands. Phytal nematodes from *Sargassum* had 60 annual generations

Table 17. Annual number of generations for some common nematode species. (HEIP et al. 1985a)

Taxon	Generations
Oncholaimus oxyuris	1.6
Monhystrella parelegantula	5
Rhabidits marina	10
Chromadorina germanica	13
Monhystera denticulata	15
Diplolaimelloides brucei	17

Table 18. P/B ratios calculated for some meiobenthic species. (HEIP et al. 1985b)

Taxon	P/B-ratio
Harpacticoida	
Huntemannia jadensis	3.8
Tachidius discipes	9.3
Microarthridion littorale	18.0
Paronychocamptus nanus	24.5
Nematoda	
Oncholaimus oxyuris	3–6
Paracanthonchus caecus	10.4
Monhystrella parelegantula	18.2
Chromadora nudicapitata	31.4
Ostracoda	
Cyprideis torosa	2.7

(KITO 1982 in HEIP et al. 1985a) and *Monhystera disjuncta* in cultures can reach up to 70 annual generations (VRANKEN and HEIP 1986).

This high variability within the same taxonomical group renders calculations of the P/B for "nematodes" (viz. for the bulk of "meiofauna") which base on just one ratio, not too meaningful (Table 18; HEIP et al. 1985b).

For the harpacticoid *Parastenhelia megarostrum*, HICKS (1985) found seven generations and a P/B-ratio of 15. This is close to the P/B-value of 18 given by FLEEGER and PALMER (1982) for another common harpacticoid, *Microarthridion littorale*. GERLACH (1971, 1978) based his estimates of quantitative energy fluxes on an arbitrary overall P/B-ratio for all meiofauna of 9. This value is close to 8.4, calculated by WARWICK and PRICE (1979) on the basis of rather detailed meiofauna data and the formula given by McNEILL and LAWTON (1970, see below). However, as pointed out above, using generalized values one should keep in mind the extreme variability within the different meiofauna groups, even within related species and also habitats. This can be underlined by comparing deep-sea or stygobiotic meiofauna with annual ratios close to one and the ciliates from the shallow sublittoral of the Baltic Sea with a P/B of 260 (SICH 1990).

A method which solved the problems of estimating generation time was suggested by HEIP (1976). He calculated the density and biomass of eliminated individuals vs. average standing stock. Using field and laboratory studies he resulted in a generalized P/B-ratio of about 15 for a typical harpacticoid, but emphasized the possibility of lower values for other meiofauna. For comparison with macrobenthos, GRAY (1984) mentioned as corresponding values of larger benthic animals 0.1 and 5.5.

According to BREY (1987) the P/B-value for macrobenthic ecosystems can be roughly estimated applying only weight figures without the problematical generation time. Summarizing numerous data sets from benthic studies, he provided the formula

$$\log P/B = -0.2455 - 0.1663 \log B - 0.2019 \log W,$$

where B is the mean annual biomass of a species and W the mean individual weight. Similarly, BREY (1987) calculated the mean annual production of a benthic species as

$$\log P = -0.2578 + 0.9905 \cdot \log W.$$

9.2.2.4 Respiration

Rather than using biomass and generation time, calculation of production is often derived from animal respiration with subsequent conversion into production of carbon or caloric values. The high respiration rates of meiofauna (as compared to macrofauna) given by GERLACH (1971) are rough estimates, but they indicate that the metabolic rate of meiobenthos is about five times larger than that of macrobenthos, i.e., meiofauna consumes five times more energy than the equivalent of macrofauna would do.

Today, the oxygen consumption of single species can be measured fairly reliably even for meiobenthos in polarographic microrespirometers (GNAIGER 1983, 1991) under variable environmental conditions (temperature, salinity). However, extrapolation from these values to the community or field situation has often been criticized as yielding unnatural figures. In other approaches, community respiration under in-situ conditions has been recorded with bell-jar apparatus (PFANNKUCHE and LOCHTE 1990; PFANNKUCHE 1993).

All the sophisticated respiration techniques are methodologically hampered by problems of low reproducibility and require careful calibration and confirmation by replicate runs. Respiration of the test animals often interferes with that of developing bacteria in the test chamber and has to be adjusted through "blank runs." The serious impact of methodological errors in the measurement of respiration has been compiled by MALAN and McLACHLAN (1991). In addition to the methodological problems, the experimenting scientist faces considerable problems of interpretation. As with other processes in living organisms, respiration depends on individual metabolism with all its regular and irregular changes. Intermittent phases of an anaerobic metabolism are surprisingly common in meiobenthos which confirms the statements of SIKORA and SIKORA (1982) and REVSBECH and JØRGENSEN (1986) about the frequent anoxia to which meiobenthos is exposed (see Chap. 8.5). The respiration rate will become modified by the structure of the substrate but also by population density and excretion of metabolites (see also BOADEN 1989a) which have to be considered when interpreting values of oxygen consumption. Moreover, and almost needless to point out, respiration rates vary considerably within species belonging to one taxonomic group (Table 19) rendering extrapolations from single-species respiration experiments for calculation of whole-group production problematical.

Table 19. Respiration rates ($nlO_2 \, \mu g^{-1}$ dry wt. h^{-1}) calculated for various meiobenthic groups. (HERMAN et al. 1985)

Taxon or ecological group	Respiration
Nematoda	
Non-selective deposit feeders	1.77
Epigrowth feeders	2.56
Omnivores and predators	3.78
Ostracoda	
Cyprideis torosa	1.78
Harpacticoida	
Canuella perplexa	3.72
Mesochra lilljeborgi	10.47
Tachidius discipes	12.59

Values for respiration can also be mathemetically assessed (Table 19). They are calculated from dry weight and circumvent the problems inherent in direct measurement (HEIP et al. 1982a):

$$R = a \cdot dwt \, b, \text{ where } a = a \text{ species-specific constant value}$$

transformed:

$$\log R [nlO_2 \cdot h^{-1}] = \log a + b \cdot \log dwt, \text{ where } b = 0.75; \text{ "three quarters rule".}$$

Calculating with $a = 0.42$ for nematode respiration, HEIP et al. (1982a) resulted in a straight regression line (logarithmic scales) when relating respiration to body weight.

From values for respiration, production can be calculated using formulae suggested by McNEILL and LAWTON (1970):

$$\log P [\text{kcal} \, m^{-2} yr^{-1}] = 0.8262 \cdot \log R - 0.0948$$

or by HUMPHREYS (1979):

$$\log P [\text{kcal} \, m^{-2} yr^{-1}] = 1.069 \cdot \log R - 0.601.$$

Using conversion factors, the resulting energy values can be expressed as weight units:

$$1 \, g [\text{ash free dry wt.}] = 5.6 \, \text{kcal or } 23 \, \text{kJ (WINBERG 1971), or, emphasizing}$$
nematodes:

$$1 \, g [\text{ash free dry wt.}] = 6.1 \, \text{kcal (SIKORA et al. 1977) or } 5.3 \, \text{kcal (CECCHERELLI}$$
and MISTRI 1991)

respectively, and

$$1 \, g [\text{ash free dry wt.}] = 25 \, \text{kJ (SCHIEMER 1982).}$$

It remains to be shown how reliable these mathematically assessed figures prove to be.

9.2.3   The Energetic Divergence Between Meiofauna and Macrofauna

Compared to macrofauna, meiofauna usually has a low standing stock; but even in tidal flats and in the shallow subtidal, where macrofauna is relatively rich, the higher turn-over rate of meiobenthos generates a high production frequently exceeding that of macrofauna. In extreme biotopes such as beaches, meiofauna is even more clearly favoured (Table 20). In the upper reaches of a beach and in estuaries, the contribution of micro- and meiofauna to the total metabolism can even go up to 97%.

Meiofauna (with nematodes and harpacticoids dominating) have been calculated to consume twice the amount of carbon and to produce four times as much as the ambient macrofauna with only half its biomass (WARWICK et al. 1979). Nematodes have a turnover rate 21 times higher than that of the macrobenthos in the same biotope. FENCHEL (1978) underlines the resulting productive importance of the small benthos when he states that in the sandy sublittoral the micro- and meiofauna contribute 40% of the total benthic metabolism.

Generalizing, one can estimate that in an average benthic marine biotope, meiofauna produces about one quarter of the total energetic budget, to which microfauna and microphytobenthos add at least the same portion (MUNRO et al.

Table 20. Comparison of biomass and production for macrofauna vs. meiofauna. The three subtables represent values from different biotopes and have differing units of measurement. (**A** from KOOP and GRIFFITHS 1982; **B** from KUIPERS et al. 1981; **C** from GERLACH 1978)

A Exposed beach, South Africa	Biomass (g dry wt m^{-1})	Production (g dry wt m^{-1} yr^{-1})	P/B ratio	Productive share (%)
Macrofauna	241	603	2.5	3
Meiofauna	227	2267	10	10
Bacteria	663	19,892	30	87
B Tidal flats, Netherlands	(g C m^{-2})	(g C m^{-2} yr^{-1})		
Macrofauna	27	27	1	29.3
Meiofauna + microfauna	1	65	65	70.7
C Sublittoral silty sand, generalized	(g dry wt m^{-2})	(g dry m^{-2} yr^{-1})		
Macrofauna (only deposit feeders)	6	12	2	9.4
Meiofauna	1.5	11	10	8.7
Microfauna + bacteria	5	104	20	82

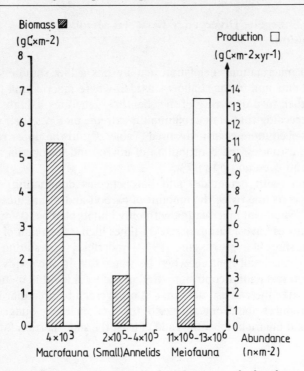

Fig. 98. The inverse relation of biomass and production in the benthos – a comparison of macro- and meiobenthos. (Based on data from WARWICK et al. 1979)

1978). For the slowly growing benthic fauna, the fixation of energy is the dominating characteristic, for the smaller organisms, the production and turnover of energy is the domain. This contention imparts to the more conspicuous macrofauna only a moderate productive significance, an aspect often not adequately considered in ecology textbooks.

The data given by WARWICK et al. (1979) for a tidal mudflat (Fig. 98), should be taken as a general example for the inverse relation in the energy budget of meiofauna and macrofauna.

More detailed reading: GERLACH (1978); WARWICK et al. (1979); HEIP et al. (1982a, 1985b); FELLER and WARWICK (1988)

9.3   The Position of Meiofauna in the Benthic Ecosystem

Energy flux diagrams have repeatedly been designed not only to illustrate, but to quantify (in cal or J) the energetic connections of meiofauna with other faunal compartments in the benthic system (WARWICK et al. 1979). However, this approach is severely hampered by the trophic selectivity characteristic for many meiofauna. Our limited knowledge of trophic relations and life history data (see above) requires generalizing data obtained from a few studies and summarizing them in a reductionistic way for large benthic groups like "nematodes" or even "meiofauna". Using this approach, meiofauna has been found linked to almost all trophic compartments (compare ELMGREN 1978), a result of little enlightening value.

For principal reasons, WATLING (1991) sharply criticizes bulk measurements such as organic carbon and nitrogen as indicators of nutritional value of a faunal compartment or a sediment. Arguing that much of this organic matter is bound into humic polymers and not available to the animals, he suggests techniques analyzing the digestible fractions to be much more meaningful. A powerful tool to better assess transfer of energy through meiofauna might be fluorescence analysis of gut contents and amounts of reserve substances such as lipids in meiobenthic animals. Even more promising and universally applicable will be future work on food relationships and energetic transfer using molecular biology and genetic probes which directly mark the organisms involved.

In any case, rash generalizations are questionable considering the nutritional selectively even of closely related meiofauna species. This is confirmed by the laborious assessment of singular trophic relations (WATZIN 1985; see Fig. 32, Chap. 2.2.4). Another method which can evidence the detailed trophic relations

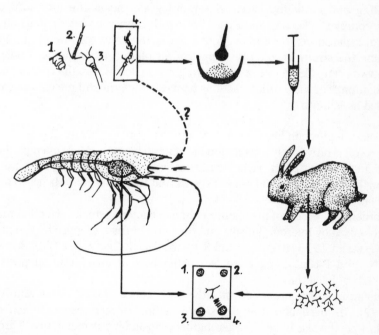

Fig. 99. The assessment of predator-prey relations with a serological test (After FELLER and WARWICK 1988). For further explanation see text

between macro- and meiobenthic species is based on a modified serological antigen-antibody reaction (Fig. 99; FELLER et al. 1979; FELLER and WARWICK 1988). This elegant, although tedious method uses the highly specific precipitative reaction of antigens from the stomach content of a putative predator with antibodies developed from a mixture of possible meiobenthic prey organisms (in rabbit blood serum). Recently, this method has been refined to an extent to even allow quantification of different food items in a heterogeneous diet (FELLER et al. 1990).

A direct proof and quantification of trophic interactions is the labelling with radioactive isotope tracers such as ^{14}C-carbon dioxide/^{3}H-thymidine or ^{14}C-acetate/^{3}H-thymidine (MONTAGNA 1983; MONTAGNA and BAUER 1988). With this method it was ascertained that nematode communities regulate bacterial numbers to a limited degree only, consuming just 3.6% of their biomass.

The value of both methods lies in their possible application to animals taken from the field or kept under largely natural conditions. Thus, we can hope to overcome the artificial situation which is prone to bias in many quantitative meiofauna experiments.

9.3.1 Links of Meiofauna to "the Small Food Web"

The close links between microorganisms, detritus and meiofauna with mutually facilitating and inhibiting effects closely integrate meiofauna into a "detrital trophic complex" (TENORE and RICE 1980) with numerous interactions and no clear-cut separations. Data from MONTAGNA (1984) and NILSSON et al. (1991) underline the strong regulatory impact of meiofauna on bacteria. Studies by ESCARAVAGE et al. (1989) show the bearing role of detritus on the energy budget of meiofauna. The various interactions between bacteria and meiofauna can be compiled as follows:

1. Meiofauna preys on bacteria, but, due to the immense bacterial productivity, the result is only rarely a reduction of bacterial biomass. More often predation has a stabilizing effect, removing senescent cells and maintaining the cell growth in an exponential phase (YINGST and RHOADS 1980; MEYER-REIL and FAUBEL 1980; MONTAGNA 1984).
2. "Gardening" effects of meiofauna enhance bacterial stocks. Excreted mucus is colonized by microorganisms and they are, in turn, cropped by meiofauna (HYLLEBERG 1975; RIEMANN and SCHRAGE 1978; GERLACH 1978). Secretion of N- and P-containing dissolved metabolites of meiofauna supports the growth of bacteria.
3. Meiofauna bioturbation activates the geochemical fluxes (ALLER and ALLER 1992). In particular the diffusion rate for oxygen becomes activated, enlarging the oxic habitats of aerobic bacteria and many meiofauna. Through the minute mucus-stabilized burrows and trails of nematodes, and the digging

and burrowing activity of ostracodes and harpacticoids, a considerably thicker oxic layer could be recorded in experimental sediments.

4. Mechanical break-down of detrital particles by meiofauna supports the bacterial decomposition (TENORE et al. 1977).

5. Decaying meiofauna serves as food for bacteria.

6. Specialized meiofauna serves as symbiotic partner for bacteria (see Chap. 8.5).

Mainly nematodes are considered to be linked into the detritus/bacteria-based food chain. But the contribution of bacteria to the diet of e.g. nematodes, the main component of meiofauna, is far from being unambiguous (see Chap. 9.3). Using tracer techniques, it was calculated that bacterivorous nematodes alone could probably not greatly stimulate bacterial growth (HERMAN and VRANKEN 1988). However, this interpretation might not be generally valid since nematodes lack important digestive enzymes. This lack may account for a considerable divergence between estimates of ingested and digested bacterial biomass and may result in contradicting calculations on their regulatory role. The extremely high bacterial turnover rates rapidly supplement the amount being devored by meiobenthos, so that the bacterial standing stock is reduced only to a very small percentage by meiofauna (3% of bacterial stock per h, see MONTAGNA 1984; < 1%, see EPSTEIN and SHIARIS 1992). ALONGI (1990a) calculated for sediment bacteria a production of 45 mg to 1.7 g carbon per m^2 and day! Further refinement of tracer methods (MONTAGNA and BAUER 1988; DOBBS et al. 1989) will hopefully yield more detailed data on the bacterial-meiobenthos trophic interactions and the energy transport through the food web.

The second main component of meiobenthos beside nematodes, harpacticoids, are mostly phytobenthos-based (see RUDNICK 1989). The high trophic relevance of microphytobenthos for meiofauna has been stressed by BLANCHARD (1991) while MONTAGNA et al. (1983) and NILSSON et al. (1991) concluded from their work a limited significance only of diatoms as nutritional source for meiobenthos (1% per h removed).

The grazing impact of benthic ciliates can be considerable, though divergent in the different trophic groups. It was calculated by EPSTEIN et al. (1992) to amount to 11% of the bacterial production, 10% of the total primary production and 93% of the daily dinoflagellate production, while only 6% of the produced diatoms was ingested. In addition to the components mentioned above, MONTAGNA (unpubl.) underlined the role of protozoa mediating in a "bacterial food loop" between bacteria and meiobenthos. This general view is supported by studies of WALTERS and MORIARTY (in press), who generalize that meiofauna mainly feeds on protozoa rather than bacteria. A high, but species-specific predatory pressure of meiofauna on ciliates is also reported by EPSTEIN and GALLAGHER (1992) in experiments from a sandy tidal flat on the White Sea (Russia). However, there is no unambiguous agreement regarding the trophic role of protozoa for meiofauna. Their impact on bacteria is assumed to be of subordinate rank compared to that of meiobenthic animals (WARWICK 1987).

9.3.2   Links of Meiofauna to the Macrobenthos

Compiling the vast literature on meiofauna-macrofauna interrelations there emerge two contrasting contentions. According to the earlier "dead-end hypothesis" meiofauna maintains close energetic links only to the detritus/bacteria complex discussed above (Chap. 9.3.1). The resulting energy, fixed by meiobenthic organisms, is usually not transmitted to a large extent to higher trophic levels (McINTYRE 1969), viz. to the macrofauna. Hence, meiofauna is often considered to represent a dead end in the food web, their stocks being only little regulated by predation of macrofauna. According to this conception, deposit feeding macrobenthos lives mainly on microorganisms and only 1/5 of its trophic needs are covered by consumption of meiofauna (GERLACH 1978).

Later, COULL and BELL (1979) argued that these values cannot be generalized and should be restricted to sandy sediments only. Here, the endobenthic life style would considerably reduce predation pressure by macrobenthos. Most meiobenthic biomass from sandy substrates will indeed be directly remineralized in a short-cut destruent cycle rather than transferred to macrofauna. For sand meiofauna, despite its considerable energetic turnover, this principal model of a rather isolated biotic element has been recently underlined by McLACHLAN and ROMER (1990). In contrast, in muds, the meiobenthos is richer and more concentrated at the superficial layers and frequently has an epibenthic life style. Here, the resulting higher exposure to predation renders meiobenthos important as a food source for macrobenthos.

Numerous studies support this alternative and point to strong mutual links between meio- and macrofauna. These exceed the merely nutritional aspect with meiobenthic animals serving as food for macrobenthos. Feeding rates of macrobenthic deposit feeders have been experimentally found to increase in the presence of meiofauna (TENORE and RICE 1980). Breaking down macrofaunal faecal pellets mechanically, meiofauna enhances the bacterial decomposition rate and, thus, helps to regenerate the sediment as the trophic basis for macrobenthos. On the other hand, as pointed out in Chap. 2.2.5, bioturbation and tube building of macrofauna increase the chemical fluxes and the structural complexity of the sediment. The filter currents of many macrobenthic species enrich bacterial biomass. Enrichment with oxygen caused by the locomotory activities of juvenile flatfish, shrimps and clams may lead to an enhanced meiofauna development in the eutrophicated, oxygen-deficient lagoonal system (CASTEL 1992). While these macrofauna activities contribute to produce a richer meiofauna, they support the contention that there is little direct competition or predatory pressure between the two benthic groups.

There is, however, also a direct negative impact exerted by permanent meiofauna on (juvenile) macrofauna (BELL and COULL 1980). The mechanisms of these inhibitory interactions are often not yet fully understood, but may be a combination of adverse sediment changes and competition. Macrobenthic animals develop very often from tiny planktonic larvae. After settling they become temporary meiofauna and are then directly exposed to the feeding pressure of permanent meiofauna (ELMGREN 1978). Meiofauna are known to be very effective food utilizers (WARWICK 1989; see Chap. 9.1.2). For meiofauna-sized macrofauna there has been postulated to exist a "meiofauna bottleneck" (BELL and COULL 1980) which was held to considerably reduce the macrobenthic offspring. The range of its impact being yet debated (ZOBRIST and COULL 1992), it is probably mostly based on predation and less so on disturbance and competition. With increasing growth, the situation becomes reversed and many macrobenthic species start to feed on meiofauna (see REISE 1979; BELL 1980; WATZIN 1985, 1986; TIPTON and BELL 1988; GEE 1989; COULL 1990; SERVICE et al. 1992).

What is the extent to which meiofauna directly serves as food for macrofauna? Particularly in the upper sediment layers and in muddy bottoms, meiofauna is intensively devoured by a multitude of smaller macrofauna such as small and juvenile fish, early ontogenetic stages of shrimps, crabs, polychaetes and ophiurids. The selectivity of this diet on meiobenthos is underlined by recent studies of FLEEGER (pers. comm.): With increasing size of a larval fish (spot, *Leiostomus*), the composition of its diet shifted from planktonic to meiobenthic copepods. Here, the initially preferred harpacticoid species were later replaced by other species and became then supplemented by nematodes.

According to the "packaging model" (LACKEY 1936), it is energetically much more parsimonious to ingest one meiofauna organism than numerous dispersed microorganisms of the same energetic value (SIKORA et al. 1977). This concentration of trophic energy in one larger meiobenthic particle could even compensate for the loss of total energy usually attributed to passing over from one trophic level to the next (approximating one order of magnitude). In limnic ecosystems, the few analyses of energy fluxes which consider meiobenthos point out that

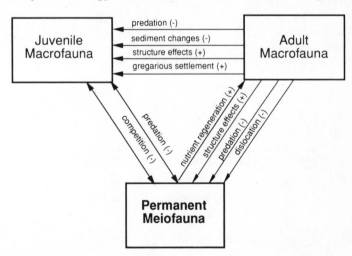

Fig. 100. Schematic compilation of positive and negative interactions between meiofauna, temporary meiofauna (juvenile macrofauna), and macrofauna. (After BELL and COULL 1980)

half of the overall energy passes through the meiobenthos, but is then transferred almost completely (80%) to predacious macrobenthic larvae of tanypodid midges, and, thus, ultimately exported from the system (STRAYER and LIKENS 1986).

Summarizing the numerous organismic interactions of meiobenthos in aquatic sediments, it results that these links are often much more differenciated and go beyond just straightforward predator-prey relations (Fig. 100). This renders their study fairly complicated. Meiofauna are as tightly linked to the macrofauna as they are to the complex of detritus and microorganisms.

9.3.3   Meiobenthos as an Integrative Benthic Complex

Meiobenthos is recognized as attaining a significant position first of all within the "small food web" (Fig. 101; KUIPERS et al. 1981).

Beside the small food web, there are numerous ties to the macrobenthos which put meiobenthos in a central position in the entire benthic food web. Judging from biomass only, the significance of meiobenthos is admittedly low, but its contribution to energy flow is convincing (Fig. 102), a fact often disregarded in ecology textbooks.

Admittedly, many exact and experimentally proven data are still lacking to unify the often controversial results and contentions about meiofauna as an ecological complex. The construction of ecosystem models clearly reveals the links of meiobenthos to other compartments of the system, but also the large gaps which still exist. At present, these models evidence mostly qualitative aspects rather than representing quantitatively reliable bases for the calculation of energy fluxes.

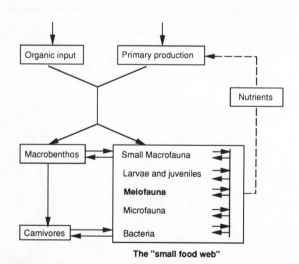

Fig. 101. The small food web, integrating meiofauna in a substantial compartment of the benthic energy flux. (After KUIPERS et al. 1981)

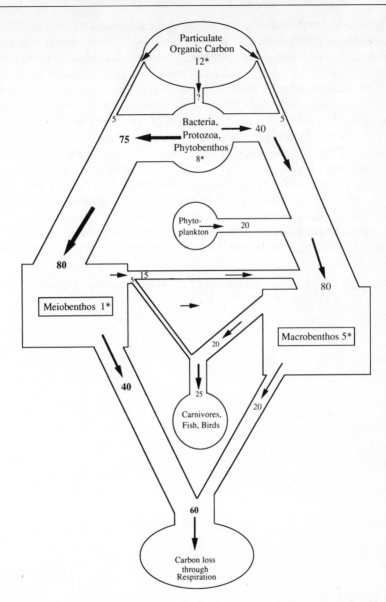

Fig. 102. The role and position of meiobenthos in a compilatory energy flow diagram. (After PLATT 1981): *numbers* reflect relative rates of carbon production (g C m^{-2}yr^{-1}), *numbers with asterisks* represent biomass values (g C m^{-2})

Future research will lead to numerous corrections in the quantitative relations. These refinements will not only refer to marginal aspects of energetic pathways. This was documented by recent results that benthic energy fluxes are based at least for 50% on sulphate reduction and that this process can readily be maintained also under oxic conditions (JØRGENSEN and BAK 1991). Another parameter that could largely change our view on oceanic transports of energy fixed in particles is the possible role of "marine snow" (Chap. 7.2 and 8.3). For future calculations of energy flow we need more reliable data on the life history of the (meiobenthic) animals. More scientific information and fewer hypotheses must be the platform for any generalization or even quantitative calculation of large-scale energy budgets. Once these data are provided, a more professional computation and computer-assisted modelling than usually practised today in the average institution will yield more generally valid and, consequently, more relevant cognitions on meiobenthos and its ecological impact. Only then will the role of meiofauna be realistically assessed in ecosystem models of predictive value.

More detailed reading: TENORE (1977); COULL and BELL (1979); TIETJEN (1980); WARWICK (1987); COULL (1990)

CHAPTER 10

A Retrospect on Meiobenthology and an Outlook to Future Research

Summarizing the short history of (marine) meiofaunal studies, meiobenthology can be largely grouped in periods with dominant research "trends". There are, of course, many singular publications that do not really fit into this gross categorization of meiofauna research, but a majority of papers seems to follow certain lines which are defined by the scientific standard of the time and by contentions about fields of major importance.

Last century until the end of the 1950s. Taxonomic and faunistic papers; microscope studies; establishment of the special character of many meiobenthic animals; attempts to create meiobenthos "coenoses"

1960 into the 1970s. Descriptive ecology of the meiobenthic habitat; analysis of abiotic factors (restricted to the technically more simple parameters) and their distributional impact; first experimental work on meiobenthic animals.

70s into the beginning of the 1980s. Experimental work, tolerance-preference tests; first studies on "difficult" abiotic parameters like the oxygen/sulphide complex and its impact on the distribution; biotic and trophic factors (predator-prey interactions); ultrastructural studies with respect to phylogenetic implications.

1980s. Assessment and conservation of meiofauna biodiversity on the basis of taxonomic studies; manipulative ecology helping to understand the role of meiofauna in benthic ecosystems; calculations on production and energetic interactions; work on diversity-relevant factors (habitat complexity); meiofauna in special biotopes like the deep-sea and sulphidic environments; impact of natural and anthropogenic disturbance on the diversity of meiofauna, recolonization and dispersive mechanisms of meiofauna, pollution studies using meiofauna as a tool.

The cognition of COULL and PALMER (1984) that "the easy part of meiofaunal research is done" holds still valid today and can be generalized. There are many unsolved problems in meiobenthology, not only of academic interest, but also of relevance for the more general fields of biology:

– to assess reliable data on life history and population dynamics of relevant species for calculation of meiobenthic production.
– to explain small-scale distributional patterns of meiofauna as a reaction to

structural complexity and micro-stratification of some dominant abiotic factors such as oxygen and hydrogen sulphide,

– to ascertain the impact of microbial processes on meiofauna, combined with further studies on the decomposition of organic matter,

– to further refine (e.g. by multifactorial experimental designs or in microcosms) the impact of complicated biotic factors like competition, trophic relations, facilitation and amensalism in our ecological studies. "Manipulative ecology" could be the appropriate method. Many data of this discipline point to the relevance of biotic factors about which we still know very little,

– to study the behavioural pattern and micro-ecological niche of meiobenthic animals to achieve a better understanding of the "normal" function, adaptive value and ecological role of structures and organisms. This would include meticulous investigations into the abiotic factor system, e.g the local hydro-dynamic fluxes, but also analyses of the structural/behavioral constraints of food uptake,

– to solve the problems of a thiobios and realize its ecological potential,

– to learn more about the ecology of freshwater meiofauna which attains partic-ular importance through global problems of drinking water supply.

– to analyze the reaction of meiofauna to natural and anthropogenic disturbance. This would help to better assess consequences of pollution and to follow the pathways or organism recolonization,

– to increase our understanding of the (eco-)physiology of meiobenthos,

– to understand the dispersive processes in meiofauna which would bring us further to solve the notorious problem of meiofaunal zoogeographical dis-tribution.

– to use morphology and ultrastructure as tools to understand the diversity, the function, the taxonomic relationships and the origin of meiobenthic forms,

– to reveal phylogenetic links which, in many cases, seem to focus on animals of meiobenthic size bridging the gaps between large taxa. A powerful tool will be application of RNA and protein fingerprinting techniques. New data on evolution would also increase our understanding of the role of refuge biotopes like the deep-sea, caves, continental groundwater.

For these goals, we need innovative, often experimental studies using modern methods (molecular biology, serology, manipulative ecology; histochemistry, ultrastructure, microtechnology). Attempts to transfer methods and terms from marine to freshwater meiobenthology and to compare pertinent studies (PALMER 1990b) would eventually connect the marine and freshwater realm as two related fields and strengthen the combined impact of their research.

Collecting more reliable data about the biology and ecology of meiobenthic animals, meiobenthology could reach a predictive power high enough to be of value for the solution of general ecological problems. Special questions have to be answered first in the laboratory under controlled conditions, but all laboratory work has to be cross-checked with corresponding results from the field.

A better knowledge would draw the attention of a wider scientific public to meiobenthos not only as a scientifically most rewarding animal group, but also as an exemplary and convenient scientific tool. The advantages of meiofauna, if restricted to higher taxonomic levels, as opposed to macrofauna for monitoring and pollution problems have to be emphasized more clearly, and adequate guides for identification should be created for this purpose. Thus, the inherent psychological problem of using animals of microscopical size, unknown to the wider public, could be overcome. All these steps could lead meiobenthology out of the somewhat isolated "corner" where specialists discuss their problems in some distance from the biological and ecological mainstream.

▓   *More detailed reading*: COULL and PALMER (1984); COULL and GIERE (1988)

References

Åkesson B (1973) Reproduction and larval morphology of five *Ophryotrocha* species (Polychaeta, Dorvilleidae). Zool Scr 2: 145–155

Åkesson B (1975) Bioassay studies with polychaetes of the genus *Ophryotrocha* as test animals. In: Koeman JH, Strik JJ (eds) Sublethal effects of toxic chemicals on aquatic animals. Proc Swedish-Netherlands Symp, Wageningen, The Netherlands, Sept 2–5, pp 121–135

Åkesson B (1977) Parasite-host relationships and phylogenetic systematics. The taxonomic position of dinophilids. Mikrofauna Meeresboden 61: 19–28

Ali A (1984) A simple and efficient sediment corer for shallow lakes. J Environ Qual 13: 63–66

Alkemade R, Wielemaker A, Hemminga MA (1992) Stimulation of decomposition of *Spartina anglica* leaves by the bacterivorous marine nematode *Diplolaimelloides bruciei* (Monhysteridae). J Exp Mar Biol Ecol 159: 267–278

Aller RC (1980) Relationships of tube-dwelling benthos with sediment and overlying water chemistry. In: Tenore KR, Coull BC (eds) Marine benthic dynamics. University of South Carolina Press, Columbia, SC, pp 285–308 (Belle W Baruch Libr Mar Sci, vol 11)

Aller RC, Aller JY (1992) Meiofauna and solute transport in marine muds. Limnol Oceanogr 37: 1018–1033

Aller RC, Yingst JY (1978) Biogeochemistry of tube-dwelling: a study of the sedentary polychaete *Amphitrite ornata* (Leidy). J Mar Res 36: 201–254

Aller RC, Yingst JY (1985) Effects of marine deposit-feeders, *Heteromastus filiformis*, *Macoma balthica* and *Tellina texana* and averaged sedimentary solute transport, reaction rates, and microbial distribution. J Mar Res 43: 615–645

Alongi DM (1985) Effects of physical disturbance on population dynamics and trophic interactions among microbes and meiofauna. J Mar Res 43: 351–364

Alongi DM (1987) Intertidal zonation and seasonality of meiobenthos in tropical mangrove estuaries. Mar Biol 95: 447–458

Alongi DM (1990a) Community dynamics of free-living nematodes in some tropical mangrove and sandflat habitats. Bull Mar Sci 46: 358–373

Alongi DM (1990b) The ecology of tropical soft-bottom benthic ecosystems. Oceanogr Mar Biol Annu Rev 28: 381–496

Alongi DM, Tenore KR (1985) Effect of detritus supply on trophic relationships within experimental benthic food webs. I. Meiofauna-polychaete [*Capitella capitata* (type I) Fabricius] interactions. J Exp Mar Biol Ecol 88: 153–166

Alongi DM, Tietjen JH (1980) Population growth and trophic interactions among free-living marine nematodes. In: Tenore KR, Coull BC (eds) Marine benthic dynamics. University of South Carolina Press, Columbia, SC, pp 151–166 (Belle W Baruch Libr Mar Sci, vol 11)

Anderson JG, Meadows PS (1969) Bacteria on intertidal sand grains. Hydrobiologia 33: 33–46

Anderson JG, Meadows PS (1978) Microenvironments in marine sediments. Proc R Soc Edinb B 76: 1–16

Andrew NL, Mapstone BD (1987) Sampling and the description of spatial pattern in marine ecology. Oceanogr Mar Biol Annu Rev 25: 39–90

Angelier E (1953) Recherches écologiques et biogéographiques sur la faune des sables submergés. Arch Zool Exp Gén 90: 37–161

Ankar S (1977) Digging profile and penetration of the Van Veen grab in different sediment types. Contrib Askö Lab Univ Stockholm, Sweden 16, pp 1–22

Ankar S, Elmgren R (1976) The benthic macro- and meiofauna of the Askö-Landsort area (Northern Baltic proper). A stratified sampling survey. Contr Asko Lab Univ Stockholm, Sweden 11, p 115

Archer D, Devol A (1992) Benthic oxygen fluxes on the Washington shelf and slope: a comparison of in situ microelectrode and chamber flux measurements. Limnol Oceanogr 37: 614–629

Archer D, Emerson S, Reimers C (1989) Dissolution of calcite in deep-sea sediments: pH and O_2 microelectrode results. Geochim Cosmochim Acta 53: 2831–2845

Arlt G (1988) Temporal and spatial meiofauna fluctuations in an inlet of the South West Baltic (Darss-Zingst Bodden chain) with special reference to the Harpacticoida (Copepoda, Crustacea). Int Rev Ges Hydrobiol 73: 297–308

Arlt G, Müller B, Warnack KH (1982) On the distribution of meiofauna in the Baltic sea. Int Rev Ges Hydrobiol 67: 97–111

Armonies W (1986) Plathelminth abundance in North Sea salt marshes: environmental instability causes high diversity. Helgol Meeresunters 40: 229–240

Armonies W (1988a) Active emergence of meiofauna from intertidal sediment. Mar Ecol Prog Ser 43: 151–159

Armonies W (1988b) Hydrodynamic factors affecting behaviour of intertidal meiobenthos. Ophelia 28: 183–194

Armonies W (1988c) Plathelminth fauna of North Sea salt marshes is strongly affected by salinity. In: Ax P, Ehlers U, Sopott-Ehlers B (eds) Free-living and symbiotic Plathelminthes. Fischer, Stuttgart, pp 505–509 (Fortschr Zool, vol 36)

Armonies W (1988d) Common pattern of plathelminth distribution in North Sea salt marshes and in the Baltic Sea. Arch Hydrobiol 111: 625–636

Armonies W (1989a) Meiofauna emergence from intertidal sediment measured in the field: significant contribution to nocturnal planktonic biomass in shallow waters. Helgol Meeresunters 43: 29–43

Armonies W (1989b) Semiplanktonic Plathelminthes in the Wadden Sea. Mar Biol 101: 521–528

Armonies W (1989c) Occurrence of meiofauna in Phaeocystis seafoam. Mar Ecol Prog Ser 53: 305–309

Armonies W (1990) Short-term changes of meiofaunal abundance in intertidal sediments. Helgol Meeresunters 44: 375–386

Arnaud PM, Poizat C, Salvini-Plawen L v (1986) Interstitial gastropods. In: Botosaneanu L (ed) Stygofauna Mundi. A faunistic, distributional, and ecological synthesis of the world fauna inhabiting subterranean waters (including the marine interstitial). Brill, Backhuys, Leiden, pp 153–176

Austen MC, Widbom B (1991) Changes in and slow recovery of a meiobenthic nematode assemblage following a hypoxic period in the Gullmar Fjord basin, Sweden. Mar Biol 111: 139–145

Austen MC, Warwick RM, Rosado MC (1989) Meiobenthic and macrobenthic community structure along a putative pollution gradient in southern Portugal. Mar Pollut Bull 20: 398–404

Ax P (1956) Die Gnathostomulida, eine rätselhafte Wurmgruppe aus dem Meeressand. Abh Akad Wiss Lit Mainz Math-Nat KI 8: 1–32

Ax P (1963) Die Ausbildung eines Schwanzfadens in der interstitiellen Sandfauna und die Verwertbarkeit von Lebensformcharakteren für die Verwandtschaftsforschung. (Mit Beschreibungen zweier neuer Turbellarien aus den Ordnungen Acoela und Seriata). Zool Anz 171: 51–76

Ax P (1966) Die Bedeutung der interstitiellen Sandfauna für allgemeine Probleme der Systematik, Ökologie und Biologie. Veröff Inst Meeresforsch Bremerhaven Sbd 2: 15–65

Ax P (1969) Populationsdynamik, Lebenszyklen und Fortpflanzungsbiologie der Mikrofauna des Meeressandes. Verh Dtsch Zool Ges Innsbruck 1968. Zool Anz Suppl 32: 66–113

Ax P (1985) The position of the Gnathostomulida and Platyhelminthes in the phylogenetic system of the Bilateria. In: Conway Morris S, George JD, Gibson R, Platt HM (eds) The origins and relationships of lower invertebrates. Clarendon Press, Oxford, pp 168–180 (Syst Ass, vol 28)

Ax P, Armonies W (1990) Brackish water Plathelminthes from Alaska as evidence for the existence of a boreal brackish water community with circumpolar distribution. Microfauna Mar 6: 7–110

Azovsky AI (1988) Colonization of sand "islands" by psammophilous ciliates: the effect of microhabitat size and stage of succession. Oikos 51: 48–56

Barnard JL (1969) The families and genera of marine gammaridean Amphipoda. Bull US Natl Hist Mus 271: p 535

Barnard JL, Barnard CM (1983) Freshwater Amphipoda of the world. I. Evolutionary patterns. II Handbook and bibliography. Hayfield Associates, Mt Vernon, Virginia, p 830

Barnes PO (1973) An in situ interstitial water sampler for use in unconsolidated sediments. Deep-Sea Res 20: 1125–1128

Barnett BE (1980) A physico-chemical method for the extraction of marine and estuarine benthos from clays and resistant muds. J Mar Biol Assoc UK 60: 225

Barnett PRO, Watson J, Connelly D (1984) A multiple corer for taking virtually undisturbed samples from shelf, bathyal and abyssal sediments. Oceanol Acta 7: 399–408

Bartolomaeus T, Ax P (1992) Protonephridia and Metanephridia – their relation within the Bilateria. Z Zool Syst Evolutionsforsch 30: 21–45

Bartsch I (1972) Ein Beitrag zur Systematik, Biologie und Ökologie der Halacaridae (Acari) aus dem Litoral der Nord- und Ostsee. I. Systematik und Biologie. Abh Verh Naturwiss Ver Hamb 16: 155–230

Bartsch I (1974) Ein Beitrag zur Systematik, Biologie und Ökologie der Halacaridae (Acari) aus dem Litoral der Nord- und Ostsee. II. Ökologische Analyse der Halacaridenfauna. Abh Verh Naturwiss Ver Hamb 17: 9–53

Bartsch I (1979) Halacaridae (Acari) von der Atlantikküste Nordamerikas. Beschreibung der Arten. Mikrofauna Meeresboden 79: 1–62

Bartsch I (1982) Halacaridae (Acari) von der Atlantikküste des borealen Nordamerikas. Ökologische und tiergeographische Faunenanalyse. Helgol Meeresunters 35: 13–46

Bartsch I (1989) Marine mites (Halacaridea: Acari): a geographical and ecological survey. Hydrobiologia 178: 21–42

Battaglia B (1957) Ricerche sul ciclo biologico di *Tisbe gracilis* (T. Scott), (Copepoda, Harpacticoida), studiato in condizioni di laboratorio. Arch Oceanogr Limnol 11: 29–46

Battaglia B, Beardmore JA (eds) (1978) Marine organisms. Genetics, ecology and evolution. In: NATO Advanced Research Institute, Venice 1977. Plenum, New York, 472 pp

Battaglia B, Bisol PM, Fava G (1978) Genetic variability in relation to the environment in some marine invertebrates. In: Battaglia B, Beardmore J (eds) Marine Organisms. Genetics, ecology and evolution. Plenum, New York, pp 53–70

Bell SS (1980) Meiofauna-macrofauna interactions in a high salt marsh habitat. Ecol Monogr 50: 487–505

Bell SS, Coen LD (1982) Investigations on epibenthic meiofauna I. Abundances on and repopulation of the tube-caps of *Diopatra cuprea* (Polychaeta: Onuphidae) in a subtropical system. Mar Biol 67: 303–309

Bell SS, Coull BC (1980) Experimental evidence for a model of juvenile macrofauna-meiofauna interactions. In: Tenore KR, Coull BC (eds) Marine benthic dynamics. University of South Carolina Press, Columbia, SC, pp 179–192 (Belle W Baruch Libr Mar Sci, vol 11)

Bell SS, Hicks GRF (1991) Marine landscapes and faunal recruitment: a field test with seagrasses and copepods. Mar Ecol Prog Ser 73: 61–68

Bell SS, Sherman KM (1980) A field investigation of meiofaunal dispersal: tidal resuspension and implications. Mar Ecol Prog Ser 3: 245–249

Bell SS, Hicks GRF, Walters K (1989) Experimental investigations of benthic reentry by migrating meiobenthic copepods. J Exp Mar Biol Ecol 130: 291–303

Berg CJ, Adams NL (1984) Microwave fixation of marine invertebrates. J Exp Mar Biol Ecol 74: 195–199

Berge JA, Leinaas HP, Sandoy K (1985) The solitary bryozoan, *Monobryozoon limicola* Franzén (Ctenostomata), a comparison of mesocosm and field samples from Oslofjorden, Norway. Sarsia 70: 91–94

Bergmans M, Dahms H-U, Schminke HK (1991) An r-strategist in Antarctic pack ice. Oecologia (Berl) 86: 305–309

Bernhard JM, Bowser SS (1992) Bacterial biofilms as a trophic resource for certain benthic Foraminifera. Mar Ecol Prog Ser 83: 263–272

Beukema JJ (1988) An evaluation of the ABC-method (abundance/biomass comparison) as applied to macrozoobenthic communities living on tidal flats in the Dutch Wadden Sea. Mar Biol 99: 425–433

Biddanda B, Riemann F (1992) Detrital carbon and nitrogen relations, examined with degrading cellulose. PSZNI Mar Ecol 13: 271–283

Blanchard GF (1990) Overlapping microscale dispersion patterns of meiofauna and microphytobenthos. Mar Ecol Prog Ser 68: 101–111

Blanchard GF (1991) Measurement of meiofauna grazing rates on microphytobenthos – is primary production a limiting factor? J Exp Mar Biol Ecol 147: 37–46

Blomqvist S (1985) Reliability of core sampling of soft bottom sediment – an in situ study. Sedimentology 32: 605–612

Blomqvist S (1990) Sampling performance of Ekman grabs – in situ observations and design improvements. Hydrobiologia 206: 245–254

Blomqvist S (1991) Quantitative sampling of soft-bottom sediments: problems and solutions. Mar Ecol Prog Ser 72: 295–304

Blomqvist S, Abrahamsson B (1985) An improved Kajak-type gravity core sampler for soft bottom sediments. Schweiz Z Hydrol 47: 81–84

Boaden PJS (1962) Colonization of graded sand by interstitial fauna. Cah Biol Mar 3: 245–248

Boaden PJS (1964) Grazing in the interstitial habitat: a review. In: Crisp (ed) Grazing in terrestrial and marine environments. Blackwell, Oxford, pp 299–303

Boaden PJS (1968) Water movement – a dominant factor in interstitial ecology. Sarsia 34: 125–136

Boaden PJS (1974) Three new thiobiotic Gastrotricha. Cah Biol Mar 15: 367–378

Boaden PJS (1975) Anaerobiosis, meiofauna and early metazoan evolution. Zool Scr 4: 21–24

Boaden PJS (1977) Thiobiotic facts and fancies (aspects of the distribution and evolution of anaerobic meiofauna). Mikrofauna Meeresboden 61: 45–63

Boaden PJS (1989a) Adaptation of intertidal sand meiofaunal oxygen uptake to temperature and population density. Sci Mar 53: 329–334

Boaden PJS (1989b) Meiofauna and the origins of the Metazoa. Zool J Linn Soc 96: 217–227

Boaden PJS, Erwin DG (1971) *Turbanella hyalina* versus *Protodriloides symbioticus*: a study in interstitial ecology. Vie Milieu 22: 479–492 (Suppl)

Boaden PJS, Platt HM (1971) Daily migration patterns in an intertidal meiobenthic community. Thalassia Jugosl 7: 1–12

Bock E, Wilderer PA, Freitag A (1988) Growth of *Nitrobacter* in the absence of dissolved oxygen. Water Res 22: 245–250

Bodin P (1988) Results of ecological monitoring three beaches polluted by the 'Amoco Cadiz' oil spill: Development of meiofauna from 1978–1984. Mar Ecol Prog Ser 42: 105–123

Bodin P (1991) Perturbations in the reproductive cycle of some harpacticoid copepod species further to the Amoco Cadiz oil spill. Hydrobiologia 209: 245–258

Bodin P, Boucher D (1983) Évolution à moyen terme du méiobenthos at des pigments chlorophylliens sur quelques plages polluées par la marée de l' "Amoco Cadiz". Oceanol Acta 6: 321–332

Boisseau JP (1957) Technique pour l' étude de la faune interstitielle des sables. C R Congr Soc Savent Paris Départm, Bordeaux, pp 117–119

Bongers T (1990) The maturity index: an ecological measure of environmental disturbance based on nematode species composition. Oecologia 83: 14–19

Bongers T, Haar J v de (1990) On the potential of basing an ecological typology of aquatic sediments on the nematode fauna: an example from the river Rhine. Hydrobiol Bull 24: 37–45

Bongers T, Alkemade R, Yeates GW (1991) Interpretation of disturbance-induced maturity decrease in marine nematode assemblages by means of the maturity index. Mar Ecol Prog Ser 76: 135–142

Botosaneanu L (ed) (1986a) Stygofauna Mundi. A faunistic, distributional, and ecological synthesis of the world fauna inhabiting subterranean waters (including the marine interstitial). Brill, Backhuys, Leiden, p 740

Botosaneanu L (1986b) Spelaeogriphacea. In: Botosaneanu L (ed) Stygofauna Mundi. A faunistic, distributional, and ecological synthesis of the world fauna inhabiting subterranean waters (including the marine interstitial). Brill, Backhuys, Leiden, p 493

Botosaneanu L, Holsinger JR (1991) Some aspects concerning colonization of the subterranean realm – especially of subterranean waters: a response to Rouch & Danielopol, 1987. Stygologia 6: 11–39

Bou C (1974) Les méthodes de récolte dans les eaux souterraines interstitielles. Ann Spéléol 29: 611–619

Boucher G (1980) Impact of Amoco Cadiz oil spill on intertidal and sublittoral meiofauna. Mar Pollut Bull 11: 95–100

Boucher G, Chamroux S (1976) Bacteria and meiofauna in an experimental sand ecosystem. I. Material and preliminary results. J Exp Mar Biol Ecol 4: 237–249

Bouguenec V, Giani N (1989) Biological studies upon *Enchytraeus variatus* Bougenec & Giani 1987 in breeding cultures. In: Kaster JL (ed) Aquatic oligochaete biology. Hydrobiologia 180: 151–165

Bouillon J, Grohmann PA (1990) *Pinushydra chiquitita* gen. et sp. nov. (Cnidaria, Hydrozoa, Athecata), a solitary marine mesopsammic polyp. Cah Biol Mar 31: 291–305

Bourne GC (1903) *Oligotrema psammites*; a new ascidian belonging to the family Molgulidae. Q J Microsc Sci 47: 233–272

Boutin C, Coineau N (1991) "Regression model", "modèle biphase" d'évolution et origine des microorganismes stygobies interstitiels continentaux. Rev Micropaléontol 33: 303–322

Bovée F de, Soyer J (1974) Cycle annuel quantitatif du méiobenthos des vases terrigènes côtières. Distribution verticale. Vie Milieu 24 Sér B: 141–157

Bowman TE, Iliffe TM (1985) *Mictocaris halope*, a new unusual peracaridian crustacean from marine caves on Bermuda. J Crustacean Biol 5: 58–73

Bretschko G, Klemens WE (1986) Quantitative methods and aspects in the study of the interstitial fauna of running waters. Stygologia 2: 297–316

Brey T (1987) Empirical relations between production, P/B-ratio, biomass and individual weight in macrobenthic invertebrate populations. In: Ros J (ed) Topics in marine biology. 22nd Eur Mar Biol Symp, 17–22 Aug 1987, Barcelona (Abstr)

Brinkhurst RO (1982a) Oligochaeta. In: Parker SP (ed) Synopsis and classification of living organisms, vol 1. Mc Graw Hill, New York, pp 50–61

Brinkhurst RO (1982b) Evolution in the Annelida. Can J Zool 60: 1043–1059

Brinkhurst RO (1984) The position of the Haplotaxidae in the evolution of oligochaete annelids. Hydrobiologia 115: 25–36

Brinkhurst RO (1991) A phylogenetic analysis of the Tubificinae (Oligochaeta, Tubificidae). Can J Zool 69: 392–397

Brinkhurst RO (1992) Evolutionary relationships within the Clitellata. Soil Biol Biochem 24: 1201–1205

Brown TJ, Sibert JR (1977) The food of some benthic harpacticoid copepods. J Fish Res Board Can 34: 1028–1031

Brunberg L (1964) On the nemertean fauna of Danish waters. Ophelia 1: 77–111

Bryan JR, Riley JP, LeWilliams B (1976) A Winkler procedure for making precise measurements of oxygen concentration for productivity and related studies. J Exp Mar Biol Ecol 21: 191–197

Bryant C (ed) (1991) Metazoan life without oxygen. Chapman and Hall, London, 291 pp

Buchanan JB (1971) Sediments. In: Buchanan JB, Kain JM (eds) Measurement of the physical and chemical environment. In: Holme NA, McIntyre AD (eds) Methods for the study of marine benthos. Blackwell, Oxford, pp 30–52 (IBP Handbook 16, 1st edn)

Buchanan JB (1984) Sediment Analysis. In: Holme NA, McIntyre AD (eds) Methods for the study of marine benthos. Blackwell, Oxford, pp 41–65 (IBP Handbook 16, 2nd edn)

Bunke D (1967) Zu Morphologie und Systematik der Aeolosomatidae Beddard 1895 und Potamodrilidae nov.fam. (Oligochaeta). Zool Jahrb Syst 94: 187–368

Burd BJ, Nemec A, Brinkhurst RO (1990) The development and application of analytical methods in benthic marine infaunal studies. Adv Mar Biol 26: 169–247

Butman CA (1989) Sediment-trap experiments on the importance of hydrodynamic processes in distributing settling larvae in near-bottom waters. J Exp Mar Biol Ecol 134: 37–88

Cannon LRG (1986) Turbellaria of the World: a guide to families and genera. Queensland Museum, Brisbane, 136 pp

Carey AG Jr (1992) The ice fauna in the shallow southwestern Beaufort Sea, Arctic Ocean. J Mar Syst 3: 225–236

Carey PG (1992) Marine interstitial ciliates. An illustrated key. Chapman & Hall, London, 351 pp

Carman KR, Thistle D (1985) Microbial food partitioning by three species of benthic copepods. Mar Biol 88: 143–148

Cary SC, Vetter RD, Felbeck H (1989) Habitat characterization and nutritional strategies of the endosymbiont-bearing bivalve Lucinoma aequizonata. Mar Ecol Prog Ser 55: 31–45

Castel J (1992) The meiofauna of coastal lagoon ecosystems and their importance in the food web. Vie Milieu 42: 125–135

Ceccherelli VU, Mistri M (1991) Production of the meiobenthic harpacticoid copepod Canuella perplexa. Mar Ecol Prog Ser 68: 225–234

Chandler GT (1989) Foraminifera may structure meiobenthic communities. Oecologia 81: 354–360

Chandler GT (1990) Effects of sediment-bound residues of the pyrethroid insecticide Fenvalerate on survival and reproduction of meiobenthic copepods. Mar Environ Res 29: 65–76

Chandler GT, Fleeger JW (1984) Tube-building by a marine meiobenthic harpacticoid copepod. Mar Biol 82: 15–19

Chandler GT, Fleeger JW (1987) Facilitative and inhibitory interactions among estuarine meiobenthic harpacticoid copepods. Ecology 68: 1906–1919

Chandler GT, Shirley TC, Fleeger JW (1988) The tom-tom corer: a new design of the Kajak corer for use in meiofauna sampling. Hydrobiologia 169: 129–134

Chapman PM (1986) Sediment quality criteria from the sediment quality triad: an example. Environ Toxicol Chem 5: 957–964

Child CA (1979) Shallow-water Pycnogonida of the Isthmus of Panama and the coast of Middle America. Smithson Contrib Zool 293: 1–86

Child CA (1988) Pycnogonida. In: Higgins RP, Thiel H (eds) Introduction to the study of meiofauna. Smithsonian Inst Press, Washington, DC, pp 423–424

Chitwood BG, Chitwood MB (1974) An introduction to nematology, reprint of 2nd edn + additions. University Park Press, Baltimore, 334 pp

Christensen B, O'Connor FB (1958) Pseudofertilization in the genus Lumbricillus (Enchytraeidae). Nature 181: 1085–1086

Chua KE, Brinkhurst RO (1973) Bacteria as potential nutritional resources for three sympatric species of tubificid oligochaetes. In: Stephenson LH, Colwell RR (eds) Estuarine microbial ecology. University of South Carolina Press, Columbia, pp 513–517 (Belle W Baruch Libr Mar Sci, vol 1)

Clarke KR (1993) Non-parametric multivariate analyses of changes in community structure. Aust J Ecol 18: 117–143

Cline JD (1969) Spectrophotometric determination of hydrogen sulfide in natural waters. Limnol Oceanogr 14: 454–458

Coineau Y, Haupt J, Delamare Deboutteville C (1978) Un remarquable exemple de convergence écologique: l'adaptation de *Gordialycus tuzetae* (Nematalycidae, Acariens) à la vie dans les interstices des sables fins. C R Acad Sci Paris 287: 883–886

Colacino JM, Kraus DW (1984) Hemoglobin-containing cells of *Neodasys* (Gastrotricha, Chaetonotida). II. Respiratory significance. Comp Biochem Physiol 79 A: 363–369

Condé B (1965) Présence de Palpigrades dans le milieu interstitiel littoral. C R Acad Sci Paris 261: 1898–1900

Connell JH, Slatyer RD (1977) Mechanisms of succession in natural communities and their role in community stability and organisation. Am Nat 111: 1119–1144

Conway Morris S, George JD, Gibson R, Platt HM (eds) (1985) The origins and relationships of lower invertebrates. Clarendon Press, Oxford 397 pp (Syst Ass, vol 28)

Cook PL (1963) Observations on live lunulitiform zoaria of Polyzoa. Cah Biol Mar 4: 407–413

Cook PL (1966) Some sand fauna polyzoa (Bryozoa) from Eastern Africa and the Northern Indian Ocean. Cah Biol Mar 7: 207–223

Cook PL (1988) Bryozoa. In: Higgins RP, Thiel H (eds) Introduction to the study of meiofauna. Smithsonian Inst Press, Washington, DC pp 438–443

Corliss JO (1974) The changing world of ciliate systematics: historical analysis of past efforts and a newly proposed phylogenetic scheme of classification for the protistan phylum Ciliophora. Syst Zool 23: 91–138

Corliss JO (1975) Taxonomic characterization of the suprafamilial groups in a revision of recently proposed schemes of classification for the phylum Ciliophora. Trans Am Microsc Soc 94: 224–267

Corliss JO (1979) The ciliated Protozoa: characterization, classification, and guide to the literature, 2nd edn. Pergamon Press, Oxford, 455 pp

Coull BC (1970) Shallow water meiobenthos of the Bermuda platform. Oecologia (Berl) 4: 325–357

Coull BC (1973) Estuarine meiofauna: a review: trophic relationships and microbial interactions. In: Stephenson LH, Colwell RR (eds) Estuarine microbial ecology. University of South Carolina Press, Columbia, pp 499–512 (Belle W Baruch Libr Mar Sci, vol 1)

Coull BC (1985) Long-term variability of estaurine meiobenthos: an 11 year study. Mar Ecol Prog Ser 24: 205–218

Coull BC (1990) Are members of the meiofauna food for higher trophic levels? Trans Am Microsc Soc 109: 233–246

Coull BC, Bell SS (1979) Perspectives of marine meiofaunal ecology. In: Livingston RJ (ed) Ecological processes in coastal and marine systems. Plenum, New York, pp 189–216

Coull BC, Chandler GT (1992) Pollution and meiofauna: field, laboratory, and mesocosm studies. Oceanogr Mar Biol Annu Rev 30: 191–271

Coull BC, Giere O (1988) The history of meiofaunal research. In: Higgins RP, Thiel H (eds) Introduction to the study of meiofauna. Smithsonian Inst Press. Washington, DC, pp 14–17

Coull BC, Grant J (1981) Encystment discovered in a marine copepod. Science 212: 343–344

Coull BC, Palmer MA (1984) Field experimentation in meiofaunal ecology. Hydrobiologia 118: 1–19

Coull BC, Ellison RL, Fleeger JW, Higgins RP, Hope WD, Hummon WD, Rieger RM, Sterrer WE, Thiel H (1977) Quantitative estimates of the meiofauna from the deep sea of North Carolina. Mar Biol 39: 233–240

Coull BC, Bell SS, Savory AM, Dudley BW (1979) Zonation of meiobenthic copepods in a southeastern United States salt marsh. Estuarine Coastal Mar Sci 9: 181–188

Coull BC, Hicks GRF, Wells JBJ (1981) Nematode/copepod ratios for monitoring pollution: a rebuttal. Mar Pollut Bull 12: 378–381

Craib JS (1965) A sampler for taking short undisturbed marine cores. J Cons Perm Int Explor Mer 30: 34–39

Creed EL, Coull BC (1984) Sand dollar, *Mellita quinquiesperforata* (Leske), and sea pansy, *Renilla reniformis* (Cuvier) effects on meiofaunal abundance. J Exp Mar Biol Ecol 84: 225–234

Crezeé M (1976) Solenofilomorphidae (Acoela), major component of a new turbellarian association in the sulfide system. Int Rev Ges Hydrobiol 61: 105–129

Crisp DJ (1984) Energy flow measurements. In: Holme NA, McIntyre AD (eds) Methods for the study of marine benthos. Blackwell Oxford, pp 284–372 (IBP Handbook 16)

Crisp DJ, Mwaiseje B (1989) Diversity in intertidal communities with special reference to the *Corallina officinalis* community. Sci Mar 53: 365–372

Cullen DJ (1973) Bioturbation of superficial marine sediments by interstitial meiobenthos. Nature 242: 323–324

Dahms H-U (1990) Naupliar development of Harpacticoida (Crustacea, Copepoda) and its significance for phylogenetic systematics. Microfauna Mar 6: 169–272

Dahms H-U (1991) Erster Nachweis eines Harpacticoiden (Copepoda) mit zystenloser Diapause. First indication of nonencysted diapause for Harpacticoida (Copepoda).Verh Dtsch Zool Ges 84: 442–443

Dahms H-U, Bergmans M, Schminke HK (1990) Distribution and adaptations of sea ice-inhabiting Harpacticoida (Crustacea, Copepoda) of the Weddell Sea (Antarctica). PSZNI Mar Ecol 11: 207–226

Dando PR, Southward AJ, Southward EC, Terwilliger N, Terwilliger RC (1985) Sulphur-oxydising bacteria and haemoglobin in gills of the bivalve mollusc *Myrtea spinifera*. Mar Ecol Prog Ser 23: 85–98

Danielopol DL (1976) The distribution of the fauna in the interstitial habitats of riverine sediments of the Danube and the Piesting (Austria). Int J Spéléol 8: 23–51

Danielopol DL (1980) An essay to assess the age of the freshwater interstitial ostracods of Europe. Bijdr Dierkd 50: 243–291

Danielopol DL (1989) Groundwater fauna associated with riverine aquifers. J N Am Benthol Soc 8: 18–35

Danielopol DL (1990a) On the interest of the "Cytherissa" project and on the present state of researches. Bull Inst Géol Bassin d'Aquitaine Bordeaux 47: 15–26

Danielopol DL (1990b) The origin of the anchialine cave fauna – the "deep sea" versus the "shallow water" hypothesis tested against the empirical evidence of the Thaumato-cyprididae (Ostracoda). Bijdr Dierkd 60: 137–143

Danielopol DL (1991) Spatial distribution and dispersal of interstitial Crustacea in alluvial sediments of a backwater of the Danube at Vienna. Stygologia 6: 97–110

Danielopol DL, Niederreiter R (1987) A sampling device for groundwater organisms and oxygen measurement in multi-level monitoring wells. Stygologia 3: 252–263

Danielopol DL, Niederreiter R (1990) New sampling equipment and extraction methods for meiobenthic organisms. Bull Inst Géol Bassin d'Aquitaine Bordeaux 47: 277–286

Danielopol DL, Rouch R (1991) L'adaptation des organismes au milieu aquatique souterrain. Réflexions sur l'apport des recherches écologiques récentes. Stygologia 6: 129–142

Dash MC, Cragg JB (1972) Selection of microfungi by Enchytraeidae (Oligochaeta) and other members of the soil fauna. Pedobiologia 12: 282–286

Davey JT, Watson PG, Bruce RH, Frickers PE (1990) An instrument for the monitoring and collection of the vented burrow fluids of benthic infauna in sediment microcosms and its application to the polychaetes *Hediste diversicolor* and *Arenicola marina*. J Exp Mar Biol Ecol 139: 135–149

Decho AW (1990) Microbial expolymer secretions in ocean environments: their role(s) in food webs and marine processes. Oceanogr Mar Biol Annu Rev 28: 73–153

Decho AW, Fleeger JW (1988) Ontogenetic feeding shifts in the meiobenthic harpacticoid copepod *Nitocra lacustris*. Mar Biol 97: 191–197

Deflaun MF, Mayer LM (1983) Relationships between bacteria and grain surfaces in intertidal sediments. Limnol Oceanogr 28: 873–881

Delamare Deboutteville C (1960) Biologie des eaux souterraines littorales et continentales. Herman, Paris, 740 pp (Vie Milieu Suppl 9)

De Zio S, Grimaldi P (1966) Ecological aspects of Tardigrada distribution in South Adriatic beaches. Veröff Inst Meeresforsch Bremerhaven Suppl 2: 87–94

D'Hondt J-L (1971) Gastrotricha. Oceanogr Mar Biol Annu Rev 9: 141–192

Dick MH, Buss LW (1993) What flies can tell us about worms: homeotic genes in the polychaete *Ctenodrilus serratus*. Helgol Meeresunters Sbd (in press)

Dinet A, Grassle F, Tunnicliffe V (1988) Premières observations sur la méiofauna des sites hydrothermaux de la dorsale Est-Pacifique (Guaymas, 21 °N) et de l'Explorer Ridge. In: Hydrothermalism, biology and ecology. Proc. Symp, Actes du colloque Hydrothermalisme, bioloqie et ecologie, Paris, 4–7 Nov 1985. Oceanol Acta Sp 8: 7–14

Dinet A, Sornin J-M, Sablière A, Delmas D, Feuillet-Girard M (1990) Influence de la biodéposition de bivalves filtreurs sur les peuplements méiobenthiques d'un marais maritime. Cah Biol Mar 31: 307–322

Dittmann S (1987) Die Bedeutung der Biodeposite für die Benthosgemeinschaft der Wattsedimente. Unter besonderer Berücksichtigung der Miesmuschel *Mytilus edulis* L. Diss Universität Göttingen, 182 pp

Dittmann S (1993) Towards a tropical "Königshafen" – research in tidal flats of Australia. Helgol Meeresunters (in press)

Dobbs FC, Guckert JB, Carman KR (1989) Comparison of three techniques for administering radiolabeled substrates to sediments for trophic studies: incorporation by microbes. Microb Ecol 17: 237–250

Dörjes J (1968) Zur Ökologie der Acoela (Turbellaria) in der Deutschen Bucht. Helgol Wiss Meeresunters 18: 78–115

Dole-Olivier MJ, Marmonier P (1992) Patch distribution of interstitial communities: prevailing factors. Freshwater Biol 27: 177–191

Doty MS (1971) Measurement of water movement in reference to benthic algal growth. Bot Mar 14: 32–35

Dragesco J (1960) Ciliés mésopsammiques littoraux. Systématique, morphologie, écologie. Trav Stat Biol Roscoff 12: 1–356

Duffy JE, Tyler S (1984) Quantitative differences in mitochondrial ultrastructure of a thiobiotic and oxybiotic turbellarian. Mar Biol 83: 95–102

Dujardin F (1851) Sur un petit animal marin, l'Echinodère, formant un type intermédiaire entre les Crustacés et les Vers. Ann Sci Nat Zool Sér 3, 15: 158–160

Dye AH (1978) An ecophysiological study of the meiofauna of the Swartskops estuary. 1. The sampling sites: Physical and chemical features. Zool Afr 13: 1–18

Dye AH, Lasiak TA (1986) Microbenthos, meiobenthos and fiddler crabs: trophic interactions in a tropical mangrove sediment. Mar Ecol Prog Ser 32: 259–264

Eckman JE (1979) Small-scale patterns and processes in a soft-substratum intertidal community. J Mar Res 37: 437–456

Eckman JE (1983) Hydrodynamic processes affecting benthic recruitment. Limnol Oceanogr 28: 241–257

Eckman JE (1985) Flow disruption by an animal-tube mimic affects sediment bacterial colonization. J Mar Res 43: 419–435

Eckman JE (1990) A model of passive settlement by planktonic larvae onto bottoms of differing roughness. Limnol Oceanogr 35: 887–901

Eckman JE, Thistle D (1991) Effects of flow about a biologically produced structure on harpacticoid copepods in San-Diego trough. Deep-Sea Res 38: 1397–1416

Eckman JE, Nowell ARM, Jumars PA (1981) Sediment destabilization by animal tubes. J Mar Res 39: 361–374

Edmonds SJ (1982) A sipunculan reported to be "interstitial" from the Netherland Antilles. Bijdr Dierkd 52: 228–230

Ehlers U (1985) Phylogenetic relationships within the Platyhelminthes. In: Conway Morris S, George JD, Gibson R, Platt HM (eds) The origins and relationships of lower invertebrates. Clarendon Press, Oxford, pp 143–158 (Syst Ass, vol 28)

Eleftheriou A, Nicholson MD (1975) The effects of exposure on beach fauna. Cah Biol Mar 16: 695–710

Ellison RL (1984) Foraminifera and meiofauna on an intertidal mudflat, Cornwall England: populations, respiration and secondary production, and energy budget. Hydrobiologia 109: 131–147

Elmgren R (1978) Structure and dynamics of Baltic benthos communities, with particular reference to the relationship between macro- and meiofauna. Kieler Meeresforsch Sbd 4: 1–22

Elmgren R, Frithsen JB (1982) The use of experimental ecosystems for evaluating the environmental impact of pollutants: a comparison of an oil spill in the Baltic Sea and two long-term, low-level oil addition experiments in mesocosms. In: Grice GD, Reeve MR (eds) Marine mesocosms. Biological and chemical research in experimental ecosystems. Springer, Berlin Heidelberg New York, pp 153–165

Elmgren R, Radziejewska T (1989) Recommendations for quantititative benthic meiofauna studies in the Baltic. Balt Mar Biol Publ 12, Szczecin, Poland, 20 pp

Elmgren R, Hansson S, Larsson U, Sundelin B, Boehm PD (1983) The "Tsesis" oil spill: acute and long-term impact on the benthos. Mar Biol 73: 51–65

Elofson O (1941) Zur Kenntnis der marinen Ostracoden Schwedens. Zool Bidr Upps 19: 217–534

Epstein SS, Gallagher ED (1992) Evidence for facilitation and inhibition of ciliate population growth by meiofauna and macrofauna on a temperate zone sandflat. J Exp Mar Biol Ecol 155: 27–39

Epstein SS, Shiaris MP (1992) Rates of microbenthic and meiobenthic bacterivory in a temperate muddy tidal flat community. Appl Environ Microbiol 58: 2426–2431

Epstein SS, Burkovsky IV, Shiaris MP (1992) Ciliate grazing on bacteria, flagellates, and microalgae in a temperate zone sandy tidal flat: ingestion rates and food niche partitioning. J Exp Mar Biol Ecol 165: 103–123

Erséus C (1980) Specific and generic criteria in marine Oligochaeta, with special emphasis on Tubificidae. In: Brinkhurst RO, Cook DG (eds) Aquatic oligochaete biology. Plenum Press, London, pp 9–24

Erséus C (1984a) Taxonomy and phylogeny of the gutless Phallodrilinae (Oligochaeta, Tubificidae), with description of one new genus and twenty-two new species. Zool Scr 13: 239–272

Erséus C (1984b) Aspects of the phylogeny of the marine Tubificidae. Hydrobiologia 115: 37–44

Erséus C (1987) Phylogenetic analysis of the aquatic Oligochaeta under the principle of parsimony. Hydrobiologia 155: 75–89

Erséus C (1990a) Cladistic analysis of the subfamilies within the Tubificidae (Oligochaeta). Zool Scr 19: 57–63

Erséus C (1990b) The marine Tubificidae (Oligochaeta) of the barrier reef ecosystems at Carrie Bow Cay, Belize, and other parts of the Caribbean Sea, with descriptions of twenty-seven new species and revision of *Heterodrilus*, *Thalassodrilides* and *Smithsonidrilus*. Zool Scr 19: 243–303

Escaravage V, García ME, Castel J (1989) The distribution of meiofauna and its contribution to detritic pathways in tidal flats (Arcachon Bay, France). Sci Mar 53: 551–559

Eskin RA, Coull BC (1984) A priori determination of valid control sites: an example using marine meiobenthic nematodes. Mar Environ Res 12: 161–172

Farke H, Riemann F (1980) Dissolved organic carbon in littoral sediments: concentrations and available amounts demonstrated by the percolation method. Veröff Inst Meeresforsch Bremerhaven 18: 235–244

Farris RA, O'Leary DJ (1985) Applications of videomicroscopy to the study of interstitial fauna. Int Rev Ges Hydrobiol 70: 891–895

Faubel A (1976a) Interstitielle Acoela (Turbellaria) aus dem Litoral der nordfriesischen Inseln Sylt und Amrum (Nordsee). Mitt Hamb Zool Mus Inst 73: 17–56

Faubel A (1976b) Populationsdynamik und Lebenszyklen interstitieller Acoela und Macrostomida (Turbellaria). Mikrofauna Meeresboden 56: 1–107

Faubel A (1982) Determination of individual meiofauna dry weight values in relation to definite size classes. Cah Biol Mar 23: 339–345

Faubel A (1984) On the abundance and activity pattern of zoobenthos inhabiting a tropical reef area, Cebu, Philippines. Coral Reefs 3: 205–213

Fauchald K (1974) Polychaete phylogeny: a problem in protostome evolution. Syst Zool 23: 493–506

Fauré-Fremiet E (1950) Ecologie des Ciliés psammophiles littoraux. Bull Biol Fr Belg 84: 35–75

Fegley SR (1987) Experimental variation of near-bottom current speeds and its effects on depth distribution of sand-living meiofauna. Mar Biol 95: 183–192

Fegley SR (1988) A comparison of meiofaunal settlement onto the sediment surface and recolonization of defaunated sandy sediment. J Exp Mar Biol Ecol 123: 97–113

Feller RJ (1982) Antigenic similarities among estuarine soft-bottom benthic taxa. Oecologia (Berl) 52: 305–310

Feller RJ, Warwick RM (1988) Energetics. In: Higgins RP, Thiel H (eds) Introduction to the study of meiofauna. Smithsonian Inst Press, Washington, DC, pp 181–196

Feller RJ, Taghon GL, Gallagher ED, Kenny GE, Jumars PA (1979) Immunological methods for food web analysis in a soft-bottom benthic community. Mar Biol 54: 61–74

Feller RJ, Coull BC, Hentschel BT (1990) Meiobenthic copepods: tracers of where juvenile *Leiostomus xanthurus* (Pisces) feed? Can J Fish Aquat Sci 47: 1913–1919

Fenchel T (1967) The ecology of marine microbenthos. I. The quantitative importance of ciliates as compared with metazoans in various types of sediments. Ophelia 4: 121–137

Fenchel T (1968a) The ecology of marine microbenthos. II. The food of marine benthic ciliates. Ophelia 5: 73–121

Fenchel T (1968b) The ecology of marine microbenthos. III. The reproductive potential of ciliates. Ophelia 5: 123–136

Fenchel T (1969) The ecology of marine microbenthos. IV. Structure and function of the benthic ecosystem, its chemical and physical factors and the microfauna communities with special reference to the ciliate Protozoa. Ophelia 6: 1–182

Fenchel T (1970) Studies on the decomposition of organic detritus derived from the turtle grass *Thalassia testudinum*. Limnol Oceanogr 15: 14–20

Fenchel TM (1978) The ecology of micro- and meiobenthos. Annu Rev Ecol Syst 9: 99–121

Fenchel T (1992) What can ecologists learn from microbes: life beneath a square centimetre of sediment surface. Funct Ecol 6: 499–507

Fenchel T, Finlay BJ (1989) *Kentrophoros*: a mouthless ciliate with a symbiotic kitchen garden. Ophelia 30: 75–93

Fenchel T, Finlay BJ (1991) The biology of free-living anaerobic ciliates. Eur J Protistol 26: 201–215

Fenchel TM, Riedl RJ (1970) The sulfide system: a new biotic community underneath the oxidized layer of marine sand bottoms. Mar Biol 7: 255–268

Fenchel T, Straarup BJ (1971) Vertical distribution of photosynthetic pigments and the penetration of light in marine sediments. Oikos 22: 172–182

Fenchel T, Perry T, Thane A (1977) Anaerobiosis and symbiosis with bacteria in free-living ciliates. J Protozool 24: 154–163

Ferguson JC (1982) A comparative study of the net metabolic benefits derived from the uptake and release of free amino acids by marine invertebrates. Biol Bull (Woods Hole) 162: 1–17

Ferris VR, Ferris JM (1979) Thread worms (Nematoda). In: Hart CW, Fuller SL (eds) Pollution ecology of estuarine invertebrates. Academic Press, New York, pp 1–33

Field JG, Clarke KR, Warwick RM (1982) A practical strategy for analysing multispecies distribution patterns. Mar Ecol Prog Ser 8: 37–52

Findlay RH, King GM, Watling L (1989) Efficacy of phospholipid analysis in determining microbial biomass in sediment. Appl Environ Microbiol 55: 2888–2893

Findlay SEG (1981) Small-scale spatial distribution of meiofauna on a mud- and sandflat. Estuarine Coastal Shelf Sci 12: 471–484

Finlay BJ, Span ASW, Harman JMP (1983) Nitrate respiration in primitive eukaryotes. Nature 303: 333–336

Fisher CR (1990) Chemoautotrophic and methanotrophic symbioses in marine invertebrates. Rev Aquat Sci 2: 399–436

Fleeger JW, Palmer MA (1982) Secondary production of the estuarine, meiobenthic copepod *Microarthridion littorale*. Mar Ecol Prog Ser 7: 157–162

Fleeger JW, Shirley TC, Ziemann DA (1989) Meiofaunal responses to sedimentation from an Alaskan spring bloom. I. Major taxa. Mar Ecol Prog Ser 57: 137–145

Fleeger JW, Palmer MA, Moser EB (1990) On the scale of aggregation of meiobenthic copepods on a tidal mudflat. PSZNI Mar Ecol 11: 227–237

Forster S, Graf G (1992) Continuously measured changes in the redox potential influenced by oxygen penetrating from burrows of *Callianassa subterranea*. Hydrobiologia 235/236: 527–532

Fossing H, Jørgensen BB (1990) Oxidation and reduction of radiolabeled inorganic sulfur compounds in an estuarine sediment, Kysing Fjord, Denmark. Geochim Cosmochim Acta 54: 2731–2742

Fox CA, Powell EN (1986) Meiofauna and the sulfide system: the effects of oxygen and sulfide on the adenylate pool of three turbellarians and a gastrotrich. Comp Biochem Physiol 85 A: 37–44

Fox CA, Powell EN (1987) The effect of oxygen and sulfide on CO_2 production by three acoel turbellarians. Are thiobiotic meiofauna aerobic? Comp Biochem Physiol 86 A: 509–514

Foy MS, Thistle D (1991) On the vertical distribution of a benthic harpacticoid copepod: field, laboratory, and flume results. J Exp Mar Biol Ecol 153: 153–164

Franzén A (1960) Monobryozoon limicola n.sp., a ctenostomatous bryozoan from the detritus layer on soft sediments. Zool Bidr Upps 33: 135–148

Freitag A, Rudert M, Bock E (1987) Growth of Nitrobacter by dissimilatoric nitrate reduction. FEMS Microbiol Lett 48: 105–109

Frey DG (1987) The taxonomy and biogeography of the Cladocera. In: Ferró L, Frey DG (eds) Proc Cladocera Symp, Budapest 1985. Hydrobiologia 145: 5–17

Fricke H, Giere O, Stetter K, Alfredsson GA, Kristjansson JK, Stoffers P, Svavarson J (1989) Hydrothermal vent communities at the shallow subpolar Mid-Atlantic ridge. Mar Biol 102: 425–429

Frithsen JB, Rudnick DT, Elmgren (1983) A new, flow-through corer for quantitative sampling of surface sediments. Hydrobiologia 99: 75–79

Fukui M, Takii S (1990) Survival of sulfate-reducing bacteria in oxic surface sediment of a seawater lake. FEMS Microbiol Ecol 73: 317–322

Furstenberg JP, Wet AG de (1982) A comparison of two extractors for separating meiobenthic nematodes from fine-grained sediments. S Afr J Zool 17: 41–43

Gabel B (1971) Die Foraminiferen der Nordsee. Helgol Wiss Meeresunters 22: 1–65

Gabrich A, Jaros PP, Brockmeyer V (1991) Application of immunological methods for the taxonomic study of Enchytraeus (Annelida) and Tisbe (Arthropoda, Crustacea). Z Zool Syst Evolutionsforsch 29: 381–392

Gage JG, Tyler PA (1991) Deep-sea biology: a natural history of organisms at the deep-sea floor. Cambridge University Press, Cambridge, 504 pp

Gaudy R, Guérin JP (1977) Dynamique des populations de Tisbe holothuriae (Crustacea: Copepoda) en élevage sur trois régimes artificiels différents. Mar Biol 39: 137–145

Gee JM (1987) Impact of epibenthic predation on estuarine intertidal harpacticoid copepod populations. Mar Biol 96: 497–510

Gee JM (1989) An ecological and economic review of meiofauna as food for fish. Zool J Linn Soc 93: 243–261

Gerlach SA (1954) Das Supralitoral der sandigen Meeresküsten als Lebensraum einer Mikrofauna. Kieler Meeresforsch 10: 121–129

Gerlach SA (1971) On the importance of marine meiofauna for benthos communities. Oecologia (Berl) 6: 176–190

Gerlach SA (1977a) Attraction to decaying organisms as a possible cause for distribution of nematodes in a Bermuda beach. Ophelia 6: 151–166

Gerlach SA (1977b) Means of meiofauna dispersal. Mikrofauna Meeresboden 61: 89–103

Gerlach SA (1978) Food-chain relationships in subtidal silty-sand marine sediments and the role of meiofauna in stimulating bacterial growth. Oecologia (Berl) 33: 55–69

Gerlach SA, Riemann F (1973/74) The Bremerhaven checklist of aquatic Nematoda

Adenophorea excluding the Dorylaimida. Veröff Inst Meeresforsch Bremerhaven Suppl 4: 736

Gerlach SA, Schrage M (1971) Life cycles in marine meiobenthos. Experiments at various temperatures with *Monhystera disjuncta* and *Theristus pertenuis* (Nematoda). Mar Biol 9: 272–280

Gerlach SA, Hahn AE, Schrage M (1985) Size spectra of benthic biomass and metabolism. Mar Ecol Prog Ser 26: 161–173

Gerner L (1969) Nemertinen der Gattungen *Cephalothrix* und *Ototyphlonemertes* aus dem marinen Mesopsammal. Helgol Wiss Meeresunters 19: 68–110

Giard A (1904) Sur une faunule charactéristique des sables à diatomées d'Ambleteuse. C R Séances Soc Biol Paris 56: 107–165

Gibbons MJ (1991) Rocky shore meiofauna: a brief overview. Trans R Soc S Afr 47: 595–603

Gibbons MJ, Griffiths CL (1988) An improved quantitative method for estimating intertidal meiofaunal standing stock on an exposed rocky shore. S. Afr J Mar Sci 6: 55–58

Gibbs PE (1985) On the genus *Phascolion* (Sipuncula) with particular reference to the North-East Atlantic species. J Mar Biol Assoc UK 65: 311–323

Gibson GR, Parkes RJ, Herbert RA (1989) Biological availability and turnover of acetate in marine and estuarine sediments in relation to dissimilitory sulphate reduction. FEMS Microbiol Ecol 62: 303–306

Giere O (1973) Oxygen in the marine hygropsammal and the vertical microdistribution of oligochaetes. Mar Biol 21: 180–189

Giere O (1975) Population structure, food relations and ecological role of marine oligochaetes. With special reference to meiobenthic species. Mar Biol 31: 139–156

Giere O (1977) An ecophysiological approach to the microdistribution of meiobenthic Oligochaeta. I. *Phallodrilus monospermathecus* (Knöllner) (Tubificidae) from a subtropical beach at Bermuda. In: Keegan BF, O'Ceidigh P, Boaden PJS (eds) Biology of benthic organisms. Pergamon Press, Oxford pp 285–296

Giere O (1979) The impact of oil pollution on intertidal meiofauna. Field studies after the La Coruña-spill, May 1976. Cah Biol Mar 20: 231–251

Giere O (1980) Tolerance and preference reactions of marine Oligochaeta in relation to their distribution. In: Brinkhurst RO, Cook DG (eds) Aquatic oligochaete biology. Plenum Press, London pp 385–409

Giere O (1981) The gutless marine oligochaete *Phallodrilus leukodermatus*. Structural studies on an aberrant tubificid associated with bacteria. Mar Ecol Prog Ser 5: 353–357

Giere O (1992) Benthic life in sulfidic zones of the sea – ecological and structural adaptations to a toxic environment. Verh Dtsch Zool Ges 85: 77–93

Giere O, Hauschildt D (1979) Experimental studies on the life cycle and production of the littoral oligochaete *Lumbricillus lineatus*, and its response to oil pollution. In: Naylor E, Hartnoll RG (eds) Cyclic phenomena in marine plants and animals. Pergamon Press, Oxford, pp 113–122

Giere O, Langheld C (1987) Structural organization, transfer and biological fate of endosymbiotic bacteria in gutless oligochaetes. Mar Biol 93: 641–650

Giere O, Pfannkuche O (1982) Biology and ecology of marine Oligochaeta, a review. Oceanogr Mar Biol Annu Rev 20: 173–308

Giere O, Welberts H (1985) An artificial "sand system" and its application for studies on interstitial fauna. J Exp Mar Biol Ecol 88: 83–89

Giere O, Liebezeit G, Dawson R (1982) Habitat conditions and distribution pattern of the gutless oligochaete *Phallodrilus leukodermatus*. Mar Ecol Prog Ser 8: 291–299

Giere O, Eleftheriou A, Murison DJ (1988a) Abiotic factors. In: Higgins RP, Thiel H (eds) Introduction to the study of meiofauna. Smithsonian Inst Press, Washington, DC, pp 61–78

Giere O, Rhode B, Dubilier N (1988b) Structural peculiarities of the body wall of *Tubificoides benedii* (Oligochaeta) and possible relations to its life in sulphidic sediments. Zoomorphology 108: 29–39

Giere O, Conway NM, Gastrock G, Schmidt C (1991) "Regulation" of gutless annelid ecology by endosymbiotic bacteria. Mar Ecol Prog Ser 68: 287–299

Gilboa-Garber N (1971) Direct spectrophotometric determination of inorganic sulfide in biological materials and in other complex mixtures. Anal Biochem 43: 129–133

Gilmour THJ (1989) A method for studying the hydrodynamics of microscopic animals. J Exp Mar Biol Ecol 133: 189–193

Gnaiger E (1983) The Twin-Flow micro respirometer and simultaneous calorimetry. In: Gnaiger E, Forstner H (eds) Polarographic oxygen sensors. Aquatic and physiological applications. Springer, Berlin Heidelberg New York, pp 134–166

Gnaiger E (1991) Animal energetics at very low oxygen: information from calorimetry and respirometry. In: Woakes AJ, Grieshaber MK, Bridges CR (eds) Physiological strategies for gas exchange and metabolism. Cambridge University Press, Cambridge, pp 149–171

Goerke H, Ernst W (1975) ATP content of estuarine nematodes: contribution to the determination of meiofauna biomass by ATP measurements. Proc 9th Eur Mar Biol Symp, Oban 1974. Aberdeen University Press, Aberdeen, pp 683–691

Golemansky V (1978) Adaptations morphologiques de Thécamoebiens psammobiontes du psammal supralittoral des mers. Acta Protozool 17: 141–152

Gomme J (1982) Epidermal nutrient absorption in marine invertebrates: a comparative analysis. Am Zool 22: 691–708

Gooday AJ (1986) Meiofaunal foraminiferans from the bathyal Porcupine Seabight (northern Atlantic): size structure, standing stock, taxonomic composition, species diversity and vertical distribution in the sediment. Deep-Sea Res 33: 1345–1373

Gooday AJ, Turley CM (1990) Responses by benthic organisms to inputs of organic material to the ocean floor: a review. Philos Trans R Soc Lond A 331: 119–138

Gooday AJ, Levin LA, Linke P, Heeger T (1992) The role of benthic Foraminifera in deep-sea food webs and carbon cycling. In: Rowe GT, Pariente V (eds) Deep-sea food chains and the global carbon cycle. Kluwer, Dordrecht, pp 63–91

Goulden CE (1971) Environmental control of the abundance and distribution of the chydorid Cladocera. Limnol Oceanogr 16: 320–331

Gowing MM, Silver MW (1983) Origins and microenvironments of bacteria mediating fecal pellets decomposition in the sea. Mar Biol 73: 7–16

Gradinger R, Janssen HH, Weissenberger J (1993) Acoel turbellaria: a major component of sea ice meiofauna. Helgol Meeresunters (in press)

Graf G, Bengtson W, Diesner U, Schulz R, Theede H (1982) Benthic response to sedimentation of a spring phytoplankon bloom: Process and budget. Mar Biol 67: 201–208

Grainger EH (1991) Exploitation of arctic sea ice by epibenthic copepods. Mar Ecol Prog Ser 77: 119–124

Gray JS (1966) The attractive factors of intertidal sands to *Protodrilus symbioticus* Giard. J Mar Biol Assoc UK 46: 627–645

Gray JS (1971) Factors controlling population localization in polychaete worms. Vie Milieu 22: 707–722

Gray JS (1978) The structure of meiofauna communities. Sarsia 64: 265–272

Gray JS (1979) Pollution-induced changes in populations. Philos Trans R Soc Lond B 286: 545–561

Gray JS (1981) The ecology of marine sediments. An introduction to the structure and function of benthic communities. Cambridge University Press, Cambridge, 185 pp (Cambridge Studies in Modern Biology 2)

Gray JS (1984) Ökologie mariner Sedimente. Eine Einführung. Springer, Berlin Heidelberg New York, 193 pp

Greiser N, Faubel A (1988) Biotic factors. In: Higgins RP, Thiel H (eds) Introduction to the study of meiofauna. Smithsonian Inst Press, Washington, DC, pp 79–114

Grelet Y (1985) Vertical distribution of meiobenthos and estimation of nematode biomass from sediments of the Gulf of Aqaba (Jordan, Red Sea). Proc 5th Int Coral Reef Congr, Tahiti, pp 251–256

Grimaldi de Zio S, D'Addabbo Gallo M (1975) Reproductive cycle of *Batillipes pennaki* Marcus (Heterotardigrada) and observations on the morphology of the female genital apparatus. Pubbl Staz Napoli Suppl 39: 212–225

Grimaldi de Zio S, Morone de Lucia RM, D'Addabbo Gallo M (1983) Marine tardigrades ecology. Oebalia NS 9: 15–31

Grimaldi de Zio S, Morone de Lucia MR, D'Addabbo Gallo M (1984) Relazione tra morfologia ed ecologia nei tardigrada marini (Heterotardigrada-Arthrotardigrada). Cah Biol Mar 25: 67–73

Grimaldi de Zio S, D'Addabbo Gallo M, Morone de Lucia MR (1987) Adaptive radiation and phylogenesis in marine Tardigrada and the establishment of Neostygarctidae, a new family of Heterotardigrada. Boll Zool 54: 27–33

Grossmann S, Reichardt W (1991) Impact of *Arenicola marina* on bacteria in intertidal sediments. Mar Ecol Prog Ser 77: 85–94

Haarløv N, Weis-Fogh T (1953) A microscopical technique for studying the undisturbed texture of soils. Oikos 4: 44–57

Hadzi J (1956) Das Kleinsein und Kleinwerden im Tierreiche. Ein weiterer Beitrag zu meiner Turbellarientheorie der Knidarien. In: Brunn AF (ed) 14th Int Congr Zool, Kopenhagen 1953, Proc, Danish Science Press, Copenhagen, pp 154–158

Hagermann GM, Rieger RM (1981) Dispersal of benthic meiofauna by wave and current action in Bogue South, North Carolina, USA. PSZNI Mar Ecol 2: 245–270

Hakala I (1971) A new model of the Kajak-bottom sampler, and other improvements in the zoobenthos sampling techniques. Ann Zool Fenn 8: 422–426

Håkanson L (1973) Sampling of recent sedimentary deposits: a new sampler. Naturvårdsverk Limnol Undersökn, Rapport 65, 20 p

Hall MO, Bell SS (1988) Response of small motile epifauna to complexity of epiphytic algae on seagrass blades. J Mar Res 46: 613–630

Hall SJ, Basford DJ, Robertson MR, Raffaeli DG, Tuck I (1991) Patterns of recolonization and the importance of pit-digging by the crab *Cancer pagurus* in a subtidal sand habitat. Mar Ecol Prog Ser 72: 93–102

Hamilton AL (1969) A method of separating invertebrates from sediments using longwave ultraviolet light and fluorescent dyes. J Fish Res Board Can 26: 1667–1672

Hansen LS, Blackburn TH (1992) Mineralization budgets in sediment microcosms: effect of the infauna and anoxic conditions. FEMS Microbiol Ecol 102: 33–43

Hargrave BT (1972) Oxidation-reduction potentials, oxygen concentrations and oxygen uptake of profundal sediments in an eutrophic lake. Oikos 23: 167–177

Harris RP (1972) Seasonal changes in the meiofauna population of an intertidal sand beach. J Mar Biol Assoc UK 52: 389–404

Hartmann G (1963) Zur Phylogenie und Systematik der Ostracoden. Z Zool Syst Evolutionsforsch 1: 1–154

Hartmann G (1966–1989) Ostracoda. In: Gruner H-E (ed) Bronns Klassen und Ordnungen des Tierreichs, 5, 1 Abt, Buch 2, 4 T, Lfg 1-4. Akad Verlagsgesellschaft Geest & Portig, Leipzig und VEB Gustav Fischer, Jena, 1067 p

Hartmann G (1973) Zum gegenwaertigen Stand der Erforschung der Ostracoden interstitieller Systeme. Ann Spéléol 28: 417–426

Hartmann G (1986) Biogeographie und Plattentektonik. Gondwana und die rezente Verteilung der Organismen. Naturwissenschaften 73: 471–480

Hartmann G (1988) Gibt es biologische Argumente zur Entstehung der Süderdteile? Geowissenschaften 6: 270–275

Hartmann G (1990) Antarktische benthische Ostracoden VI. Auswertung der Reise der "Polarstern" Ant. VI-2 (1. Teil, Meiofauna und Zehnerserien) sowie Versuch einer vorläufigen Auswertung aller bislang vorliegenden Daten. Mitt Hamb Zool Mus Inst 87: 191–245

Hartmann G, Kühl C (1978) Zur Variabilität der Oberflächenornamente der Schalen lebender Ostracoden-Populationen. Mitt Hamb Zool Mus Inst 75: 221–223

Hartwig E (1973a) Die Ciliaten des Gezeiten-Sandstrandes der Nordseeinsel Sylt. I. Systematik. Mikrofauna Meeresboden 18: 387–453

Hartwig E (1973b) Die Ciliaten des Gezeiten-Sandstrandes der Nordseeinsel Sylt. II. Ökologie. Mikrofauna Meeresboden 21: 1–71

Healy B, Walters K (1993) The distribution and abundance of Oligochaeta in a Spartina salt marsh, Sapelo Island, Georgia, U.S.A. Hydrobiologia (in press)

Hed J (1977) Extinction of fluorescence by crystal violet and its use to differentiate between attached and ingested microorganisms in phagocytosis. FEMS Microbiol Lett 1: 357–361

Heinzelmann C (1990) Lebensgemeinschaften stabilisieren das Flußbett. Untersuchungen zur Benthosbesiedlung und Sohlenerosion. Forsch Mitt DFG 4: 10–12

Heip C (1976) The calculation of eliminated biomass. Biol Jahrb Dodonea 44: 217–225

Heip C (1980a) The influence of competition and predation on production of meiobenthic copepods. In: Tenore KR, Coull BC (eds) Marine benthic dynamics. University of South Carolina Press, Columbia, SC, pp 167–177 (Belle W Baruch Libr Mar Sci, vol 11)

Heip C (1980b) Meiobenthos as a tool in the assessment of marine environmental quality. Rapp P-V Réun Cons Int Explor Mer 179: 182–187

Heip C, Decraemer W (1974) The diversity of nematode communities in the southern North Sea. J Mar Biol Assoc UK 54: 251–255

Heip C, Smol N, Hautekiet W (1974) A rapid method of extracting meiobenthic nematodes and copepods from mud and detritus. Mar Biol 28: 79–81

Heip C, Herman PMJ, Coomans A (1982a) The productivity of marine meiobenthos. Acad Anal K1 Wetenschappen 44: 1–20

Heip C, Vincx M, Smol N, Vranken G (1982b) The systematics and ecology of free-living marine nematodes. Helminthol Abstr Ser B 51: 1–31

Heip C, Vincx M, Vranken G (1985a) The ecology of marine nematodes. Oceanogr Mar Biol Annu Rev 23: 399–489

Heip C, Herman PMJ, Smol N, Brussel D van, Vranken G (1985b) Energy flow through the meiobenthos. In: Heip C, Polk P (eds) Benthic studies of the Southern Bight of the North Sea and its adjacent continental estuaries. Biological processes and translocations. Ministry of Scientific Policy, Brussels, Belgium, pp 11-40 (Concerted Actions Oceanography 3)

Heip C, Warwick RM, Carr MR, Herman PMJ, Huys R, Smol N, Holsbeke K van (1988) Analysis of community attributes of the benthic meiofauna of Frierfjord/Langesundfjord. Mar Ecol Prog Ser 46: 171–180

Helder W, Bakker JF (1985) Shipboard comparison of macro- and minielectrodes for measuring oxygen distribution in marine sediments. With addition: technical description of manufacturing and application of needle- and microoxygen electrodes, in sediments. Limnol Oceanogr 30: 1106–1108

Hellwig-Armonies M (1988) High abundance of Plathelminthes in a North Sea salt marsh creek. Progr Zool 36: 499–504

Hennig HF-KO, Eagle GA, Fielder L, Fricke A, Gledhill WJ, Greenwood PJ, Orren MJ (1983a) Ratio and population density of psammolitoral meiofauna as a perturbation indicator of sandy beaches in South Africa. Environ Monit Assess 3: 45–60

Henning HF-KO, Fricke AH, Martin CT (1983b) The effects of meiofauna and bacteria on nutrient cycles in sandy beaches. In: McLachlan A, Erasmus T (eds) Sandy beaches as ecosystems. Junk, The Hague, pp 235–247

Herman PMJ, Heip C (1988) On the use of meiofauna in ecological monitoring: Who needs taxonomy? Mar Pollut Bull 19: 665–668

Herman PMJ, Vranken G (1988) Studies of the life-history and energetics of marine and brackish-water nematodes. II. Production, respiration and food uptake by *Monhystera disjuncta*. Oecologia 77: 457–463

Herman RL, Dahms HU (1992) Meiofauna communities along a depth transect off Halley Bay (Weddell Sea-Antarctica). Polar Biol 12: 313-320

Herman R, Vincx M, Heip C (1985) Meiofauna of the Belgian coastal waters: Spatial and temporal variability and productivity. In: Heip C, Polk P (eds) Benthic Studies of the Southern Bight of the North Sea and its adjacent continental estuaries. Biological processes and translocations. Ministry of Scientific Policy, Brussels, Belgium, pp 41–63 (Concerted Actions Oceanography 3)

Hicks GRF (1985) Meiofauna associated with rocky shore algae. In: Moore, PG, Seed R (eds) The ecology of rocky coasts. Hodder and Stoughton, London, pp 36–64

Hicks GRF (1986) Distribution and behaviour of meiofaunal copepods inside and outside seagrass beds. Mar Ecol Prog Ser 31: 159–170

Hicks GRF (1988) Sediment rafting: a novel mechanism for the small-scale dispersal of intertidal estuarine meiofauna. Mar Ecol Prog Ser 48: 69–80

Hicks GRF (1989) Does epibenthic structure negatively affect meiofauna? J Exp Mar Biol Ecol 133: 39–55

Hicks GRF (1991) Monitoring with meiofauna: a compelling option for evaluating environmental stress in tidal inlets. Water Qual Centre Publ 21: 387–391

Hicks GRF, Coull BC (1983) The ecology of marine meiobenthic harpacticoid copepods. Oceanogr Mar Biol Annu Rev 21: 67–175

Higgins RP (1964) A method for meiobenthic invertebrate collection. Am Zool 4: 291

Higgins RP (1981) Kinorhyncha. In: Parker SP (ed) Synopsis and Classification of Living Organisms, vol 1. McGraw Hill, New York, pp 873–877

Higgins RP (1986) Redescription of *Echinoderes pilosus* (Kinorhyncha: Cyclorhagida). Proc Biol Soc Wash 99: 399–405

Higgins RP (1988) Kinorhyncha. In: Higgins RP, Thiel H (eds) Introduction to the study of meiofauna. Smithsonian Inst Press, Washington, DC, pp 328–331

Higgins RP, Kristensen RM (1986) New Loricifera from southeastern United States coastal waters. Smithson Contrib Zool 438: 1–70

Higgins RP, Storch V (1989) Ultrastructural observations of the larva of *Tubiluchus corallicola* (Priapulida). Helgol Meeresunters 43: 1–11

Higgins RP, Storch V (1991) Evidence for direct development in *Meiopriapulus fijiensis* (Priapulida). Trans Am Microsc Soc 110: 37–46

Higgins RP, Thiel H (eds) (1988) Introduction to the study of meiofauna. Smithsonian Inst Press, Washington, DC, 488 pp

Hines ME, Jones GE (1985) Microbial biogeochemistry and bioturbation in the sediment of Great Bay, New Hampshire. Estuarine Coastal Shelf Sci 20: 729–742

Hockin DC (1982a) Experimental insular zoogeography: some tests of the equilibrium theory using meiobenthic harpacticoid copepods. J Biogeogr 9: 487–498

Hockin DC (1982b) The spatial population structure of a harpacticoid copepod community in spring. Hydrobiologia 96: 201–209

Hofker J (1977) The Foraminifera of Dutch tidal flats and salt marshes. Neth J Sea Res 11: 223–296

Hogue EW, Miller CB (1981) Effects of sediment microtopography on small-scale spatial distributions of meiobenthic nematodes. J Exp Mar Biol Ecol 53: 181–191

Holme NA, McIntyre AD (eds) (1984) Methods for the study of marine benthos, 2nd edn. Blackwell, Oxford, 387 pp (IBP Handbook 16)

Holopainen IJ, Paasivirta L (1977) Abundance and biomass of the meiobenthos in the oligotrophic and mesohumic lake Pääjärvi, southern Finland. Ann Zool Fenn 14: 124–134

Holopainen IJ, Sarvala J (1975) Efficiencies of two corers in sampling softbottom invertebrates. Ann Zool Fenn 12: 280-284

Hopper BE, Meyers SP (1966) Aspects of the life cycles of marine nematodes. Helgol Wiss Meeresunters 13: 444–449

Howes BL, Wakeham SG (1985) Effects of a sampling technique on measurements of porewater constituents in salt marsh sediments. Limnol Oceanogr 30: 221–227

Hulings NC, Gray JS (1976) Physical factors controlling abundance of meiofauna on tidal and atidal beaches. Mar Biol 34: 77–83

Hummon WD (1971) The marine and brackish-water Gastrotricha in perspective. In: Hulings NC (ed) Proc. 1st International Conference on Meiofauna, Tunesia. Smithson Contrib Zool 76: 21–23

Hummon WD (1972) Dispersion of Gastrotricha in a marine beach of the San Juan Archipelago, Washington. Mar Biol 16: 349–355

Hummon WD (1974) Respiratory and osmoregulatory physiology of a meiobenthic marine gastrotrich, *Turbanella ocellata* Hummon 1974. Cah Biol Mar 16: 255–268

Hummon WD (1989) The fetch-energy index: an a priori estimator of coastal exposure, applied to littoral marine Gastrotricha of the British Isles. In: Ryland JS, Tyler PA (eds) Reproduction, genetics and distributions of marine organisms. Olsen & Olsen, Fredensborg, pp 387–393

Humphreys WF (1979) Production and respiration in animal populations. J Anim Ecol 48: 427–453

Hurlbert SH (1984) Pseudoreplication and the design of ecological field experiments. Ecol Monogr 54: 187–211

Huston M (1979) A general hypothesis of species diversity. Am Nat 113: 81–101

Huys R, Herman PMJ, Heip CHR, Soetaert K (1992) The meiobenthos of the North Sea: density, biomass trends and distribution of copepod communities. ICES J Mar Sci 49: 23–44

Hylleberg J (1975) Selective feeding by *Abarenicola pacifica* with notes on *Abarenicola vagabunda* and a concept of gardening in lugworms. Ophelia 14: 113–137

Hylleberg J, Henriksen K (1980) The central role of bioturbation in sediment mineralization and element re-cycling. Ophelia 1: 1–16

Iharos G (1975) Summary of the results of forty years of research on Tardigrada. Mem Ist Ital Idrobiol Suppl 32: 159–169

Iliffe TM (1990) Crevicular dispersal of marine cave faunas. Mém Biospéol 17: 93–96

Iliffe TM, Wilkens H, Parzefall J, Williams D (1984) Marine lava cave fauna: composition, biogeography and origins. Science 225: 309–311

Ivester MS, Coull BC (1977) Niche fractionation studies of two sympatric species of *Enhydrosoma*. Mikrofauna Meeresboden 61: 131–145

Jackson D (1986) A manually operated core-sampler suitable for use on fine-particulate sediments. Estuarine Coastal Shelf Sci 23: 419–422

Jahnke RA (1988) A simple, reliable, and inexpensive pore-water sampler. Limol Oceanogr 33: 483–487

Jansson B-O (1966a) Microdistribution of factors and fauna in marine sandy beaches. Veröff Inst Meeresforsch Bremerhaven Sbd 2: 77–86

Jansson B-O (1966b) On the ecology of *Derocheilocaris remanei* Delamare and Chappuis (Crustacea, Mystacocarida). Vie Milieu 17: 143–186

Jansson B-O (1967a) Diurnal and annual variations of temperature and salinity of interstitial water in sandy beaches. Ophelia 4: 173–201

Jansson B-O (1967b) The significance of grain size and pore water content for the interstitial fauna of sandy beaches. Oikos 18: 311–322

Jenkins RJF (1991) The early environment. In: Bryant C (ed) Metazoan life without oxygen. Chapman and Hall, London, pp 38–64

Jennings JB, Hick AJ (1990) Differences in the distribution, mitochondrial content and probable roles of haemoglobin-containing parenchymal cells in four species of ento-symbiotic turbellarians (Rhabdocoela: Umagillidae and Pterastericolidae). Ophelia 31: 163–175

Jensen P (1981) Species distribution and a microhabitat theory for marine mud dwelling Comesomatidae (Nematoda) in European waters. Cah Biol Mar 22: 231–241

Jensen P (1982) A new meiofauna sample splitter. Ann Zool Fenn 19: 233–236

Jensen P (1983) Meiofaunal abundance and vertical zonation in a sublittoral soft bottom, with a test of the Haps corer. Mar Biol 74: 319–326

Jensen P (1984) Ecology of benthic and epiphytic nematodes in brackish waters. Hydrobiologia 108: 201–217

Jensen P (1986) Nematode fauna in the sulphide-rich brine seep and adjacent bottoms of the East Flower Garden, NW Gulf of Mexico. IV. Ecological aspects. Mar Biol 92: 489–503

Jensen P (1987a) Feeding ecology of free-living aquatic nematodes. Mar Ecol Prog Ser 35: 187–196

Jensen P (1987b) Differences in microhabitat, abundance, biomass and body size between oxybiotic and thiobiotic free-living marine nematodes. Oecologia (Berl) 71: 564–567

Jensen P, Emrich R, Weber K (1992) Brominated metabolites and reduced numbers of meiofauna organisms in the burrow wall lining of the deep-sea enteropneust *Stereobalanus canadensis*. Deep-Sea Res 39: 1247–1253

Joint IR, Gee JM, Warwick RM (1982) Determination of fine-scale vertical distribution of microbes and meiofauna in an intertidal sediment. Mar Biol 72: 157–164

Jones RW, Charnock MA (1985) "Morphogroups" of agglutinating Foraminifera. Their life positions and feeding habits and potential applicability in (palaeo)ecological studies. Rev Paléobiol 4: 311–320

Jonge VN, Bouwmann LA (1977) A simple density separation technique for quantitative isolation of meiobenthos using the colloidal silica Ludox-TM. Mar Biol 42: 143–148

Jørgensen BB, Revsbech NP (1985) Diffuse boundary layers and the oxygen uptake of sediments and detritus. Limnol Oceanogr 30: 111–122

Jørgensen BB (1977) Bacterial sulfate reduction within reduced microniches of oxidized marine sediments. Mar Biol 41: 7–17

Jørgensen BB (1988) Ecology of the sulphur cycle: oxidative pathways in sediments. In: Cole JA, Ferguson SJ (eds) The nitrogen and sulphur cycles. 42nd Symp Soc Gen Microbiol, Univ of Southampton. Cambridge University Press, Cambridge, pp 65–98

Jørgensen BB (1990) A thiosulfate shunt in the sulfur cycle of marine sediments. Science 249: 152–154

Jørgensen BB, Bak F (1991) Pathways and microbiology of thiosulfate transformations and sulfate reduction in a marine sediment (Kattegat, Denmark). Appl Environ Microbiol 57: 847–856

Jørgensen CB (1976) August Pütter, August Krogh, and the modern ideas on the use of dissolved organic matter in aquatic environments. Biol Rev 51: 291–328

Jørgensen NOG, Mopper K, Lindroth P (1980) Occurrence, origin, and assimilation of free amino acids in an estuarine environment. Ophelia Suppl 1: 179–192

Jørgensen NOG, Lindroth P, Mopper K (1981) Extraction and distribution of free amino acids and ammonium in sediment interstitial waters from the Limfjord, Denmark, Oceanol Acta 4: 465–474

Josefson AB, Widbom B (1988) Differential response of benthic macrofauna and meiofauna to hypoxia in the Gullmar Fjord basin. Mar Biol 100: 31–40

Jouin C (1992) The ultrastructure of a gutless annelid, *Parenterodrilus* gen. nov. *taenioides* (= *Astomus taenioides*) (Polychaeta, Protodrilidae). Can J Zool 70: 1833–1848

Jouk PEH, Martens PM, Schockaert ER (1988) Horizontal distribution of the Plathelminthes in a sandy beach of the Belgian coast. In: Ax P, Ehlers U, Sopott-Ehlers B (eds) Free-living and symbiotic Plathelminthes. Fischer, Stuttgart, pp 481–487 (Fortschr Zool 36)

Kanneworff E, Nicolaisen W (1973) The "Haps"; a frame-supported bottom corer. Ophelia 10: 119–129

Karaman SL (1935) Die Fauna unterirdischer Gewässer Jugoslawiens. Verh Int Ver Theor Angew Limnol 7: 46–53

Karl DM, La Rock PA (1975) Adenosine triphosphate measurements in soil and marine sediments. J Fish Res Board Can 32: 599–607

Kemp PF (1988) Bacterivory by benthic ciliates: significance as a carbon source and impact on sediment bacteria. Mar Ecol Prog Ser 49: 163–169

Kemp PF (1994) Benthic Microbial Ecology. In: Kennish MJ, Luth PL (eds) Marine Science Book Series. CRC Press, Boca Raton

Keppner EJ, Tarjan AC (1989) Illustrated key to the genera of free-living marine nematodes of the order Enoplida. NOAA Tech Rep N2F77, US Dep Commerce, 26 pp

King GM (1986) Inhibition of microbial activity in marine sediments by a bromphenol from a hemichordate. Nature 323: 257–259

Kinne O (1964) Non-genetic adaptation to temperature and salinity. Helgol Wiss Meeresunters 9: 433–458

Kirsteuer E (1976) Notes on adult morphology and larval development of *Tubiluchus corallicola* (Priapulida), based on in vivo and scanning electron microscopic examinations of specimens from Bermuda. Zool Scr 5: 239–255

Kirsteuer E (1977) Remarks on taxonomy and geographic distribution of the genus *Ototyphlonemertes* Diesing (Nemertina, Monostilifera), Mikrofauna Meeresboden 61: 167–181

Kisielewski J (1987) Two new interesting genera of Gastrotricha (Macrodasyoida and Chaetonotida) from the Brazilian freshwater psammon. Hydrobiologia 153: 23–30

Kisielewski J (1990) Origin and phylogenetic significance of freshwater psammic Gastrotricha. Stygologia 5: 87–92

Koelmel R (1974) Ein neuer "Meiofaunastecher" zur quantitativen Probennahme in Weichböden. Mar Biol 25: 163–169

Koop K, Griffiths CL (1982) The relative significance of bacteria, meio- and macrofauna on an exposed sandy beach. Mar Biol 66: 295–300

Kovalevsky A (1901) Les Hédylides, études anatomiques. Mém Acad Sci St Petersburg (Sci Math Phys Nat) 12: 1–32

Kowarc VA (1990) Production of a harpacticoid copepod from the meiofaunal community of a second order mountain stream. Stygologia 5: 25–32

Kraus MG, Found BW (1975) Preliminary observation on the salinity and temperature tolerances and salinity preferences of *Derocheilocaris typica* Pennak and Zinn 1943. Cah Biol Mar 16: 751–762

Kristensen E (1984) Life cycle, growth and production in estuarine populations of the polychaetes *Nereis virens* and *N. diversicolor*. Holarct Ecol 7: 249–256

Kristensen E, Blackburn TH (1987) The fate of organic carbon and nitrogen in experimental marine systems: Influence of bioturbation and anoxia. J Mar Res 45: 231–257

Kristensen E, Jensen MH, Andersen TK (1985) The impact of polychaete (*Nereis virens* Sars) burrow on nitrification and nitrate reduction in estuarine sediment. J Exp Mar Biol Ecol 85: 75–91

Kristensen RM (1982) The first record of cyclomorphosis in Tardigrada based on a new genus and species from Arctic meiobenthos. Z Zool Syst Evolutionsforsch 20: 249–270

Kristensen RM (1983) Loricifera, a new phylum with Aschelminthes characters from the meiobenthos. Z Zool Syst Evolutionsforsch 21: 161–180

Kristensen RM (1984) Nyt dyr-korsetdyret-opdaget. Nat Verden 1984: 357–367

Kristensen RM (1991a) Loricifera. In: Harrison FW, Ruppert EE (eds) Microscopic Anatomy of Invertebrates, vol 4: Aschelminthes. Wiley-Liss, New York, pp 351–375

Kristensen RM (1991b) Loricifera – a general biological and phylogenetic overview. Verh Dtsch Zool Ges 84: 231–246

Kristensen RM, Higgins RP (1984) A new family of Arthrotardigrada (Tardigrada: Heterotardigrada) from the Atlantic coast of Florida, U.S.A. Trans Am Microsc Soc 103: 295–311

Kristensen RM, Higgins RP (1991) Kinorhyncha. In: Harrison FW, Ruppert EE (eds) Microscopic anatomy of invertebrates, vol 4: Aschelminthes. Wiley-Liss, New York, pp 377–404

Krogh A, Spärck R (1936) On a new bottom sampler for investigation of the microfauna of the sea bottom. K Dan Vidensk Selsk Skr 13: 1–12

Krumbein WC (1939) Graphic presentation and statistical analysis of sedimentary data. In: Trask PD (ed) Recent marine sediments. A symposium (reprinted 1955 with review of advances since 1939). The American Association of Petroleum Geologists, Tulsa, Oklahoma. Murby & Co, London, pp 558–591

Kruskal JB, Wish M (1978) Multidimensional scaling. Sage Publications, Beverley Hills, USA, 93 pp

Kuipers BR, Wilde PAWJ de, Creutzberg F (1981) Energy flow in a tidal flat ecosystem. Mar Ecol Prog Ser 5: 215–221

Kunz H (1935) Zur Ökologie der Copepoden Schleswig-Holsteins und der Kieler Bucht. Schr Naturwiss Ver Schleswig-Holstein 21: 84–132

Kurdziel JP, Bell SS (1992) Emergence and dispersal of phytal-dwelling meiobenthic copepods. J Exp Mar Biol Ecol 163: 43–64

LaBarbera M, Vogel S (1976) An inexpensive thermistor flowmeter for aquatic biology. Limnol Oceanogr 21: 750–756

Lackey JB (1936) Occurrence and distribution of the marine protozoan species in the Woods Hole area. Biol Bull (Woods Hole) 70: 264–278

Lafon M, Durbec A (1990) Essai de description biologique des interactions entre eau de surface et eau souterraine: vulnérabilité d'un aquifère à la pollution d'un fleuve. Ann Limnol 26: 119–129

Lambshead PJD (1984) The nematode/copepod ratio. Some anomalous results from the Firsth of Clyde. Mar Pollut Bull 15: 256–259

Lambshead PJD (1986) Sub-catastrophic sewage and industrial waste contamination as revealed by marine nematode faunal analysis. Mar Ecol Prog Ser 29: 247–259

Lambshead PJD, Gooday AJ (1990) The impact of seasonally deposited phytodetritus of epifaunal and shallow infaunal benthic foraminiferal populations in the bathyal northeast Atlantic: the assemblage response. Deep-Sea Res 37: 1263–1283

Lambshead PJD, Paterson GLJ (1986) Ecological cladistics – an investigation of numerical cladistics as a method for analysing ecological data. J Nat Hist 20: 895–909

Lambshead PJD, Platt HM, Shaw KM (1983) The detection of differences among assemblages of marine benthic species based on an assessment of dominance and diversity. J Nat Hist 17: 859–874

Lang K (1948) Monographie der Harpacticoidea, Bd 1 und 2. Håkan Ohlsson, Lund, 1682 pp

Lee JJ (1980a) A conceptual model of marine detrital decomposition and the organisms associated with the process. Adv Aquat Microbiol 2: 257–291

Lee JJ (1980b) Nutrition and physiology of the Foraminifera. In: Levandowsky M, Hutner SH (eds) Biochemistry and physiology of Protozoa, 2nd edn, vol 3. Academic Press, New York, pp 43–66

Levin LA (1991) Interactions between metazoans and large agglutinating protozoans: implications for the community structure of deep-sea bottoms. Am Zool 31: 886–900

Levinton JS (1983) The latitudinal compensation hypothesis: Growth data and a model of latitudinal growth differentiation based upon energy budgets. I. Interspecific comparison of Ophryotrocha (Polychaeta: Dorvilleidae). Biol Bull (Woods Hole) 165: 686–698

Liebezeit G, Felbeck H, Dawson R, Giere O (1983) Transepidermal uptake of dissolved carbohydrates by the gutless marine oligochaete Phallodrilus leukodermatus (Annelida). Océanis 9: 205–211

Lombardi J, Ruppert EE (1982) Functional morphology of locomotion in Derocheilocaris typica (Crustacea, Mystacocarida). Zoomorphology 100: 1–10

Lorenzen S (1981) Entwurf eines phylogenetischen Systems der freilebenden Nematoden. Veröff Inst Meeresforsch Bremerhaven Suppl 7: 472

Lorenzen S (1986) Nematoda: Interstitial nematodes from marine, brackish and hypersaline environments. In: Botosaneanu L (ed) Stygofauna Mundi. A faunistic, distributional, and ecological synthesis of the world fauna inhabiting subterranean waters (including the marine interstitial). Brill, Backhuys, Leiden, pp 133–142

Lorenzen S, Prein M, Valentin C (1987) Mass aggregations of the free-living marine nematode Pontonema vulgare (Oncholaimidae) in organically polluted fjords. Mar Ecol Prog Ser 37: 27–34

Luckenbach MW (1986) Sediment stability around animal tubes: the roles of hydrodynamic processes and biotic activity. Limnol Oceanogr 31: 779–787

Malan DE, McLachlan A (1991) In situ benthic oxygen fluxes in a nearshore coastal marine system: a new approach to quantify the effect of wave action. Mar Ecol Prog Ser 73: 69–81

Mangum C (1991) Precambrian oxygen levels, the sulfide biosystem, and the origin of the Metazoa. J Exp Zool 260: 33–42

Marcotte BM (1983) The imperatives of copepod diversity: perception, cognition, competition and predation. In: Schram FR (ed) Crustacean Phylogeny. Balkema, Rotterdam, pp 47–72

Marcotte BM (1984) Behaviourally defined ecological resources and speciation in Tisbe (Copepoda: Harpacticoida). J Crustacean Biol 4: 404–416

Marcotte BM (1986) Sedimentary particle sizes and the ecological grain of food resources for meiobenthic copepods. Estuarine Coastal Shelf Sci 23: 423–427

Mare MF (1942) A study of a marine benthic community with special reference to the micro-organisms. J Mar Biol Assoc UK 25: 517–554

Marion AF, Kovalevsky AO (1886) Organisation du Lepidomenia hystrix, nouveau type de Solénogastre. CR Hebd Séances Acad Sci Paris 103: 757–759

Martens PM, Schockaert ER (1986) The importance of turbellarians in the marine meiobenthos: a review. Hydrobiologia 132: 295–303

Mason WT, Yevich PP (1967) The use of Phloxine B and Rose Bengal stains to facilate sorting benthic samples. Trans Am Microsc Soc 86: 221–223

Mayer LM, Rossi PM (1982) Specific surface areas in coastal sediments: relationships with other textural factors. Mar Geol 45: 241–252

McDowell EM (1978) Fixation and processing. In: Trump BF, Jones RT (eds) Diagnostic electron microscopy, vol 1. Wiley, New York, pp 113–139

McIntyre AD (1968) The meiofauna and macrofauna of some tropical beaches. J Zool Lond 156: 377–392

McIntyre AD (1969) Ecology of marine meiobenthos. Biol Rev 44: 245–290

McIntyre AD (1977) Effects of pollution on inshore benthos. In: Coull BC (ed) Ecology of marine benthos. University of South Carolina Press, Columbia, pp 301–318

McIntyre AD, Warwick RM (1984) Meiofauna techniques. In: Holme NA, McIntyre AD (eds) Methods for the study of marine benthos. Blackwell, Oxford, pp 217–244 (IBP Handbook 16)

McIntyre AD, Elliott JM, Ellis DV (1984) Introduction: design of sampling programmes. In: Holme NA, McIntyre AD (eds) Methods for the study of marine benthos. Blackwell, Oxford, pp 1–26 (IBP Handbook 16)

McLachlan A (1980) The definition of sandy beaches in relation to exposure: a simple rating system. S Afr J Sci 76: 137–138

McLachlan A (1989) Water filtration by dissipative beaches. Limnol Oceanogr 34: 774–779

McLachlan A, Romer G (1990) Trophic relations in a high energy beach and surf-zone ecosystem. In: Barnes M, Gibson RN (eds) Trophic relations in the marine environment. Proc 24th Eur Mar Biol Symp. Aberdeen University Press, Aberdeen, pp 356–371

McLachlan A, Erasmus T, Furstenberg JP (1977) Migrations of sandy beach meiofauna. Zool Afr 12: 256–277

McLachlan A, Dye AH, Ryst P van der (1979) Vertical gradients in the fauna and oxidation of two exposed sandy beaches. S Afr J Zool 14: 43–49

McLachlan A, Woolridge T, Dye AH (1981) The ecology of sandy beaches in southern Africa. S Afr J Zool 16: 219–231

McNeill S, Lawton JH (1970) Annual production and respiration in animal populations. Nature 225: 472–474

Meadows PS (1986) Biological activity and seabed sediment structure. Nature 323: 207

Meadows PS, Anderson JG (1966) Micro-organisms attached to marine and freshwater sand grains. Nature 212: 1059–1060

Meadows PS, Anderson JG (1968) Micro-organisms attached to marine sand grains. J Mar Biol Assoc UK 48: 161–175

Meadows PS, Tait J (1985) Bioturbation, geotechnics and microbiology at the sediment-water interface in deep-sea sediments. In: Gibbs PE (ed) Proc 19th Eur Mar Biol Symp. Cambridge University Press, Cambridge, pp 191–199

Meadows PS, Tait J (1989) Modification of sediment permeability and shear strength by two burrowing invertebrates. Mar Biol 101: 75–82

Meadows PS, Tait J, Hussain SA (1990) Effects of estuarine infauna on sediment stability and particle sedimentation. Hydrobiologia 190: 263–266

Meineke T, Westheide W (1979) Gezeitenabhängige Wanderungen der Interstitialfauna in einem Gezeitenstrand der Insel Sylt (Nordsee). Mikrofauna Meeresboden 75: 203–236

Menker D, Ax P (1970) Zur Morphologie von Arenadiplosoma migrans n.g. n.sp., einer vagilen Ascidien-Kolonie aus den Mesopsammal der Nordsee (Tunicata, Ascidiacea). Z Morphol Tiere 66: 323–336

Meyer HA, Bell SS (1989) Response of harpacticoid copepods to detrital accumulation on seagrass blades: a field experiment with Metis holothuriae (Edwards). J Exp Mar Biol Ecol 132: 141–149

Meyer-Reil L-A (1987) Biomass and activity of benthic bacteria. In: Rumohr J, Walger E, Zeitzschel B (eds) Lecture notes on coastal and estuarine studies, vol 13: Seawater-sediment interactions. Springer, Berlin Heidelberg New York, pp 93–110

Meyer-Reil L-A, Faubel A (1980) Uptake of organic matter by meiofauna organisms and interrelationships with bacteria. Mar Ecol Prog Ser 3: 251–256

Meyer-Reil L-A, Dawson R, Liebezeit G, Tiedge H (1978) Fluctuations and interactions of bacterial activity in sandy sediments and overlying waters. Mar Biol 48: 161–171

Meyers MB, Fossing H, Powell EN (1987) Microdistribution of interstitial meiofauna, oxygen and sulfide gradients, and the tubes of macro-infauna. Mar Ecol Prog Ser 35: 223–241

Meyers MB, Powell EN, Fossing H (1988) Movement of oxybiotic and thiobiotic meiofauna in response to changes in pore-water oxygen and sulfide gradients around macro-infaunal tubes. Mar Biol 98: 395–414

Michelson AR, Jacobson ME, Scranton MI, Mackin JE (1989) Modelling the distribution of acetate in anoxic estuarine sediments. Limnol Oceanogr 34: 747–757

Monniot C, Monniot F (1984) Nouvelles Sorberacea (Tunicata) profondes de l'Atlantique Sud et de l'Océan Indien. Cah Biol Mar 25: 197–215

Monniot C, Monniot F (1990) Revision of the class Sorberacea (benthic turnicates) with descriptions of 7 new species. Zool J Linn Soc 99: 239–290

Monniot C, Monniot F, Gaill F (1975) Les Sorberacea: une nouvelle classe de Tuniciers. Arch Zool Exp Gén 116: 77–122

Monniot F (1962) Recherches sur le graviers à Amphioxus de la région de Banyuls-sur-Mer. Vie Milieu 13: 231–322

Monniot F (1965) Ascidies interstitielles des côtes d'Europe. Mém Mus Natl Hist Nat Paris Sér A 35: 1–154

Monniot F (1966) Un Palpigrade interstitiel *Leptokoenenia scurra* n sp. Rev Ecol Biol Sol 3: 41–64

Monniot F (1971) Les Ascidies littorales et profondes des sédiments meubles. In: Hulings NC (ed) Proc 1st International Conference on Meiofauna, Tunesia. Smithson Contrib Zool 76: 119–126

Monniot F, Monniot C (1988) Tunicata. In: Higgins RP, Thiel H (eds) Introduction to the study of meiofauna. Smithsonian Inst Press, Washington, DC, pp 461–464

Monod T (1940) Thermosbaenacea. In: Schellenberg A (ed) Bronns Klassen und Ordnungen des Tierreichs, Bd 5, 1 Abt, 4 T, Buch 4. Akad Verlagsgesellschaft Geest & Portig, Leipzig und VEB Gustav Fischer, Jena, pp 1–24

Montagna PA (1983) Live controls for radioisotope food chain experiments using meiofauna. Mar Ecol Prog Ser 12: 43–46

Montagna PA (1984) In situ measurement of meiobenthic grazing rates on sediment bacteria and edaphic diatoms. Mar Ecol Prog Ser 18: 119–130

Montagna PA (1989) Meiofaunal-microbial interactions in food chains and nutrient cycling. 7th Int Meiofauna Conf, Vienna 1989 (Abstr)

Montagna PA, Bauer JE (1988) Partitioning radiolabeled thymidine uptake by bacteria and meiofauna using metabolic blocks and poisons in benthic feeding studies. Mar Biol 98: 101–110

Montagna PA, Coull BC, Herring TL, Dudley BW (1983) The relationship between abundances of meiofauna and their suspected microbial food (diatoms and bacteria). Estuarine Coastal Shelf Sci 17: 381–394

Montagna PA, Bauer JE, Hardin D, Spies RB (1989) Vertical distribution of microbial and meiofauna populations in sediments of a natural coastal hydrocarbon seep. J Mar Res 47: 657–680

Moore CG, Bett BJ (1989) The use of meiofauna in marine pollution impact assessment. Zool J Linn Soc 96: 263–280

Moore CG, Stevenson JM (1991) The occurrence of intersexuality in harpacticoid copepods and its relationship with pollution. Mar Pollut Bull 22: 72–74

Moore HB (1931) The muds of the Clyde Sea area. III. Chemical and physical conditions; rate and nature of sedimentation; and fauna. J Mar Biol Assoc UK 17: 325–358

Moore HB, Neill RG (1930) An instrument for sampling marine muds. J Mar Biol Assoc UK 16: 589–594

Moriarty DJW (1980) Measurement of bacterial biomass in sandy sediments. Proc Int Symp Environ Biogeochem (ISEB) 4, pp 131–138

Morill AC, Powell EN, Bidigare RR, Shick JM (1988) Adaptations to life in the sulfide system: a comparison of oxygen detoxifying enzymes in thiobiotic and detoxifying enzymes in thiobiotic and oxybiotic meiofauna (and freshwater planarians). J Comp Phys B 158: 335–344

Morris JT, Coull BC (1992) Population dynamics, numerical production, and potential predation impact on a meiobenthic copepod. Can J Fish Aquat Sci 49: 609–616

Morse MP (1981) *Meiopriapulus fijiensis* n.sp.: an interstitial priapulid from coarse sand in Fiji. Trans Am Microsc Soc 100: 239–252

Mortensen T (1925) An apparatus for catching the micro-fauna of the sea bottom. Vidensk Medd Dan Naturhist Foren København 80: 445–451

Müller KJ, Walossek D (1985) Skaracarida, a new order of Crustacea from the Upper Cambrian of Västergötland, Sweden, Fossils Strata 17: 1–65

Müller KJ, Walossek D (1991) Ein Blick durch das <Orsten>-Fenster in die Arthropodenwelt vor 500 Millionen Jahren. Verh Dtsch Zool Ges 84: 281–294

Müller U, Ax P (1971) Gnathostomulida von der Nordseeinsel Sylt mit Beobachtungen zur Lebensweise und Entwicklung von Gnathostomula paradoxa Ax. Mikrofauna Meeresboden 9: 1–41

Mullineaux LS (1987) Organisms living on manganese nodules and crusts: distribution and abundance at three North Pacific sites. Deep-Sea Res 34: 165–184

Munro ALS, Wells JBJ, McIntyre AD (1978) Energy flow in the flora and meiofauna of sandy beaches. Proc R Soc Edinb B 76: 297–315

Murrell MC, Fleeger JW (1989) Meiofauna abundance on the Gulf of Mexico continental shelf affected by hypoxia. Cont Shelf Res 9: 1049–1062

Muus B (1964) A new quantitative sampler for the meiobenthos. Ophelia 1: 209–216

Muus B (1968) A field method for measuring "exposure" by means of plaster balls. A preliminary account. Sarsia 34: 61–68

Neel JK (1948) A limnological investigation of the psammon in Douglas Lake, Michigan, with special reference to shoal and shoreline dynamics. Trans Am Microsc Soc 67: 1–53

Nehring S, Jensen P, Lorenzen S (1990) Tube-dwelling nematodes: tube construction and possible ecological effects on sediment-water interfaces. Mar Ecol Prog Ser 64: 123–128

Nelson AL, Coull BC (1989) Selection of meiobenthic prey by juvenile spot (Pisces): an experimental study. Mar Ecol Prog Ser 53: 51–57

Nicholas WL (1984) The biology of free-living nematodes, 2nd edn. Clarendon Press, Oxford, 251 pp

Nicholas WL, Goodchild DJ, Stewart A (1987) The mineral composition of intracellular inclusions in nematodes from thiobiotic mangrove mud-flats. Nematologica 33: 167–179

Nicholas WL, Elek JA, Stewart AC, Marples TG (1991) The nematode fauna of a temperate Australian mangrove mudflat – its population density, diversity and distribution. Hydrobiologia 209: 13–28

Nicholls AG (1935) Copepods from the interstitial fauna of a sandy beach. J Mar Biol Assoc UK 20: 379–405

Nichols JA (1979) A simple flotation technique for separating meiobenthic nematodes from fine-graded sediments. Trans Am Microsc Soc 98: 127–130

Nilson P, Jönsson B, Lindström-Swanberg I, Sundbäk K (1991) Response of a marine shallow-water sediment system to an increased load of inorganic nutrients. Mar Ecol Prog Ser 71: 275–290

Nodot C (1978) Cycle biologiques de quelques espèces de copépodes harpacticoides psammiques. Téthys 8: 241–248

Noldt U, Wehrenberg C (1984) Quantitative extraction of living Plathelminthes from marine sands. Mar Ecol Prog Ser 20: 193–201

Noodt W (1965) Natürliches System und Biogeographie der Syncarida (Crustacea, Malacostraca). Gewäss Abwäss 37/38: 77–186

Noodt W (1971) Ecology of the Copepoda. In: Hulings NC (ed) Proc 1st International Conference on Meiofauna, Tunesia. Smithson Contrib Zool 76: 97–102

Nordheim H von (1989) Six new species of Protodrilus (Annelida Polychaeta) from Europe and New Zealand, with a concise presentation of the genus. Zool Scr 18: 245–268

Norenburg JL (1988) Remarks on marine interstitial nemertines and key to the species. Hydrobiologia 156: 87–92

Norenburg JL, Morse MP (1983) Systematic implications of *Euphysa ruthae* n.sp. (Athecata: Corymorphidae), a psammophilic solitary hydroid with unusual morphogenesis. Trans Am Microsc Soc 102: 1–17

Novak R (1989) Ecology of nematodes in the Mediterranean seagrass *Posidonia oceanica* (L.) Delile. 1. General part and faunistics of the nematode community. PSZNI Mar Ecol 10: 335–363

Nuß B, Trimkowski V (1984) Physikalische Mikroanalysen an kristolloiden Einschlüssen bei *Tobrilus gracilis* (Nematoda, Enoplida). Veröff Inst Meeresforsch Bremerh 20: 17–27

Ockelmann KW (1964) An improved detritus-sledge for collecting meiobenthos. Ophelia 1: 217–222

'Olafsson E (1991) Intertidal meiofauna of four sandy beaches in Iceland. Ophelia 33: 55–65

'Olafsson E (1992) Small-scale distribution of marine meiobenthos – the effects of decaying macrofauna. Oecologia 90: 37–42

'Olafsson E, Moore CG (1990) Control of meiobenthic abundance by macroepifauna in a subtidal muddy habitat. Mar Ecol Prog Ser 65: 241–249

'Olafsson E, Moore CG, Bett BJ (1990) The impact of *Melinna palmata* Grube, a tube-building polychaete, on meiofaunal community structure in a soft-bottom subtidal habitat. Estuarine Coastal Shelf Sci 31: 883–893

Orghidan T (1955) Ein neuer Lebensraum des unterirdischen Wassers: der hyporheische Biotop. Arch Hydrobiol 55: 392–414

Osenga GA, Coull BC (1983) *Spartina alterniflora* Loisel root structure and meiofaunal abundance. J Exp Mar Biol Ecol 67: 221–225

Ott JA (1972) Determination of fauna boundaries of nematodes in an intertidal sand flat. Int Rev Ges Hydrobiol 57: 645–663

Ott JA, Novak R (1989) Living at an interface: meiofauna at the oxygen/sulfide boundary of marine sediments. In: Ryland JS, Tyler PA (eds) Reproduction, genetics and distributions of marine organisms. 23rd European Marine Biology Symp. Olsen & Olsen, Fredensborg, Denmark, pp 415–422

Ott J, Novak R, Schiemer F, Hentschel U, Nebelsick M, Polz M (1991) Tackling the sulfide gradient: a novel strategy involving marine nematodes and chemoautotrophic ecto-symbionts. PSZNI Mar Ecol 12: 261–279

Page HG (1955) Phi-millimeter conversion table. J Sediment Petrol 25: 285–292

Palmer JP, Round TE (1967) Persistent vertical migration rhythms in benthic microflora. VI. The tidal and diurnal nature of the rhythm in the diatom *Hantzschia virgata*. Biol Bull (Woods Hole) 132: 44–55

Palmer MA (1988) Dispersal of marine meiofauna: a review and conceptual model explaining passive transport and active emergence with implications for recruitment. Mar Ecol Prog Ser 48: 81–91

Palmer MA (1990a) Temporal and spatial dynamics of meiofauna within the hyporheic zone of Goose Creek, Virginia. J N Am Benthol Soc 9: 17–25

Palmer MA (1990b) Understanding the movement dynamics of a stream-dwelling meiofauna community using marine analogs. Stygologia 5: 67–74

Palmer MA (1992) Incorporating lotic meiofauna into our understanding of faunal transport processes. Limnol Oceanogr 37: 329–341

Palmer MA, Gust G (1985) Dispersal of meiofauna in a turbulent tidal creek. J Mar Res 43: 179–210

Palmer MA, Bely AE, Berg KE (1992) Response of invertebrates to lotic disturbance: a test of the hyporheic refuge hypothesis. Oecologia 89: 182–194

Parker RH (1975) The study of benthic communities. Elsevier, Amsterdam, 279 pp

Parry GD (1981) The meanings of r- and K-selection. Oecologia (Berl) 48: 260–264

Patterson DJ, Larsen J, Corliss JO (1988) The ecology of the heterotrophic flagellates and ciliates living in marine sediments. Prog Protist 3: 185–277

Peck LS, Uglow RF (1990) Two methods for the assessment of the oxygen content of small volumes of seawater. J Exp Mar Biol Ecol 141: 53–62

Pennak RW (1939) The microscopic fauna of the sandy beaches. Problems of lake biology. Publ Am Assoc Adv Sci 10: 94–106

Pennak RW (1940) Ecology of the microscopic Metazoa inhabiting the sandy beaches of some Wisconsin lakes. Ecol Monogr 10: 537–615

Pennak RW (1951) Comparative ecology of the interstitial fauna of fresh-water and marine beaches. L'Année Biol, Sér 3, 27: 217–248

Pennak RW (1988) Ecology of the freshwater meiofauna. In: Higgins RP, Thiel H (eds) Introduction to the study of meiofauna. Smithsonian Inst Press, Washington, DC, pp 39–60

Pennak RW, Ward JV (1985) Bathynellacea (Crustacea: Syncarida) in the United States, and a new species from the phreatic zone of a Colorado mountain stream. Trans Am Microsc Soc 104: 209–215

Pennak RW, Ward JV (1986) Interstitial fauna communities of the hyporheic and adjacent groundwater biotopes of a Colorado mountain stream. Arch Hydrobiol Suppl 74: 356–396

Pennak RW, Zinn DJ (1943) Mystacocarida, a new order of Crustacea from intertidal beaches in Massachussetts and Connecticut. Smithson Misc Collect 103: 1–11

Petersen CGJ (1913) Havets bonitering. II. Om havbundens dyresamfund og om disses betydning for den marine zoogeografi. Beret Landbrugsmin Dan Biol St Copenhagen 21: 1–42, plus addendum: 1–68

Pfannenstiel H-D (1981) Endocrine control of sexual differentiation in the protandric polychaete, *Ophryotrocha puerilis*. In: Clark WH, Adams TS (eds) Advances in invertebrate reproduction. Elsevier, North Holland 332 p

Pfannkuche O (1985) The deep-sea meiofauna of the Porcupine Seabight and abyssal plain (NE Atlantic): population structure, distribution and standing stocks. Oceanol Acta 8: 343–353

Pfannkuche O (1992) Organic carbon flux through the benthic community in the temperate abyssal northeast Atlantic. In: Rowe GT, Pariente V (eds) Deep-sea food chains and the global carbon cycle. Kluwer, Dordrecht, pp 183–198

Pfannkuche O (1993) Benthic response to the sedimentation of particulate organic matter at the BIOTRANS station, 47°N, 20°W. Deep-Sea Res II, 40: 135-149

Pfannkuche O, Lochte K (1990) Metabolismus und Energiefluß im Benthal. In: Biotrans. Biologischer Vertikaltransport und Energiehaushalt in der bodennahen Wasserschicht der Tiefsee. Ber Zentr Meeres-Klimaforsch Univ Hamb 10: 130–154

Pfannkuche O, Thiel H (1987) Meiobenthic stocks and benthic activity on the NE-Svalbard shelf and in the Nansen Basin. Polar Biol 7: 253–266

Pfannkuche O, Thiel H (1988) Sample processing. In: Higgins RP, Thiel H (eds) Introduction to the study of meiofauna. Smithsonian Inst Press, Washington, DC, pp 134–145

Pinckney J, Zingmark RG (1991) Effects of tidal stage and sun angles on intertidal benthic microalgal productivity. Mar Ecol Prog Ser 76: 81–89

Platonova TA, Gal'tsova VV (1985) Nematodes and their role in the meiobenthos (Nematody i ikh rol'v meiobenthose). Studies on marine fauna (Issledavanie fauny morei). Published for the Smithsonian Institution Libraries. Amerind Publ Co Pvt Ltd, New Delhi, 1985 (Nauka Publishers, Leningrad 1975), 366 pp

Platt HM (1981) Meiofaunal dynamics and the origin of the metazoa. In: Greenwood PH (ed) The evolving biosphere. Cambridge University Press, Cambridge, pp 207–216

Platt HM, Warwick RM (1980) The significance of free-living nematodes to the littoral ecosystem. In: Price JH, Irvine DEG, Farnham WF (eds) The shore environment. 2. Ecosystems. Academic Press, New York, pp 729–759 (Syst Ass Spec, vol 17)

Platt HM, Warwick RM (1983) Freeliving marine nematodes. Pt 1. British enoplids. Pictorial key to world genera and notes for the identification of British species. Cambridge University Press, Cambridge, 307 pp (Synopses of the British Fauna, vol 28)

Platt HM, Warwick RM (1988) Freeliving marine nematodes. Pt 2. British chromadorids. Pictorial key to world genera and notes for the identification of British species. Brill, Backhuys, Leiden, 502 pp (Synopses of the British Fauna, vol 38)

Platt HM, Warwick RM Freeliving marine nematodes. Pt 3. British monhysterids. Pictorial key to world genera and notes for the identification of British species. Brill, Backhuys, Leiden (Synopses of the British Fauna) (in prep.)

Platt HM, Shaw KM, Lambshead PJD (1984) Nematode species abundance patterns and their use in the detection of environmental perturbations. Hydrobiologia 118: 59–66

Poizat C (1985) Interstitial opisthobranch gastropods as indicator organisms in sublittoral sandy habitats. Stygologia 1: 26–42

Pollock LW (1970) Distribution and dynamics of interstitial Tardigrada at Woods Hole, Massachusetts, U.S.A. Ophelia 7: 145–165

Polz MF, Felbeck H, Novak R, Nebelsick M, Ott JA (1992) Chemoautotrophic, sulfuroxidizing symbiotic bacteria on marine nematodes: morphological and biochemical characterization. Microb Ecol 24: 313–329

Por FD, Bromley HJ (1974) Morphology and anatomy of *Maccabaeus tentaculatus* (Priapulida: Seticoronaria). J Zool (Lond) 173: 173–197

Por FD, Masry D (1968) Survival of a nematode and an oligochaete species in the anaerobic benthal of Lake Tiberias. Oikos 19: 388–391

Potel P, Reise K (1987) Gastrotricha Macrodasyida of intertidal and subtidal sandy sediments in the northern Wadden Sea. Microfauna Mar 3: 363–376

Powell EN (1989) Oxygen, sulfide and diffusion: why thiobiotic meiofauna must be sulfide-insensitive first-order respirers. J Mar Res 47: 887–932

Powell EN, Bright TJ (1981) A thiobios does exist – gnathostomulid domination of the Canyon community at the East Flower Garden brine seep. Int Rev Ges Hydrobiol 66: 675–683

Powell EN, Bright TJ, Woods A, Gittings S (1983) Meiofauna and the thiobios in the East Flower Garden brine seep. Mar Biol 73: 269–283

Powell EN, Bright TJ, Brooks JM (1986) The effect of sulfide and an increased food supply on the meiofauna and macrofauna at the East Flower Garden brine seep. Helgol Meeresunters 40: 57–83

Powell MA, Arp AJ (1989) Hydrogen sulfide oxidation by abundant nonhemoglobin heme compounds in marine invertebrates from sulfide-rich habitats. J Exp Zool 249: 121–132

Powell MA, Somero GN (1986) Hydrogen sulfide oxidation is coupled to oxidative phosphorylation in mitochondria of *Solemya reidi*. Science 233: 563–566

Pugh PJA, King PE (1985a) Vertical distribution and substrate association of the British Halacaridae. J Nat Hist 19: 961–968

Pugh PJA, King PE (1985b) Feeding in intertidal Acari. J Exp Mar Biol Ecol 94: 269–280

Purschke G (1988) Pharynx. In: Westheide W, Hermans CO (eds) The ultrastructure of Polychaeta. Microfauna Mar 4: 177–197

Raffaeli D (1987) The behaviour of the nematode/copepod ratio in organic pollution studies. Mar Environ Res 23: 135–152

Raffaeli DG, Mason CF (1981) Pollution monitoring with meiofauna, using the ratio of nematodes to copepods. Mar Pollut Bull 12: 158–163

Reeburgh WS (1967) An improved interstitial water sampler. Limnol Oceanogr 12: 163–165

Reeburgh WS, Erickson RE (1982) A dipstick sampler for rapid, continuous chemical profiles in sediments. Limnol Oceanogr 27: 556–559

Reichardt W (1989) Microbiological aspects of bioturbation. Sci Mar 53: 301–306

Reid DM (1932–33) Salinity interchange between salt water in sand overflowing fresh water at low tide. J Mar Biol Assoc UK 18: 299–306

Reise K (1979) Moderate predation on meiofauna by the macrobenthos of the wadden sea. Helgol Wiss Meeresunters 32: 453–465

Reise K (1981a) High abundance of small zoobenthos around biogenic structures in tidal sediments of the wadden sea. Helgol Meeresunters 34: 413–425

Reise K (1981b) Gnathostomulida abundant alongside polychaete burrows. Mar Ecol Prog Ser 6: 329–333

Reise K (1984) Free-living Plathelminthes (Turbellaria) of a marine sand flat: An ecological study. Microfauna Mar 1: 1–62

Reise K (1985) Tidal flat ecology. An experimental approach to species interactions. Ecol Stud 54. Springer, Berlin Heidelberg New York, 191 pp

Reise K (1987a) Spatial niches and long-term performance in meiobenthic Plathelminthes of an intertidal lugworm flat. Mar Ecol Prog Ser 38: 1–11

Reise K (1987b) Experimental analysis of processes between species on marine tidal flats. Ecol Stud 61. Springer, Berlin Heidelberg New York, pp 391–400

Reise K (1988) Plathelminth diversity in littoral sediments around the island of Sylt in the North Sea. Fortschr Zool 36: 469–480

Reise K, Ax P (1979) A meiofaunal "thiobios" limited to the anaerobic sulfide system of marine sand does not exist. Mar Biol 54: 225–237

Remane A (1927) Halammohydra, ein eigenartiges Hydrozoon der Nord- und Ostsee. Z Morphol Ökol Tiere 7: 643–677

Remane A (1932) Archiannelida. In: Grimpe G, Wagler E (eds) Die Tierwelt der Nord- und Ostsee. Akademische Verlagsgesellschaft, Leipzig, VIa 1, pp 1–36

Remane A (1933) Verteilung und Organisation der benthonischen Mikrofauna der Kieler Bucht. Wiss Meeresunters, Abt Kiel, NF 21, pp 161–221

Remane A (1934) Die Brackwasserfauna. Zool Anz Suppl 7: 34–74

Remane A (1936a) Gastrotricha und Kinorhyncha. In: Bronns Klassen und Ordnungen des Tierreichs, Bd4 Vermes, 2 Abt, 1 B, 4d, 2 T, Akademische Verlagsgesellschaft Geest & Portig, Leipzig, 385 p

Remane A (1936b) Monobryozoon ambulans n.g. n.sp., ein eigenartiges Bryozoon des Meeressandes. Zool Anz 113: 161–167

Remane A (1940) Einführung in die zoologische Ökologie der Nord- und Ostsee. In: Grimpe G, Wagler E (eds) Die Tierwelt der Nord- und Ostsee. Akademische Verlagsgesellschaft Geest & Portig, Leipzig, I a, 238 p

Remane A (1949) Die psammobionten Rotatorien der Nord- und Ostsee. Kiel Meeresforsch 6: 59–67

Remane A (1952a) Die Besiedlung des Sandbodens im Meere und die Bedeutung der

Lebensformtypen für die Ökologie. Verh Dtsch Zool Ges Wilhelmshaven 1951. Zool Anz Suppl 16: 327–359

Remane A (1952b) Die Grundlagen des natürlichen Systems, der vergleichenden Anatomie und der Phylogenetik. Theoretische Morphologie und Systematik I. Akademische Verlagsgesellschaft Geest & Portig, Leipzig, 400 pp

Remane A (1956) Die Cephalocarida, eine weitere neue Ordnung (Unterklasse) der Krebse. Mikrokosmos 45: 227–230

Remane A (1959) Regionale Verschiedenheit der Lebenwesen gegenüber dem Salzgehalt und ihre Bedeutung für die Brackwasser-Einteilung. Arch Oceanogr Limnol Suppl 11: 35–46

Remmert H (1992) Ökologie. Ein Lehrbuch, 5th edn. Springer, Berlin, Heidelberg New York, 363 pp

Renaud-Mornant JC (1982) Species diversity in marine Tardigrada. In: Nelson D (ed) Proc 3rd Int Symp on Tardigrada. East Tennessee State University Press, Johnson City, USA, pp 149–178

Renaud-Mornant J, Gourbault N (1980) Survie de la méiofauna après l'échouement de l' "Amoco Cadiz" (Chenal de Morlaix, Grève de Roscoff). Bull Mus Natl Hist Nat Paris 4e sér, 2 Sect A: 759–772

Renaud-Mornant JC, Salvat B, Bossy C (1971) Macrobenthos and meiobenthos from the closed lagoon of a Polynesian atoll. Maturei Vavao (Tuamotu). Biotropica 3: 36–55

Revsbech NP, Jørgensen BB (1986) Microelectrodes: their use in microbial ecology. Adv Microb Ecol 9: 293–352

Revsbech NP, Ward DM (1983) Oxygen microelectrode that is insensitive to medium chemical composition: use in an acid microbial mat dominated by *Cyanidium caldarium*. Appl Environ Microbiol 45: 755–759

Revsbech NP, Sørensen J, Blackburn TH, Lomholt JP (1980) Distribution of oxygen in marine sediments measured with microelectrodes. Limnol Oceanogr 25: 403–411

Rey JR, Shaffer J, Kain T, Stahl R, Crossman R (1992) Sulfide variation in the pore and surface waters of artificial salt-marsh ditches and a natural tidal creek. Estuaries 15: 257–269

Rhoads DC, Young DK (1970) The influence of deposit-feeding organisms on sediment stability and community trophic structure. J Mar Res 28: 150–178

Rhoads DC, Aller RC, Goldhaber MB (1977) The influence of colonizing benthos on physical properties and chemical diagenesis of the estuarine seafloor. In: Coull BC (ed) Ecology of marine benthos. University of South Carolina Press, Columbia, pp 113–138 (Belle W Baruch Libr Mar Sci, vol 6)

Rhoads DC, Yingst JY, Ullman WJ (1978) Seafloor stability in central Long Island Sound: Part I. Temporal changes in erodibility of fine-grained sediment. In: Wiley ML (ed) Estuarine interactions. Academic Press, New York, pp 221–244

Rhumbler L (1938) Foraminiferen aus dem Meeressand von Helgoland, gesammelt von A. Remane (Kiel). Kiel Meeresforsch 2: 157–222

Riddle MJ (1989) Bite profiles of some benthic grab samplers. Estuarine Coastal Shelf Sci 29: 285–292

Riebesell U (1992) The formation of large marine snow and its sustained residence in surface waters. Limnol Oceanogr 37: 63–76

Riedl RJ (1969) Gnathostomulida from America. Science 163: 445–452

Riedl RJ (1971) How much seawater passes through sandy beaches? Int Rev Ges Hydrobiol 56: 923–946

Riedl R (1983) Stamm: Nemertini (Schnurwürmer). In: Fauna und Flora des Mittelmeeres.

Ein systematischer Meeresführer für Biologen und Naturfreunde, 3rd edn. Parey, Hamburg, pp 217–225

Riedl RJ, Machan R (1972) Hydrodynamic patterns in lotic intertidal sands and their bioclimatological implications. Mar Biol 13: 179–209

Riedl RJ, Ott JA (1970) A suction-corer to yield electric potentials in coastal sediment layers. Senckenb Marit 2: 67–84

Rieger R (1976) Monociliated epidermal cells in Gastrotricha: significance for concepts of early metazoan evolution. Z Zool Syst Evolutionsforsch 14: 198–226

Rieger RM (1980) A new group of interstitial worms, Lobatocerebridae nov. fam. (Annelida) and its significance for metazoan phylogeny. Zoomorphologie 95: 41–84

Rieger RM (1981) Morphology of the Turbellaria at the ultrastructural level. Hydrobiologia 84: 213–229

Rieger RM (1985) The phylogenetic status of the acoelomate organization within the Bilateria: a histological perspective. In: Conway Morris S, George JD, Gibson R, Platt HM (eds) The origins and relationships of lower invertebrates. Clarendon Press, Oxford, pp 101–122 (Syst Ass, vol 28)

Rieger RM (1991) Neue Organisationstypen aus der Sandlückenraumfauna: die Lobatocerebriden und *Jennaria pulchra*. Verh Dtsch Zool Ges 84: 247–259

Rieger RM, Rieger GE (1975) Fine structure of the pharyngeal bulb in *Trilobodrilus* and its phylogenetic significance within Archiannelida. Tissue Cell 7: 267–279

Rieger RM, Rieger GE (1976) Fine structure of the archiannelid cuticle and remarks on the evolution of the cuticle within the Spiralia. Acta Zool (Stockh) 57: 53–68

Rieger RM, Ruppert E (1978) Resin embedments of quantitative meiofauna samples for ecological and structural studies.- Description and application. Mar Biol 46: 223–235

Rieger RM, Sterrer W (1975) New spicular skeletons in Turbellaria, and the occurrence of spicules in marine meiofauna. Z Zool Syst Evolutionsforsch 13: 207–248

Rieger RM, Tyler S (1979) The homology theorem in ultrastructure research. Am Zool 19: 655–664

Rieger R, Ruppert E, Rieger GE, Schoepfer-Sterrer C (1974) On the fine structure of gastrotrichs with description of *Chordodasys antennatus* sp.n. Zool Scr 3: 219–237

Riemann F (1966) Die interstitielle Fauna im Elbe-Aestuar. Verbreitung und Systematik. Arch Hydrobiol Suppl 31 (Elbe-Aestuar 3) 1/2: 1–279

Riemann F (1970) Das Kiemenlückensystem von Krebsen als Lebensraum der Meiofauna, mit Beschreibung freilebender Nematoden aus karibischen amphibisch lebenden Decapoden. Veröff Inst Meeresforsch Bremerhaven 12: 413–428

Riemann F (1985) Eisen und Mangan in pazifischen Tiefsee-Rhizopoden und Beziehungen zur Manganknollen-Genese. Int Rev Ges Hydrobiol 70: 165–172

Riemann F, Schrage M (1978) The mucus-trap hypothesis on feeding of aquatic nematodes and implications for biodegradation and sediment texture. Oecologia (Berl) 34: 75–88

Rieper-Kirchner M (1989) Microbial degradation of North Sea macroalgae: field and laboratory studies. Bot Mar 32: 241–252

Riser NW (1984) General observations on the intertidal interstitial fauna of New Zealand. Tane 30: 239–250

Romeyn K, Bouwman LA (1983) Food selection and consumption by estuarine nematodes. Hydrobiol Bull 17: 103–109

Rossi L (1961) Morfologia e riproduzione vegetative di un Madreporario nuovo per il Mediterraneo. Boll Zool 28: 261–272

Rudnick DT (1989) Time lags between the deposition and meiobenthic assimilation of phytodetritus. Mar Ecol Prog Ser 50: 231–240

Runnegar B (1991) Oxygen and the early evolution of the Metazoa In: Bryant C (ed) Metazoan life without oxygen. Chapman & Hall, London, pp 65–87

Ruppert EE (1972) An efficient, quantitative method for sampling the meiobenthos. Limnol Oceanogr 17: 629–631

Ruppert EE (1977) Zoogeography and speciation in marine Gastrotricha. Mikrofauna Meeresboden 61: 231–251

Ruppert EE (1982) Comparative ultrastructure of the gastrotrich pharynx and the evolution of myoepithelial foreguts in Aschelminthes. Zoomorphology 99: 181–220

Rutledge PA, Fleeger JW (1988) Laboratory studies on core sampling with application to subtidal meiobenthos collection. Limnol Oceanogr 33: 274–279

Rutledge PA, Fleeger JW (1993) Abundance and seasonality of meiofauna, including harpacticoid copepod species, associated with *Spartina alterniflora* stems. Estuaries (in press)

Ruttner-Kolisko A (1961) Biotop und Biozönose des Sandufers einiger österreichischer Flüsse. Verh Int Ver Limnol 14: 362–368

Ruttner-Kolisko A (1962) Porenraum und kapillare Wasserströmung im Limnopsammal, ein Beispiel für die Bedeutung verlangsamter Strömung. Schweiz Z Hydrol 24: 444–458

Saager PM, Sweerts JP, Ellermeijer HJ (1990) A simple pore water sampler for coarse, sandy sediments of low porosity. Limnol Oceanogr 35: 747–750

Salvat B (1964) Les conditions hydrodynamiques interstitielles des sediments meubles intertidaux et la répartition verticale de la fauna endogée. CR Hebd Séanc Acad Sci Paris 259: 1576–1579

Salvat B, Renaud-Mornant JC (1969) Etude écologique du macrobenthos et du méiobenthos d'un fond sableux du Lagon de Mururoa (Tuamotu-Polynésie). Cah Pac 13: 159–179

Salvini-Plawen LV (1966) Zur Kenntnis der Cnidaria des nordadriatischen Mesopsammon. Veröff Inst Meeresforsch Bremerhaven Sbd 2: 165–186

Salvini-Plawen LV (1968) Neue Formen im marinen Mesopsammon: Kamptozoa und Aculifera. Ann Naturhist Mus Wien 72: 231–272

Salvini-Plawen LV (1972) Zur Taxonomie und Ökologie mediterraner Holothuroidea-Apoda. Helgol Wiss Meeresunters 23: 459–466

Salvini-Plawen L (1985) New interstitial Solenogastres (Mollusca). Stygologia 1: 101–108

Sanders HL (1958) Benthic studies in Buzzards Bay. I. Animal-sediment relationships. Limnol Oceanogr 3: 245–258

Sanders HL (1959) The significance of the Cephalocarida in crustacean phylogeny. In: Hewer HR, Riley ND (eds) Proc 15th Int Congr Zool London 1958, pp 337–340

Sanders HL, Hessler RR, Garner SP (1985) *Hirsutia bathyalis*, a new unusual deep-sea benthic peracaridan crustacean from the tropical Atlantic. J Crustacean Biol 5: 30–57

Sandulli R, Nicola M de (1991) Responses of meiobenthic communities along a gradient of sewage pollution. Mar Pollut Bull 22: 463–467

Sassuchin DN, Kabanov NM, Neizvestnova ES (1927) K izuceniju mikroskopiceskogo nasalenija nanosnych peskov v rusle reki Oki. Ueber die mikroskopische Planzen- und Tierwelt der Sandfläche des Okaufers bei Murom. Russ Gidrobiol Zh 6: 59–83 (in Russian and German)

Sayles PL, Mangelsdorf PC, Wilson TRS, Hume DN (1976) A sampler for the in situ collection of marine sedimentary pore waters. Deep-Sea Res 23: 259–264

Scherer B (1984) Meiofauna der "Sulfidschicht" im Wattboden der Nordsee. Struktur und Abundanz. Diss, Universität Göttingen, 171 pp

Scherer B (1985) Annual dynamics of a meiofauna community from the "sulfide layer" of a North Sea sand flat. Microfauna Mar 2: 117–161

Schiemer F (1982) Food dependence and energetics of free-living nematodes. I. Respiration, growth and reproduction of *Caenorhabditis briggsae* at different levels of food supply. Oecologia (Berl) 54: 108–121

Schiemer F, Duncan A (1974) The oxygen consumption of a freshwater benthic nematode, *Tobrilus gracilis* (Bastian). Oecologia (Berl) 15: 121–126

Schiemer F, Novak R, Ott J (1990) Metabolic studies on thiobiotic free-living nematodes and their symbiotic microorganisms. Mar Biol 106: 129–137

Schmidt C (1989) Zur Ökologie des sulfidtoleranten Nematoden-Genus *Tobrilus* Andrássy 1959 aus dem Elbwatt. Diplom-Arbeit, Universität Hamburg (unveröff.), 145 pp + Anhang

Schmidt P, Teuchert G (1969) Quantitative Untersuchungen zur Ökologie der Gastrotrichen im Gezeiten-Sandstrand der Insel Sylt. Mar Biol 4: 4–23

Schminke HK (1981) Adaptation of Bathynellacea (Crustacea, Syncarida) to life in the interstitial ("Zoea theory"). Int Rev Ges Hydrobiol 66: 575–637

Schminke HK (1986) Syncarida. In: Botosaneanu, L. (ed) Stygofauna Mundi. A faunistic, distributional, and ecological synthesis of the world fauna inhabiting subterranean waters (including the marine interstitial). Brill, Backhuys, Leiden, pp 389–404

Schulz E (1950) *Psammohydra nanna*, ein neues solitäres Hydrozoon in der westlichen Beltsee. Kiel Meeresforsch 7: 122–137

Schulz E (1963) Über die Tardigraden. Zool Anz 171: 3–12

Schuster R (1962) Das marine Litoral als Lebensraum terrestrischer Kleinarthropoden. Int Rev Ges Hydrobiol 47: 359–412

Schuster R (1965) Die Ökologie der terrestrischen Kleinfauna des Meeresstrandes. Verh Dtsch Zool Ges Kiel 1964: 492–521

Schuster R (1979) Soil mites in the marine environment. Rec Adv Acarol 1: 593–602

Schuster RO, Nelson D, Grigarick A, Christenberry D (1980) Systematic criteria of the Eutardigrada. Trans Am Microsc Soc 99: 284–303

Schwinghamer P (1981a) Characteristic size distributions of integral benthic communities. Can J Fish Aquat Sci 38: 1255–1263

Schwinghamer P (1981b) Extraction of living meiofauna from marine sediments by centrifugation in a silica sol-Sorbitol mixture. Can J Fish Aquat Sci 38: 476–478

Schwinghamer P (1983) Generating ecological hypotheses from biomass spectra using causal analysis: a benthic example. Mar Ecol Prog Ser 13: 151–166

Schwoerbel J (1961a) Über die Lebensbedingungen und die Besiedlung des hyporheischen Lebensraumes. Arch Hydrobiol Suppl 25: 182–214

Schwoerbel J (1961b) Subterrane Wassermilben (Acari: Hydrachnellae, Porohalacaridae und Stygotrombiidae), ihre Ökologie und Bedeutung für die Abgrenzung eines aquatischen Lebensraumes zwischen Oberfläche und Grundwasser. Arch Hydrobiol Suppl 25: 242–306

Schwoerbel J (1967) Das hyporheische Interstitial als Grenzbiotop zwischen oberirdischem und subterranem Ökosystem und seine Bedeutung für die Primär-Evolution von Kleinsthöhlenbewohnern. Arch Hydrobiol Suppl 33: 1–62

Sepers ABJ (1977) The utilisation of dissolved organic compounds in aquatic environments. Hydrobiologia 52: 39–54

Serban M (1960) La néotenie, et le problème de la taille chez les Copépodes. Crustaceana 1: 77–83

Service SK, Bell SS (1987) Density-influenced active dispersal of harpacticoid copepods. J Exp Mar Biol Ecol 114: 49–62

Service SK, Feller RJ, Coull BC, Woods R (1992) Predation effect of three fish species and a shrimp on macrobenthos and meiobenthos in microcosms. Estuarine Coastal Shelf Sci 34: 277–293

Shanks AI, Edmonson EW (1990) The vertical flux of metazoans (holoplankton, meiofauna, and larval invertebrates) due to their association with marine snow. Limnol Oceanogr 35: 455–463

Shannon CE (1949) The mathematical theory of communication. In: Shannon CE, Weaver W (eds) The mathematical theory of communication. University of Illinois Press, Urbana, pp 3–91

Sherman KM, Coull BC (1980) The response of meiofauna to sediment disturbance. J Exp Mar Biol Ecol 46: 59–71

Shiells GM, Anderson KJ (1985) Pollution monitoring using nematode/copepod ratio. A practical application. Mar Pollut Bull 16: 62–68

Shirayama Y, Horikoshi M (1989) Comparison of the benthic size structure between sublittoral, upperslope and deep-sea areas of the western Pacific. Int Rev Ges Hydrobiol 74: 1–13

Shirayama Y, Ohta S (1990) Meiofauna in a cold-seep community off Hatsushima, Central Japan. J Oceanogr Soc Jpn 46: 118–124

Shirayama Y, Swinbanks DD (1986) Oxygen profile in deep-sea calcareous sediment calculated on the basis of measured respiration rates of deep-sea meiobenthos and its relevance to manganese diagenesis. La Mer 24: 75–80

Sich H (1990) Die benthische Ciliatenfauna bei Gabelsflach (Kieler Bucht) und deren Beeinflussung durch Bakterien. Eine Studie über Menge, Biomasse, Produktion, Bakterieningestion und Ultrastruktur von Mikroorganismen. Diss, Universität Kiel, Ber Inst Meeresk 191, 215 pp

Siewing R (ed) (1985) Lehrbuch der Zoologie. Bd 2 Systematik 3rd edn. Fischer Stuttgart, 1107 pp

Sikora JP, Sikora WB, Erkenbrecher CW, Coull BC (1977) Significance of ATP, carbon, and caloric content of meiobenthic nematodes in partitioning benthic biomass. Mar Biol 44: 7–14

Sikora WB, Sikora JP (1982) Ecological implication of the vertical distribution of meiofauna in salt marsh sediments. In: Kennedy VS (ed) Estuarine comparisons. Academic Press, New York, pp 269–282

Smith LD, Coull BC (1987) Juvenile spot (Pisces) and grass shrimp predation on meiobenthos in muddy and sandy substrata. J Exp Mar Biol Ecol 105: 123–136

Smith W, McIntyre AD (1954) A spring-loaded bottom sampler. J Mar Biol Assoc UK 33: 257–264

Soetaert K, Heip C (1989) The size structure of nematode assemblages along a Mediterranean deep-sea transect. Deep-Sea Res A 36: 93–102

Somero GN, Childress JJ, Anderson AE (1989) Transport metabolism, and detoxification of hydrogen sulfide in animals from sulfide-rich marine environments. Rev Aquat Sci 1: 591–614

Sopott B (1973) Jahreszeitliche Verteilung und Lebenszyklen der Proseriata (Turbellaria) eines Sandstrandes der Nordseeinsel Sylt. Mikrofauna Meeresboden 15: 253–358

Sopott-Ehlers B (1989) Coelogynopora visurgis nov. spec. (Proseriata) und andere freilebende Plathelminthes mariner Herkunft aus Ufersanden der Weser. Microfauna Mar 5: 87–93

St John J, Jones GP, Sale PF (1989) Distribution and abundance of soft-sediment meiofauna and a predatory goby in a coral reef lagoon. Coral Reefs 8: 51–58

Steele JH, Baird IE (1968) Production ecology of a sandy beach. Limnol Oceanogr 13: 14–25

Stephens GC (1982) Recent progress in the study of "Die Ernährung der Wassertiere und der Stoffhaushalt der Gewässer". Am Zool 22: 611–619

Sterrer W (1971) Gnathostomulida: problems and procedures. In: Hulings NC (ed) Proc. 1st International Conference on Meiofauna, Tunesia. Smithson Contrib Zool 76: 9–15

Sterrer W (1972) Systematics and evolution within the Gnathostomulida. Syst Zool 21: 151–173

Sterrer W (1973) Plate tectonics as a mechanism for dispersal and speciation in interstitial sand fauna. Neth J Sea Res 7: 200–222

Sterrer W, Ax P (eds) (1977) The meiofauna species in time and space. Proceedings of a workshop symposium, Bermuda Biological Station, 1975. Mikrofauna Meeresboden 61, 316 pp

Sterrer W, Rieger R (1974) Retronectidae – a new cosmopolitan marine family of Catenulida (Turbellaria). In: Riser NW, Morse MP (eds) Biology of the Turbellaria. McGraw Hill, New York, pp 63–92

Sterrer W, Mainitz M, Rieger RM (1985) Gnathostomulida: enigmatic as ever. In: Conway Morris S, George JD, Gibson R, Platt HM (eds) The origins and relationships of lower invertebrates. Clarendon Press, Oxford, pp 181–199 (Syst Ass, vol 28)

Stewart MG (1979) Absorption of dissolved organic nutrients by marine invertebrates. Oceanogr Mar Biol Annu Rev 17: 163–192

Stock JH (1976) A new genus and two new species of the crustacean order Thermosbaenacea from the West Indies (with full bibliography and biogeographic notes). Bijdr Dierkd 46: 47–70

Stock JH (1980) Regression model evolution as exemplified by the genus *Pseudoniphargus* (Amphipoda). Bijdr Dierkd 50: 104–144

Stock JH (1986) Deep sea origin of cave faunas, an unlikely supposition. Stygologia 2: 105–111

Strayer D (1985) The benthic micrometazoans of Mirror Lake, New Hampshire. Arch Hydrobiol Suppl 72: 287–426

Strayer D (1986) The size structure of a lacustrine zoobenthic community. Oecologia (Berl) 69: 513–516

Strayer DL (1991) Perspectives on the size structure of lacustrine zoobenthos, its causes, and its consequences. J N Am Benthol Soc 10: 210–221

Strayer D, Likens GE (1986) An energy budget for the zoobenthos of Mirror Lake, New Hampshire. Ecology 67: 303–313

Sun B, Fleeger JW (1991) Spatial and temporal patterns of dispersion in meiobenthic copepods. Mar Ecol Prog Ser 71: 1–11

Swedmark B (1964) The interstitial fauna of marine sand. Biol Rev 39: 1–42

Swedmark B (1967) *Gwynia capsula* (Jeffreys), an articulate brachiopod with brood protection. Nature 213: 1151–1152

Swedmark B (1968) The biology of interstitial Mollusca. Symp Zool Soc Lond 22: 135–149

Swedmark B (1971) A review of Gastropoda, Brachiopoda, and Echinodermata in marine meiobenthos. In: Hulings NC (ed) Proc 1st International Conference on Meiofauna, Tunesia. Smithson Contrib Zool 76: 41–45

Swedmark B, Teissier G (1958) *Armorhydra janowiczi*, n.g., n.sp., Hydroméduse benthique. C R Acad Sci Paris 247: 133–135

Tabacchi E (1990) A sampler for interstitial fauna in alluvial rivers. Regul Riv Res Manag 5: 177–182

Taylor WR (1964) Light and photosynthesis in intertidal benthic diatoms. Helgol Wiss Meeresunters 10: 29–37

Taylor WR, Gebelein CD (1966) Plant pigments and light penetration in intertidal sediments. Helgol Wiss Meeresunters 13: 229–237

Teal JM, Kanwisher J (1961) Gas exchange in a Georgia salt marsh. Limnol Oceanogr 6: 388–399

Tempel D, Westheide W (1980) Uptake and incorporation of dissolved amino acids by the interstitial Turbellaria and Polychaeta and their dependence on temperature and salinity. Mar Ecol Prog Ser 3: 41–50

Tenore KR (1977) Utilization of aged detritus derived from different sources by the polychaete *Capitella capitata*. Mar Biol 44: 51–55

Tenore KR, Rice DL (1980) A review of trophic factors affecting secondary production of deposit-feeders. In: Tenore KR, Coull BC (eds) Marine benthic dynamics. University of South Carolina Press, Columbia, SC, pp 325–340 (Belle W Baruch Libr Mar Sci, vol 11)

Tenore KR, Tietjen JH, Lee JH (1977) Effect of meiofauna on incorporation of aged eelgrass, *Zostera marina*, detritus by the polychaete *Nephthys incisa*. J Fish Res Board Can 34: 563–567

Tenore KR, Cammen L, Findlay SEG, Phillips N (1982) Perspective of research on detritus: do factors controlling the availability of detritus to macroconsumers depend on its source? J Mar Res 40: 473–490

Teuchert G (1977) The ultrastructure of the marine gastrotrich *Turbanella cornuta* Remane (Macrodasyoidea) and its functional and phylogenetical importance. Zoomorphology 88: 189–246

Teuchert G (1978) Strukturanalyse von Bewegungsformen bei Gastrotrichen. Zool Jahrb Anat 99: 12–22

Teuchert G, Lappe A (1980) Zum sogenannten "Pseudocoel" der Nemathelminthes. – Ein Vergleich der Leibeshöhlen von mehreren Gastrotrichen. Zool Jahrb Anat 103: 424–438

Thiel H (1966) Quantitative Untersuchungen über die Meiofauna des Tiefseebodens. Veröff Inst Meeresforsch Bremerhaven Sbd 2: 131–147

Thiel H (1972a) Meiofauna und Struktur der benthischen Lebensgemeinschaft des Iberischen Tiefseebeckens. "Meteor" Forsch-Ergeb D 12: 36–51

Thiel H (1972b) Die Bedeutung der Meiofauna in küstenfernen benthischen Lebensgemeinschaften verschiedener geographischer Regionen. Verh Dtsch Zool Ges 65: 37–42

Thiel H (1975) The size structure of the deep-sea benthos. Int Rev Ges Hydrobiol 60: 575–606

Thiel H, Pfannkuche O, Schriever G, Lochte K, Gooday AJ, Hemleben C, Mantoura RFG, Turley CM, Patching JW, Rieman F (1988/1989) Phytodetritus on the deep-sea floor in a central oceanic region of the Northeast Atlantic. Biol Oceanogr 6: 203–239

Thistle D (1988) A temporal difference in harpacticoid-copepod abundance at a deep-sea site: caused by benthic storms? Deep-Sea Res 35: 1015–1020

Thistle D, Eckman JE (1990) The effect of a biologically produced structure on the benthic copepods of a deep-sea site. Deep-Sea Res 37: 541–554

Thomsen L (1991) Treatment and splitting of samples for bacteria and meiofauna biomass determinations by means of a semi-automatic image analysis system. Mar Ecol Prog Ser 71: 301–306

Tiemann H (1975) Zur Eidonomie, Anatomie, Systematik und Biologie der Gattung *Porcellidium* Claus, 1860 (Copepoda, Harpacticoida). Diss Universität Hamburg 103 pp

Tietjen JH (1980) Microbial-meiofaunal interrelationships: a review. Microbiology 1980: 335–338

Tietjen JH (1989) Ecology of deep-sea nematodes from the Puerto Rico Trench area and Hatteras abyssal plain. Deep-Sea Res 36: 1579–1594

Tietjen JH (1992) Abundance and biomass of metazoan meiobenthos in the deep-sea. In: Rowe GT, Pariente V (eds) Deep-sea food chains and the global cycle. Kluwer, Leiden, pp 45–62

Tietjen JH, Lee JJ (1977) Feeding behaviour of marine nematodes. In: Coull BC (ed) Ecology of marine benthos. University of South Carolina Press, Columbia, pp 21–35 (Belle W Baruch Libr Mar Sci, vol 6)

Tipton K, Bell SS (1988) Foraging patterns of two syngnathid fishes: importance of harpacticoid copepods. Mar Ecol Prog Ser 47: 31–43

Turner PN (1990) Some interstitial Rotifera from a Florida, USA, beach. Trans Am Microsc Soc 109: 417–421

Tyler S (1977) Ultrastructure and systematics: an example from turbellarian adhesive organs. Mikrofauna Meeresboden 61: 271–286

Tzschaschel G (1979) Marine Rotatoria aus dem Interstitial der Nordseeinsel Sylt. Mikrofauna Meeresboden 71: 1–64

Tzschaschel G (1980) Verteilung, Abundanzdynamik und Biologie mariner interstitieller Rotatoria. Mikrofauna Meeresboden 81: 1–56

Uhlig G (1964) Eine einfache Methode zur Extraktion der vagilen mesopsammalen Mikrofauna. Helgol Wiss Meeresunters 11: 178–185

Uhlig G, Heimberg SHH (1981) A new versatile compression chamber for examination of living microorganisms. Helgol Meeresunters 34: 251–256

Uhlig G, Thiel H, Gray JS (eds) (1973) The quantitative separation of meiofauna. A comparison of methods. Helgol Wiss Meeresunters 25: 173–195

Van Damme D, Heip C, Willems KA (1984) Influence of pollution on the harpacticoid copepods of two North Sea estuaries. Hydrobiologia 112: 143–160

Van de Velde MC, Coomans A (1989) A putative new hydrostatic skeletal function for the epidermis in monhysterids (Nematoda). Tissue Cell 21: 525–534

Van der Land J (1970) Systematics, zoogeography and ecology of the Priapulida. Zool Verh 112: 1–118 + 5 pl

Van Gemerden H, Tughan CS, Wit R de, Herbert RA (1989) Laminated microbial ecosystems on sheltered beaches in Scapa Flow, Orkney Islands. FEMS Microbiol Ecol 62: 87–102

Van Harten D (1992) Hydrothermal vent Ostracoda and faunal association in the deep-sea. Deep-Sea Res 39: 1067–1070

Vanreusel A (1990) Ecology of the free-living marine nematodes from the Voordelta (Southern Bight of the North Sea). 1. Species composition and structure of the nematode communities. Cah Biol Mar 31: 439–462

Vanreusel A (1991) Ecology of the free-living marine nematodes from the Voordelta (Southern Bight of the North Sea). 2. Habitat preferences of the dominant species. Nematologica 37: 343–359

Viets K (1927) Die Halacaridae der Nordsee. Z Wiss Zool 130: 83–173

Vincx M, Meire P, Heip C (1990) The distribution of nematode communities in the Southern Bight of the North Sea. Cah Biol Mar 31: 107–129

Vismann B (1991) Sulfide tolerance: physiological mechanisms and ecological implications. Ophelia 34: 1–28

Visscher PT, Beukema J, van Gemerden H (1991) In situ characterization of sediments. Measurement of oxygen and sulfide profiles with a novel combined needle electrode. Limnol Oceanogr 36: 1476–1479

Voigt M, Koste W (1978) Rotatoria (Überordnung Monogononta). Die Rädertiere Mitteleuropas. Ein Bestimmungswerk, 2nd edn. Borntraeger, Berlin, vol 1: Text, 673 pp + 63 fig; vol 2: Plates, 4761 pp + 234 pl

Vranken G, Heip (1986) The productivity of marine nematodes. Ophelia 26: 429–442

Vranken G, Herman PMJ, Heip C (1988) Studies of the life-history and energetics of marine and brackish-water nematodes. I. Demography of *Monhystera disjuncta* at different temperature and feeding conditions. Oecologia 77: 296–301

Wägele J-W (1982) Zur Phylogenie der Anthuridea (Crustacea, Isopoda) mit Beiträgen zur Lebensweise, Morphologie, Anatomie und Taxonomie. Zoologica 45: 1–127

Walossek D, Müller KL (1990) Upper Cambrian stem-lineage crustaceans and their bearing upon the monophyletic origin of Crustacea and the position of *Agnostus*. Lethaia 23: 409–427

Walters K (1991) Influences of abundance, behavior, species composition, and ontogenetic stage on active emergence of meiobenthic copepods in subtropical habitats. Mar Biol 108: 207–215

Walters K, Bell SS (1986) Diel pattern of active vertical migration in seagrass meiofauna. Mar Ecol Prog Ser 34: 95–103

Walters K, Moriarty DJW (1993) The effects of complex trophic interactions on a marine microbenthic assemblage. Ecology (in press)

Ward JV, Palmer MA (1993) Distribution patterns of interstitial freshwater meiofauna over a range of spatial scales, with emphasis on alluvial river-aquifer systems. Stygologia (in press)

Ward JV, Voelz NJ (1990) Gradient analysis of interstitial meiofauna along a longitudinal stream profile. Stygologia 5: 93–100

Warwick RM (1981) The nematode/copepod ratio and its use in pollution ecology. Mar Pollut Bull 12: 329–333

Warwick RM (1984) Species size distribution in marine benthic communities. Oecologia (Berl) 61: 32–41

Warwick RM (1986) A new method for detecting pollution effects on marine macrobenthic communities. Mar Biol 92: 557–562

Warwick RM (1987) Meiofauna – their role in marine detrital systems. In: Moriarty DJW, Pullin RSV (eds) Detritus and microbial ecology in aquaculture. ICLARM Conf Proc 14: 282–295

Warwick RM (1988) Analysis of community attributes of the macrobenthos of Frierfjord/Langesundfjord at taxonomic levels higher than species. Mar Ecol Prog Ser 46: 167–170

Warwick RM (1989) The role of meiofauna in the marine ecosystem: evolutionary considerations. Zool J Linn Soc 96: 229–241

Warwick RM (1993) Environmental impact studies on marine communities: pragmatical considerations. Aust J Ecol 18: 63–80

Warwick RM, Clarke KR (1991) A comparison of some methods for analysing changes in benthic community structure. J Mar Biol Assoc UK 71: 225–244

Warwick RM, Price R (1979) Ecological and metabolic studies on free-living nematodes from an estuarine mud-flat. Estuarine Coastal Mar Sci 9: 257–271

Warwick RM, Join IR, Radford PJ (1979) Secondary production of the benthos in an estuarine environment. In: Jeffries RL, Davy AJ (eds) Ecological processes in coastal environments. Blackwell, Oxford, pp 429–450

Warwick RM, Collins NR, Gee JM, George CL (1986a) Species size distributions of benthic and pelagic Metazoa: evidence for interaction? Mar Ecol Prog Ser 34: 63–68

Warwick RM, Gee JM, Berge JA, Ambrose W Jr (1986b) Effects of the feeding activity of the polychaete *Streblosoma bairdi* (Malmgren) on meiofaunal abundance and community structure. Sarsia 71: 11–16

Warwick RM, Clarke KR, Gee GM (1990b) The effect of disturbance by soldier crabs *Mictyris platycheles* H. Milne Edwards on meiobenthic community structure. J Exp Mar Biol Ecol 135: 19–33

Warwick RM, Platt HM, Clarke KR, Agard J, Gobin J (1990a) Analysis of macrobenthic and meiobenthic community structure in relation to pollution and disturbance in Hamilton Harbour, Bermuda. J Exp Mar Biol Ecol 138: 119–142

Wasmund N (1989) Live algae in deep sediment layers. Int Rev Ges Hydrobiol 74: 589–597

Watling L (1991) The sedimentary milieu and its consequences for resident organisms. Am Zool 31: 789–796

Watzin MC (1985) Interactions among temporary and permanent meiofauna: observations on the feeding and behaviour of selected taxa. Biol Bull (Woods Hole) 169: 397–416

Watzin MC (1986) Larval settlement into marine soft-sediment systems: interactions with the meiofauna. J Exp Mar Biol Ecol 98: 65–113

Webb DG, Parsons TR (1991) Impact of predation-disturbance by large epifauna on sediment-dwelling harpacticoid copepods: Field experiments in a subtidal seagrass bed. Mar Biol 109: 485–491

Webster IT (1992) Wave enhancement of solute exchange within empty burrows. Limnol Oceanogr 37: 630–643

Weinstein F (1961) *Psammostyela delamarei* n.g. n.sp., Ascidie interstitielle des sables à Amphioxus. CR Acad Sci Paris, Sér D 252: 1843–1844

Weise W, Rheinheimer G (1978) Scanning electron microscopy and epifluorescence investigations of bacterial colonization of marine sand sediments. Microb Ecol 4: 175–188

Weissenberger J, Dieckmann G, Gradinger R, Spindler M (1992) Sea ice – a cast technique to examine and analyze brine pockets and channel structure. Limnol Oceanogr 37: 179–183

Wells JBJ (1971) A brief review of methods of sampling the meiobenthos. In: Hulings NC (ed) Proc 1st International Conference on Meiofauna, Tunesia. Smithson Contrib Zool 76: 183–186

Wells JBJ (1976) Keys to aid in the identification of marine harpacticoid copepods. University of Aberdeen, Aberdeen, 215 p + Amendment Bulletin Numbers 1–5 in Zool Publ Victoria Univ Wellington (1978–1985)

Wells JBJ (1986) Copepoda: Marine-interstitial Harpacticoida. In: Botosaneanu L (ed) Stygofauna Mundi. A faunistic, distributional, and ecological synthesis of the world fauna inhabiting subterranean waters (including the marine interstitial). Brill, Backhuys, Leiden, pp 356–381

Wells JBJ (1988) Copepoda. In: Higgins RP, Thiel H (eds) Introduction to the study of meiofauna. Smithsonian Inst Press, Washington, DC, pp 380–388

Wesenberg-Lund C (1939) Biologie der Süsswassertiere. Springer, Wien 817 pp

Westheide W (1968) Zur quantitativen Verteilung von Bakterien und Hefen in einem Gezeitenstrand der Nordseeküste. Mar Biol 1: 336–347

Westheide W (1972) La faune des polychètes et des archiannélides dans les plages sableuses à ressac de la côte méditerranéenne de la Tunisie. Bull Inst Océanogr Pêche Salammbô 2: 449–468

Westheide W (1977) The geographical distribution of interstitial polychaetes. Mikrofauna Meeresboden 61: 287–302

Westheide W (1979) *Hesionides riegerorum* n.sp., a new interstitial freshwater polychaete from the United States. Int Rev Ges Hydrobiol 64: 273–280

Westheide W (1984) The concept of reproduction in polychaetes with small body size: adaptations in interstitial species. Fortschr Zool 29: 265–287

Westheide W (1985) The systematic position of the Dinophilidae and the archiannelid problem. In: Conway Morris S, George JD, Gibson R, Platt HM (eds) The origins and relationships of lower invertebrates. Clarendon Press, Oxford, pp 310–326 (Syst Ass, vol 28)

Westheide W (1986) The nephridia of the interstitial polychaete *Hesionides arenaria* and their phylogenetic significance (Polychaeta, Hesionidae). Zoomorphology 106: 35–43

Westheide W (1987a) Progenesis as a principle in meiofauna evolution. J Nat Hist 21: 843–854

Westheide W (1987b) The interstitial polychaete *Hesionides pettiboneae* n.sp. (Hesionidae) from the US east coast and its transatlantic relationships. Biol Soc Wash Bull 7: 131–139

Westheide W (1988) Polychaeta. In: Higgins RP, Thiel H (eds) Introduction to the study of meiofauna. Smithsonian Inst Press, Washington, DC London, pp 332–344

Westheide W (1990) Polychaetes: interstitial families, Key and notes for the identification of the species. In: Kermack DH, Barnes RSK (eds) Synopses of the British fauna, vol 44. Universial Book Services/Dr W Backhuys, Oegstgeest, the Netherlands, 152 pp

Westheide W (1991) The meiofauna of the Galápagos. A review. In: James MJ (ed) Galápagos Marine Invertebrates. Taxonomy, biogeography, and evolution in Darwin's islands. Plenum Press, New York London, pp 37–73

Westheide W, Schminke HK (1991) A joint project. Solving problems of zoosystematics with biochemical and morphological methods. Preface. Z Zool Syst Evolutionsforsch 29: 321–322

Westphalen D (1993) Stromatolithoid microbial nodules from Bermuda – a special microhabitat for meiofauna. Mar Biol (in press)

Wharton DA (1986) A functional biology of nematodes. Croonhelm, London 192 pp

Whiteside MK, Williams JB, White CP (1978) Seasonal abundance and pattern of chydorid Cladocera in mud and vegetative habitats. Ecology 59: 1177–1188

Whitfield M (1974) Thermodynamic limitations on the use of the platinum electrode in Eh measurements. Limnol Oceanogr 19: 857–865

Whitlatch RB, Zajac RN (1985) Biotic interactions among estuarine infaunal opportunistic species. Mar Ecol Prog Ser 21: 299–311

Widbom B (1984) Determination of average individual dry weights and ash-free dry weights in different sieve fractions of marine meiofauna. Mar Biol 84: 101–108

Wiederholm T, Erikson L (1977) Effects of alcohol preservation on the weight of some benthic invertebrates. Zoon 5: 29–31

Wieser W (1953) Die Beziehung zwischen Mundhöhlengestalt, Ernährungsweise und Vorkommen bei freilebenden marinen Nematoden. Ark Zool 2: 439–484

Wieser W (1959) The effect of grain size on the distribution of small invertebrates inhabiting the beaches of Puget Sound. Limnol Oceanogr 4: 181–194

Wieser W (1960) Benthic studies in Buzzards Bay. II. The meiofauna. Limnol Oceanogr 5: 121–137

Wieser W (1975) The meiofauna as a tool in the study of habitat heterogeneity: Ecophysiological aspects. A review. Cah Biol Mar 16: 647–670

Wieser W, Ott J, Schiemer F, Gnaiger E (1974): An ecophysiological study of some meiofauna species inhabiting a sandy beach at Bermuda. Mar Biol 26: 235–248

Wilkens H, Parzefall J, Iliffe TM (1986) Origin and age of the marine stygofauna of Lanzarote, Canary Islands. Mitt Hamb Zool Mus Inst 83: 223–230

Williams DD (1989) Towards a biological and chemical definition of the hyporheic zone in two Canadian rivers. Freshwater Biol 22: 189–208

Williams DD, Williams NE (1974) A counterstaining technique for use in sorting benthic samples. Limnol Oceanogr 19: 152–154

Williams PL, Dusenbery DB (1990) Aquatic toxicity testing using the nematode, *Caenorhabditis elegans*. Environ Tox Chem 9: 1285–1290

Williams R (1971) A technique for measuring the interstitial voids of a sediment based on epoxy resin impregnation. In: Hulings NC (ed) Proc. 1st International Conference on Meiofauna, Tunesia. Smithson Contrib Zool 76: 199–205

Williams R (1972) The abundance and biomass of the interstitial fauna of a graded series of shell gravels in relation to available space. J Anim Ecol 41: 623–646

Williams-Howze J, Coull BC (1992) Are temperature and photoperiod necessary cues for encystment in the marine benthic harpacticoid copepod *Heteropsyllus nunni* Coull? Biol Bull (Woods Hole) 182: 109–116

Williams-Howze J, Fleeger JW (1987) Pore pattern: a possible indicator of tube-building in *Stenhelia* and *Pseudostenhelia* (Copepoda: Harpacticoida). J Crustacean Biol 7: 148–157

Williamson FA, Palframan KR (1989) An improved method for collecting and staining microorganisms for enumeration by fluorescence light microscopy. J Microsc 154: 267–272

Wilson TRS (1978) Evidence for denitrification in aerobic pelagic sediments. Nature 274: 354–356

Wilson WH (1991) Competition and predation in marine soft-sediment communities. Annu Rev Ecol Syst 21: 221–241

Winberg GG (ed) (1971) Methods of estimation of production of marine animals. Academic Press, London, 175 pp

Wiszniewski J (1934) Recherches écologiques sur le psammon et spécialement sur les rotifères psammiques (in Polish and French). Arch Hydrobiol Rybactwa 8: 149–271

Woodin SA, Jakson JBC (1979) Interphyletic competition among marine benthos. J Mar Res 34: 25–41

Wormald AP (1976) Effects of a spill of marine diesel oil on the meiofauna of a sandy beach at Picnic Bay, Hong Kong, Environ Pollut 11: 117–130

Wright JC, Westh P, Ramlov H (1982) Cryptobiosis in Tardigrada. Biol Rev Cambr Philos Soc. 67: 1–30

Yingst JY (1978) Patterns of micro- and meiofaunal abundance in marine sediments, measured with adenosine triphosphate assay. Mar Biol 47: 41–54

Yingst JY, Rhoads DC (1980) The role of bioturbation in the enhancement of bacterial growth in marine sediments. In: Tenore KR, Coull BC (eds) Marine benthic dynamics. University of South Carolina Press, Columbia, SC pp 407–421 (Belle W Baruch Libr Mar Sci, vol 11)

Zajcev JP, Ancupova LV, Vorob'eva LV, Garkavaja GP, Kulakova II, Rusnak EM (1987) Nematody v glubokovodnoj zone Cernogo Morja. Nematodes from the depth zone of the Black Sea. Dokl AN USSR, Ser B, Geol Chim Biol Nauki 11: 77–79 (in Russian)

Zelinka C (1978) Monographie der Echinodera, Engelmann, Leipzig, 396 pp + 4 pl

Zimmermann CF, Price MT, Montgomery TR (1978) A comparison of ceramic and Teflon in situ samplers for nutrient pore water determinations. Estuarine Coastal Mar Sci 7: 93–97

Zinn DS, Found BW Kraus (1982) A bibliography of the Mystacocarida. Crustaceana 43: 270–274

Zobrist EC, Coull BC (1992) Meiobenthic interactions with macrobenthic larvae and juveniles: an experimental assessment of the meiofaunal bottleneck. Mar Ecol Prog Ser 88: 1–8

Glossary

(abbreviations: *syn.*: synonym; *contr.*: contrary term; *subst.*: substantive; *adj.*: adjective)

abyssal = depth zone of the sea, mostly between 4000 and 6000 m

allopatric = living in or originating from different geographical areas

amensalism = active inhibition of a competing species through biological interactions

amphiatlantic = occurring at both side of the Atlantic Ocean

anchihaline = a coastal salt water habitat separated from the open sea, mostly used in context with marine caves

apex = the tip of a (mollusc) shell

apomorphic = morphologically derived related to the ancestral structure

aufwuchs = organisms (mostly plants) growing as a surface layer on hard substrates, *syn.* epigrowth

bell jar apparatus = experimental gadged with bell-shape jars enclosing a bottom area (mostly used for measurement of community respiration)

biofilm = thin organic layer of microorganisms and excreted mucus

bioturbation = mixing of sediment by biogenic activities, i.e. burrowing

capitate = ending with a small swelling, mostly used for tentacles

calcareous = (sediment) consisting of carbonates, mostly of biogenic origin

circum-mundane = of a worldwide distribution

cnidome = the complete set of differently shaped cnidocysts (nematocysts)

cryptic species = a group of morphologically identical species whose differences are genetic or physiological and, thus, morphologically hidden, *syn.* sibling species

cryptobiotic = a mode of passive endurance without signs of active life; *subst.* cryptobiosis

cystid = the body part of a bryozoan individual (zooid) which is more or less solid, often also encrusted (*syn.* zooecium), and into which the soft tentacular apparatus, the lophophor or polypid, can be withdrawn

desmonemes = type of nematocysts with long and thin thread coiled up in the cell

destruents = the community of organisms degrading organic matter, e.g. bacteria, fungi

diaphragm (of electrodes) = connection of internal reference electrode with ambient medium, mostly a porous ceramic disc

disjunct = with a geographically separate distribution

Ediacara = geological period preceding the Cambrian, 700 to 570 million years ago

endemic = occurring exclusively in the region investigated, not present in other regions; *subst.* endemism, endemist

ephemeral = of only very short, temporary existence

epimers = lateral plate-like extensions of the body cuticle, mostly in peracarid crustaceans

euryoecious = of a wide ecological range, ecologically fairly hard, *contr* stenoecious

eurytopic = of a wide biotopical range; *contr.* stenotopic

eutelic = condition of a fixed cell number, a cellular constancy often reached already at a juvenile stage

filiform = thread-like

fission = mode of asexual reproduction in animals by division of the body into two or several parts which become independent individuals

flotation = passive drifting of particles

generalist = ecological type of organism adapted to a wide array of habitat conditions; often also termed "colonizer" due to the ability of rapidly populating disturbed habitats after an organismic depletion. *contr.* specialist, persister

gonochoristic = unisexual, individuals having separate sexes

hapto-sessile = animals attached to a substrate, but capable of detaching their hold and slowly changing their position

hemi-sessile = slow-moving animals, temporarily fixed in their position by anchoring structures

heterogony = a regular alternation of generative modes, changing from a normal, heterosexually reproducing generation to one (or several) unisexual, parthenogenetic generations; *adj.* heterogonous

homonomous = with a repeated sequence of similar structures, e.g. with similar segments in arthropods

hydrogenosome = cellular organelles (of some protozoans) metabolically involved in the production of hydrogen, probably derived from mitochondria

hydrothermal vents = underwater, often deep-sea sites where hot water, usually rich in hydrogen sulphide, escapes under high pressure

hyperbenthic = living directly above the bottom

idioadaptation = adaptations of a high specificity for a particular organism

interstitial fauna = microscopically small fauna of motile aquatic sediments living in the interstices of sand

irrigation = the pumping of water through waving movements (mostly used for animals living in tubes or burrows)

iteroparous = reproducing several times during the life time while semelparous animals have just one reproductive period

K-strategist = ecological type of organism highly adapted to specific biotopical conditions, fully utilizing the carrying capacity of a given ecological niche; often also coined specialist, persister. *contr.* r-strategist

karstic = pertaining to limestone strata rich in subterranean waters, caves and streams

lebensformtype = the German word for a specific combination of structural and ecological adaptations to a particular habitat, e.g. to life in narrow crevices

limnogenic = of limnetic origin

marsupium = the brood pouch (here) of peracarid crustaceans formed by characteristic cuticular plates, the oostegites

meroplanktonic = living temporarily, for a certain ontogenetic phase (mostly as larvae) as plankton; *subst.* meroplankton

mesocosm = experimental simulation of a more or less large and complex natural habitat and its organisms

metasome = posterior part of body in chelicerates

mixocoelom = the body cavity of arthropods developed through fusion of primary blastocoel and secondary coelom

monotypic = a taxon which alone represents the next higher taxonomic category, e.g. a genus with a monotypic species comprises just this single species

neotenic = attaining sexual maturation already at a larval stage, somatic development retarded in comparison to sexual development; *subst.* neoteny; *syn.* progenetic

nodulus = a swelling in the setae of annelids

oostegite = cuticular plates originating at the basis of the thoracic appendages in peracarid crustaceans and forming the brood pouch

P/B value = relative index of productivity relating Production (during a given time period) to the average standing stock, the Biomass

paratomy = asexual division of body with complete subsequent regeneration of parts, a mode of asexual multiplication; see fission

parthenogenesis = development from unfertilized female gametes; *adj.* parthenogenetic

pelos = aquatic animal community living in mud or silt

peristaltic = with regular alternation of longitudinal and circular body muscles producing proceeding contractile waves

phenotypic sex determination = sex determination by external factors and not by the chromosomal combination alone

phoretic = being regularly transported by an animal of a different species; *subst.* phoresy

physical lung = structure in aquatic air-breathing animals for storage of an air bubble from which the oxygen is respired and into which dissolved oxygen is regularly diffusing from the ambient water; this type of respiration is also termed plastron respiration

phytal = an aquatic habitat of an animal community characterized and structured by plants (also used as adjective, e.g. the phytal meiofauna community)

phyton = aquatic animal community living in the plants, in the algal belt

pinger = device releasing an electronic signal upon contact, e.g. attached to grabs or nets signalling bottom contact

polypid = the soft part of a bryozoan zooecium, the lophophor (see cystid)

polyvoltine = reproductive cycle with several reproductive periods per year; similarly univoltine (one reproductive period), bivoltine (two)

progenetic = attaining sexual maturation already at a larval stage, somatic development retarded in comparison to sexual development; *subst.* progenesis; *syn.* neotenic

protandrous = condition in hermaphrodites where a functional male phase is antecedent to the subsequent female phase

protogynous = condition in hermaphrodites where a functional female phase is antecedent to the subsequent male phase (rare)

psammon = aquatic animal community living in sand

pseudofertilization = a stimulus elicited from sperm (or sexual products) not resulting in karyogamy, but required for the onset of a parthenogenetic development (in some oligochaetes)

r-strategist = ecological type of organism adapted to a wide range of biotopical conditions; with its high reproductive potential it quickly colonizes impoverished habitats after destructive events; often also coined generalist, colonizer; *contr.* K-strategist

regressive evolution = evolution leading to increasing simplification of structures or reduction of organs, notably frequent in caves or subterranean habitats

rhabdites = rod-shaped secretions in the turbellarian epidermis which dissolve rapidly when extruded

rhinophore = posterior pair of tentacles in opisthobranch molluscs with numerous chemoreceptors

semelparous = reproducing only once during life time; *contr.* iteroparous = with several reproductive periods sessile = fixed to a substrate

siliceous (also silicious) = sediment of silicate minerals, e.g. quartz

skewness = the degree of asymmetry in a statistical distribution, e.g. in the sediment grain size distribution

specialist = species of a small adaptive range well fitting in one particular narrow ecological niche; often a K-selected species; *contr.* generalist

stenoecious = ecologically of a narrow range; *contr.* euryoecious

stenothermal = with a narrow temperature range; *contr.* eurythermal

stenotopic = with a narrow biotopical range; *contr.* eurytopic

stochastic = by chance, accidental, depending on non-predictive events

stygobios = community of animals in subterranean habitats

sympatric = co-occurring in the identical area; *subst.* sympatry

(sym)plesiomorphy = the common posession of an ancestral, homologue character; *contr.* (syn)apomorphy = a derived character genuine to the taxon and not shared by the ancestral taxon; *adj.* plesiomorph, apomorph,

syncytium = cell mass or layer with numerous nuclei, but without separating cell walls; in eukaryonts secondarily derived

syntopical = occurring in the same biotope

thalassogenic = of marine origin

triradiate = pertaining to an organ with a regular tripartite and trisymmetrical structure, i.e. the pharynx musculature in Nemathelminthes

troglobitic = living in caves and corresponding subterranean spaces

troglobiont = animal living in caves

urosome = the posterior part of the abdomen in crustaceans

verrucose = warty, with small tubercles

zooid = an individual within a colonial aggregate, mostly of minute size, e.g. in bryozoans, colonial ascidians

Subject Index

Taxonomic and derived terms and author names are excluded from the following index, for information see Chapter 5 and References. Terms provided by chapter headings are usually not included here, for information see Table of Contents.

Printing: Saladruck, Berlin
Binding: Buchbinderei Lüderitz & Bauer, Berlin

DATE DUE

MAY 2 5 1999	
DEC 1 5 2005	